Practical Handbook of Earth Science

Practical Handbook of Earth Science

Jane H. Hodgkinson
Frank D. Stacey

CRC Press is an imprint of the
Taylor & Francis Group, an **informa** business

CRC Press
Taylor & Francis Group
6000 Broken Sound Parkway NW, Suite 300
Boca Raton, FL 33487-2742

Printed on acid-free paper

International Standard Book Number-13: 978-1-138-55223-4 (Hardback)
978-1-138-05444-8 (Paperback)

Library of Congress Cataloging-in-Publication Data

Names: Hodgkinson, Jane H., author. | Stacey, F. D. (Frank D.) author.
Title: Practical handbook of earth science / Jane H. Hodgkinson and Frank D. Stacey.
Description: Boca Raton : Taylor & Francis, 2017. | "A CRC title, part of the Taylor & Francis imprint, a member of the Taylor & Francis Group, the academic division of T&F Informa plc." | Includes bibliographical references.
Identifiers: LCCN 2017008776 | ISBN 9781138552234 (hardback : alk. paper) | ISBN 9781138054448 (paperback : alk. paper)
Subjects: LCSH: Earth sciences--Handbooks, manuals, etc.
Classification: LCC QE26.3 .H63 2017 | DDC 550--dc23
LC record available at https://lccn.loc.gov/2017008776

Visit the Taylor & Francis Web site at
http://www.taylorandfrancis.com

and the CRC Press Web site at
http://www.crcpress.com

Contents

SECTION III Global Geophysics

SECTION IV Major Subdivisions of the Earth: Structures and Properties

SECTION VI Rocks and Minerals

Preface

Specialised subject dictionaries, encyclopaedias and data compilations are generally multi-authored, and even multi-volumed, collections of quasi-independent components of their disciplines. There remains a need for smaller handbooks, compiled from the perspective of subject overviews but with the coherence that is more readily achieved with one or two authors. Our attempt to address this need for Earth Science originated from our colleagues' comments on several editions of a geophysics text by one of us. They were not using it as a text but as a source of data and reported that they found the appendices to be the most useful parts of the book. Although this may appear to make a case for a separate collection of such material, that would not be satisfactory. The appendices were found to be useful in the context of the textbook but would not have been effective out of context. Nevertheless, there is a message. Although simple numbers or facts sometimes satisfy enquiries, more often clarification of significance or connections with other facts are needed. We have taken the position that a stand-alone reference work is most useful if it is a partial compromise with the style of a textbook, but without the exhaustive approach of an encyclopaedia.

Our subject matter encompasses geology and geophysics, including the oceans and atmosphere, with attention to environmental implications and resources. It emphasises basic science with no coverage of exploration or instrumentation. To cover this range in a small book, some corners have been cut. A few essential references are added but we resisted the urge to include extensive citation lists, which fill large parts of some reference works. We have aimed for self-contained, concise explanations without citing data sources that may be neither readily accessible nor straightforward. For inconsistent data, as far as possible, we have checked original observations and report only those we judge to be most reliable, in some cases with our own recalculations. We hope, in this way, to present to time-constrained enquirers, who may be active in related or overlapping disciplines, the best currently available data and insights on topics encompassed by the broad term 'Earth Science'.

We appreciate the interest expressed in this project by many colleagues and thank especially James Boland, Michael Cooke, Paul Davis, John Griffiths, Micaela Grigorescu, Jonathan Hodgkinson, Cameron Huddlestone-Holmes, Mark Maxwell, Graham O'Brien, Brett Poulsen and Antonio Valero for helpful comments on a draft manuscript.

Jane H. Hodgkinson
Frank D. Stacey

Authors

Jane Helen Hodgkinson was born and brought up in London and, following her early education, began a career in merchant banking and the commodity markets. Seeking intellectual stimulation, she undertook a degree course in geology on an evening and part-time basis at Birkbeck College, London. This was a highly successful move and she graduated with first class honours in 2003, setting herself up for her second career, as a geologist. Seeing opportunities in Australia, she undertook a PhD at the Queensland University of Technology, completed in 2008, and was appointed to CSIRO, Australia's national science agency, as a research geologist. Much of her work has been concerned with the problems and opportunities presented to the mining industry by climate change and to a study of the CO_2 geosequestration potential of Queensland geology.

Frank Donald Stacey was born and educated in London, with BSc (Hons physics) 1950, PhD 1953, DSc 1968. Following appointments at the University of British Columbia, the Australian National University and the Meteorological Office research unit in Cambridge, he joined the University of Queensland, becoming Professor of Applied Physics in 1971. Since 1956, all his research has been in geophysics, initially as an experimenter in rock magnetism and subsequently in a range of other subjects, with an increasing theoretical interest. He is most widely known for his textbook, *Physics of the Earth*, which is now in its fourth edition. Retiring from the university in 1996, he joined CSIRO as honorary fellow, to continue fundamental research, especially applications of thermodynamics to the Earth.

This is the author's second joint book. The first presented a global perspective on the environment: *The Earth as a Cradle for Life: the Origin, Evolution and Future of the Environment* (World Scientific, 2013).

Section I

Notation and Units

Chapter 1

Physical Units and Constants

1.1 SI UNITS (SYSTÈME INTERNATIONAL D'UNITÉS)

There is nothing fundamental about the SI system. It is simply the result of a decision to produce a single universal system of units to replace multiple alternative systems. Modifications/additions have been introduced from time to time and the system is not without its difficulties. In the Earth sciences, the greatest difficulty is experienced in geomagnetism and rock magnetism, but practitioners in these disciplines are constrained by a consensus favouring a common system of units across all science. Some flexibility is assumed, and Table 1.1 presents some SI equivalents to the formal units with multiplying factors in Table 1.2. Thus, the geomagnetic field strength is generally expressed in millitesla, although the formal SI prescription is ampere/metre, both being SI units. Caution is required in the use of the mole as a chemical unit of material quantity because it refers to mass in grams, not the SI unit, kilograms, and is a survivor from the centimetre–gram–second system. Radiation intensity is most conveniently expressed in watts, W/unit solid angle or W/m^2 at a specified distance, without reference to its wavelength, avoiding the generally impractical SI definition of the candela. Units applied to radioactive exposure doses are not listed here. Conversions from units in common non-SI systems are listed in Section 1.2 with more details for fossil fuels in Section 24.5.1.

TABLE 1.1 SI UNITS

Quantity	Unit	Symbol	Equivalents
Distance	metre	m	
Mass	kilogram	kg	1000 g
Time	second	s	
Temperature	kelvin	K	°C + 273.15
Substance unit	mole	mol	10^{-3} kg mol
Electric current	ampere	A	C/s
Light intensity	candela	cd	
Angle	radian	rad	180°/π
Frequency	hertz	Hz	s^{-1}
Force	newton	N	kg·m/s^2
Energy	joule	J	N·m
Power	watts	W	J/s
Pressure	pascal	Pa	N/m^2
Viscosity	pascal-second	Pa s	
Electric charge	coulomb	C	A·s
Electric potential	volt	V	J/C
Electric resistance	ohm	Ω	V/A
Electric conductance	siemen	S	1/Ω
Electric resistivity	ohm-metre	Ω-m	
Electric conductivity	siemen/metre	S m^{-1}	1/(Ω-m)
Electric capacitance	farad	F	C/V
Magnetic field strength	ampere/metre	A m^{-1}	
Magnetic flux	weber	Wb	V·s
Magnetic intensity (flux density)	tesla	T	Wb/m^2
Magnetisation	ampere metre	A m	
Inductance	henry	H	V·s/A

TABLE 1.2 UNIT PREFIXES FOR MULTIPLYING FACTORS

Multiplying Factor	Prefix	Symbol
10^{18}	exa	E
10^{15}	peta	P
10^{12}	tera	T
10^{9}	giga	G
10^{6}	mega	M
10^{3}	kilo	k
10^{2}	hecto	h
10	deka	da
10^{-1}	deci	d
10^{-2}	centi	c
10^{-3}	milli	m
10^{-6}	micro	μ
10^{-9}	nano	n
10^{-12}	pico	p
10^{-15}	femto	f
10^{-18}	atto	a

Note: Some of these (such as millimetres and kilowatts) are widely used and generally familiar, but others are not and should be used sparingly or defined with their use.

1.2 UNIT CONVERSIONS: SI EQUIVALENTS OF OTHER UNITS

Length, area, angle

1 angstrom	= 10^{-10} m
1 inch	= 0.0254 m (exact)
1 foot (12 inches)	= 0.3048 m
1 yard (3 feet)	= 0.9144 m
1 chain (22 yards)	= 20.1168 m
1 furlong (220 yards)	= 201.168 m

1 statute mile (1760 yards)	= 1609.344 m (exact)
1 nautical mile	= 1852 m (originally 1 minute of latitude)
1 league (3 nautical miles)	= 5556 m
1 fathom (6 feet, water depth)	= 1.8288 m
1 astronomical unit (AU)	= $1.495978707 \times 10^{11}$ m (defined)
(≈ semi-major axis of Earth's orbit)	
1 light year	= 9.460895×10^{15} m
1 parsec	= 3.085678×10^{16} m
1 barn (nuclear cross section)	= 10^{-28} m^2
1 square (100 square feet)	= 9.2903 m^2
1 hectare	= 10^4 m^2 = 0.01 km^2
1 acre	= 4046.856 m^2
1 degree (angle)	= π/180 radian
1 arc sec (1/3600 degree)	= π/648,000 rad = $4.848... \times 10^{-6}$ rad

Volume, mass

1 litre	= 10^{-3} m^3
1 fluid ounce (Imperial)	= 0.028349523 L
1 fluid ounce (US)	= 0.029535296 L
1 acre-foot	= 1233.48 m^3
1 gallon (Imperial)	= 4.5359237 L
1 gallon (US) (231 cubic inches)	= 3.7854118 L
1 barrel (oil) (~42 US gallons)	= 158.99 L
1 barrel (oil)	= 0.146 toe = 6.113 GJ
[For other fossil fuel unit conversions see Section 24.5]	
1 tonne of oil equivalent (toe)	= 41.868 GJ
1 toe coal	= 1.428 tonnes of coal
1 grain	= 6.4798918×10^{-5} kg
1 ounce (Avoirdupois, 437.5 grains)	= 0.028349323 kg
1 dram (1/16 ounce)	= 1.7718327×10^{-3} kg
1 ounce (Troy, 480 grains)	= 0.031034768 kg
1 carat (gem stones, 1/5 gram)	= 2×10^{-4} kg
1 pound (lb, 7000 grains)	= 0.45359237 kg
1 tonne	= 1000 kg
1 ton (Imperial) (2240 lb)	= 1016.0469 kg
1 ton (US) (2000 lb)	= 907.18474 kg
1 hundredweight, hwt (1/20 ton)	
Imperial, 112 lb	= 50.802345 kg
US, 100 lb	= 45.359237 kg
1 stone (14 lb, 1/8 Imp hwt)	= 6.35029318 kg

Time, speed

1 sidereal year	$= 3.155815 \times 10^7$ s = 365.25636 days
1 tropical year (equinox to equinox)	= 365.24219 days
1 sidereal day	= 86164.091 s
1 solar day	= 86,400 s
1 km/hour	$= 0.277778$ m s^{-1}
1 mile/hour	$= 0.44704$ m s^{-1}
1 knott (nautical mile/hour)	$= 0.51444$ m s^{-1}

Force, pressure, energy

1 dyne	$= 10^{-5}$ N
1 Gal (gravity)	$= 10^{-2}$ m s^{-2} (1 mGal = 10^{-5} m s^{-2})
1 atmosphere	= 101,325 Pa
1 bar	$= 10^5$ Pa
1 psi (lb/square inch)	= 6894.8 Pa
1 dyne/cm^2	= 0.1 Pa
1 torr (1 mm of mercury)	= 133.3 Pa
1 erg	$= 10^{-7}$ J
1 electron volt (eV)	$= 1.60217657 \times 10^{-19}$ J
1 calorie, international steam	= 4.1868 J
thermochemical (USA)	= 4.184 J
1 British thermal unit (BTU)	= 1055.06 J
1 quad (quadrillion BTU) (10^{15} BTU)	= 1055 PJ
1 kilowatt-hour	$= 3.6 \times 10^6$ J
1 horsepower	= 745.7 W
1 heat flux unit [1 μcalorie/(cm^2 s)]	$= 4.1868 \times 10^{-2}$ W m^{-2}

Fluid flow

1 poise	= 0.1 Pa s
1 darcy	$= 0.987 \times 10^{-12}$ m^2 ≈ (1 μm^2) (see Section 24.8.1)
1 Sverdrup (Sv, ocean flow)	$= 10^6$ m^3 s^{-1}

Electromagnetism

1 coulomb	= 1 ampere × 1 second
1 gauss	$= 10^{-4}$ T (tesla) = 10^5 nT (gamma)
1 gamma	$= 10^{-9}$ T
1 oersted	$= 10^3/4\pi$ A m^{-1} (ampere-turn/metre)
1 gauss – cm^3 (magnetic moment)	$= 10^{-3}$ A m^2
1 e.m.u. of magnetisation	$= 10^3$ A m^{-1}
1 μS cm^{-1}	$= 10^{-4}$ S m^{-1}
1 esu, electric charge	$= 3.33564 \times 10^{-10}$ coulomb

Temperature

X °C	$= (X + 273.15)$ K
X °F	$= (5/9)(X + 459.67)$ K

Other

1 Dobson unit (ozone/unit area)	$= 2.69 \times 10^{20}$ molecules/m^2 (2.69×10^{16} ozone molecules/cm^2)

1.3 FUNDAMENTAL CONSTANTS

Mathematical constants:

$$\pi = 4(1-1/3 + 1/5-1/7 + 1/9-1/11 + ...) = 3.141592654...$$

$$e = 1 + \frac{1}{1} + \frac{1}{1\times 2} + \frac{1}{1\times 2\times 3} + \frac{1}{1\times 2\times 3\times 4} + ... = 2.718281828...$$

$$\log_e(x) \equiv \ln(x) = \ln(10) \times \log_{10}(x) = 2.302585093 \times \log_{10}(x)$$

Defined constants:

Speed of light in vacuum c	$= 2.99792458 \times 10^8$ m s^{-1}
Permeability of free space μ_0	$= 4\pi \times 10^{-7}$ H m^{-1}
Permittivity of free space ε_0	$= 1/\mu_0 c^2 = 8.8541878... \times 10^{-12}$ F m^{-1}

NOTE: A change to the SI system is planned for 2018, when μ_0 and ε_0 will no longer be defined constants, but will have uncertainties subject to the condition $(\mu_0\,\varepsilon_0) = 1/c^2$. At the same time, h, k, e and N_A will become defined constants with the best values then prevailing in the present system, which will be very close to those listed here (Newell 2014).

Physical constants: CODATA values from the NIST (National Institute of Standards and Technology) 2014 listing (with uncertainties in the last digits in parentheses):

gravitational constant, G	$= 6.67408(31) \times 10^{-11}$ m^{3}kg^{-1}s^{-2} (N m^{2}kg^{-2})
Planck constant, h	$= 6.62607040(81) \times 10^{-34}$ J s
elementary charge, e	$= 1.6021766208(98) \times 10^{-19}$ C
electron mass, m_e	$= 9.10938356(11) \times 10^{-31}$ kg
proton mass, m_p	$= 1.672621898(21) \times 10^{-27}$ kg
neutron mass, m_n	$= 1.674927471(21) \times 10^{-27}$ kg
atomic mass constant, u (^{12}C mass/12)	$= 1.660539040(20) \times 10^{-27}$ kg

Avogadro's number, $N_A = (1/u)$ $= 6.022140857(74) \times 10^{23}\ \mathrm{mol}^{-1}$
$= 6.022140857(74) \times 10^{26}\ (\mathrm{kg\ mol})^{-1}$

gas constant, R $= 8.3144598(48)\ \mathrm{J\ mol^{-1}K^{-1}}$
$= 8.3144598(48) \times 10^{3}\ \mathrm{J\ (kg\ mol)^{-1}K^{-1}}$

Boltzmann's constant, $k = R/N_A$ $= 1.38064852(79) \times 10^{-23}\ \mathrm{J\ K^{-1}}$

Stefan–Boltzmann constant,
$\sigma = 2\pi^5 k^4/15h^3c^2$ $= 5.670367(13) \times 10^{-8}\ \mathrm{Wm^{-2}K^{-4}}$

Faraday constant, $F = (e/u) = N_A\, e$ $= 9.648533289(59) \times 10^{4}\ \mathrm{C\ mol^{-1}}$
$= 9.648533289(59) \times 10^{7}\ \mathrm{C\ (kg\ mol)^{-1}}$

inverse fine structure constant,
$\alpha^{-1} = 2\mathrm{h}/\mu_0 \mathrm{c}e^2$ $= 137.035999139(31)$

Bohr magneton, $\mu_B = (eh/4\pi m_e)$ $= 9.27401000(6) \times 10^{-24}\ \mathrm{A\ m^2}$

Lorenz number, $L = (\pi k/e)^2/3$ $= 2.443003 \times 10^{-8}\ \mathrm{W\ \Omega\ K^{-2}}$

(This is a coefficient relating the electron component, κ_e, of thermal conductivity to the electrical conductivity, σ_e, of a metal by the Wiedemann–Franz law: $\kappa_e = L\sigma_e T$.)

Chapter 2

Some Shorthand Conventions

2.1 SELECTED ACRONYMS AND ABBREVIATIONS

ACC Antarctic circumpolar current (now Antarctic polar frontal zone)
AGU American Geophysical Union
ALT altitude, altimeter
AMSL above mean sea level
APW apparent polar wander
ATM atmosphere, atmospheric
AU astronomical unit (radius of Earth's orbit)
BABI basaltic achondrite best initial ratio $^{87}Sr/^{86}Sr$ (primordial)
BCC body-centred cubic
BCF billion cubic feet
BIFs banded iron formations
BIRPS British Institutions Reflection Profiling Syndicate
BGL below ground level
BLS below land surface
BSE bulk silicate Earth
C1 type 1 carbonaceous chondrite (composition)
CAIs calcium- and aluminium-rich inclusions (in meteorites)
CBM coal bed methane (CSG)
CCD carbonate compensation depth
CCNs cloud condensation nuclei
CDIAC Carbon Dioxide Information Analysis Center
CFCs chlorofluoro carbons
CMB core-mantle boundary
CME coronal mass ejection
COCORP Consortium for Continental Reflection Profiling
CODATA Committee for Data on Science and Technology
COHMAP Cooperative Holocene Mapping Project
COSPAR Committee on Space Research

CPX	clinopyroxene
CSG	coal seam gas (CBM)
D″	lowermost layer of the mantle
DTR	daily temperature range
DVI	dust veil index (volcanic)
EDAX	energy dispersive X-ray analysis
EDM	electronic distance measurement
EOS, EoS	equation of state
EPMA	electron probe microanalysis
ERI	Earthquake Research Institute (Tokyo)
ESA	European Space Agency
ESRL	Earth System Research Laboratory (NOAA)
EUG	European Union of Geosciences
FAO	Food and Agriculture Organization
FCC	face-centred cubic
GAD	geocentric axial dipole
GCM	global circulation model (atmosphere or ocean)
GISP	Greenland Ice Sheet Project (USA)
GPR	ground-penetrating radar
GPS	global positioning system
GRACE	Gravity Recovery And Climate Experiment
GRIP	Greenland Ice Core Project (Europe)
HCP	hexagonal close-packed
HFU	heat flux unit [1 calorie/(cm^2 s)]
HIMU	high μ ($^{238}U/^{204}Pb$)
HREE	heavy rare earth elements
HS	high spin (state of electron spin alignment in Fe^{2+} ions)
IAG	International Association of Geodesy
IAGA	International Association of Geomagnetism and Aeronomy
IAHS	International Association of Hydrological Sciences
IAMAP	International Association of Meteorology and Atmospheric Physics
IAPSO	International Association of Physical Sciences of the Oceans
IASPEI	International Association of Seismology and Physics of the Earth's Interior
IAVCEI	International Association of Volcanology and Chemistry of the Earth's interior
ICB	inner core boundary
ICS	International Commission on Stratigraphy
ICSU	International Council of Scientific Unions
IGRF	International Geomagnetic Reference Field

IGY International Geophysical Year (July 1957–December 1958)
ILP International Lithosphere Programme
IPCC International Panel on Climate Change
IR infra-red
IRIS Incorporated Research Institutions for Seismology
ITCZ Inter-Tropical Convergence Zone (atmospheric)
IUGG International Union of Geodesy and Geophysics
IUGS International Union of Geological Sciences
JMA Japan Meteorological Agency
JOIDES Joint Oceanographic Institutions for Deep Earth Sampling
KREEP K+REE+P-rich late solidification from magma
K-T cretaceous-tertiary (boundary)
LFG landfill gas (largely methane)
LGM last glacial maximum
LHB late heavy bombardment (of the Moon)
LID seismologically observed lithosphere
LIL large-ion lithophile ('incompatible' with mantle minerals)
LNG liquefied natural gas
LOD length-of-day
LPG liquefied petroleum gas
LREE light rare earth elements
LS low spin (state of electron spin alignment in Fe^{2+} ions)
LVZ low-velocity zone (asthenosphere)
m_b body wave magnitude (earthquake)
MHD magnetohydrodynamics
MIZ marginal ice zone
Moho Mohorovičić discontinuity (crust–mantle boundary)
MORB mid-ocean ridge basalt
M_S surface wave magnitude (earthquake)
MSL mean sea level
MT magneto-telluric (electromagnetic prospecting method)
M_W moment magnitude (earthquake)
mw magnesiowustite/ferropericlase (Mg,Fe)O
MWP medieval warm period (~900–1200 AD)
NADW North Atlantic deep water
NASA National Aeronautics and Space Administration
NDS non-linear dynamical system
NEA near–earth asteroid
NGRIP North Greenland Ice Core Project
NIST National Institute of Standards and Technology
NMR nuclear magnetic resonance

NOAA	National Oceanic and Atmospheric Administration
NRM	natural remanent magnetisation
NSIDC	National Snow and Ice Data Center
NTU	Nephelometric Turbidity Unit
ODP	ocean drilling program
OIB	ocean island basalt
OLR	outgoing long wavelength radiation
OPX	orthopyroxene
P waves	primary (compressional) waves (seismology)
pH	'power of hydrogen' (a numerical measure of acidity)
PKIKP	P waves entering the inner core
PKP	P waves entering the core
PREM	Preliminary Reference Earth Model (Dziewonski and Anderson 1981)
pv	silicate perovskite
Pyrolite	pyroxene-olivine (model of mantle composition)
QBO	quasi-biennial oscillation (stratosphere)
Qtz	quartz
REE	rare earth elements
RMS	root mean square
S waves	secondary (shear) waves (seismology)
SAR	synthetic aperture radar
SCAR	Scientific Committee on Antarctic Research
SH	horizontally polarised shear wave
sial	Si–Al crustal composition
SIO	Scripps Institution of Oceanography
sima	Si–Mg mantle composition
SKS	shear waves penetrating the core
SLR	satellite laser ranging
SNC	shergottite, nakhlite, chassignite (Martian meteorite types)
Sq	solar quiet
SSC	sudden commencement (magnetic storm)
SSSI	site of special scientific interest
SST	sea surface temperature
SV	vertically polarised shear wave
TRM	thermo-remanent magnetisation
USGS	United States Geological Survey
UTC	Coordinated Universal Time
UV	ultra-violet
VEI	volcanic explosivity index
VLBI	very long baseline interferometry

VMS volcanic massive sulphide
WHOI Woods Hole Oceanographic Institution
WMO World Meteorological Organization
WWRP World Weather Research Programme
WWSSN Worldwide Standardized Seismographic Network
XRD X-ray diffraction
XRF X-ray fluorescence spectroscopy
z term an annual cycle in apparent latitude synchronous at all observatories

2.2 THE GREEK ALPHABET

Alpha	α	Α	Nu	ν	Ν
Beta	β	Β	Xi	ξ	Ξ
Gamma	γ	Γ	Omicron	ο	Ο
Delta	δ	Δ	Pi	π	Π
Epsilon	ε	Ε	Rho	ρ	Ρ
Zeta	ζ	Ζ	Sigma	σ	Σ
Eta	η	Η	Tau	τ	Τ
Theta	θ	Θ	Upsilon	υ	Υ
Iota	ι	Ι	Phi	φ	Φ
Kappa	κ	Κ	Chi	χ	Χ
Lambda	λ	Λ	Psi	ψ	Ψ
Mu	μ	Μ	Omega	ω	Ω

Section II

The Building Blocks

Elements to Planets

Chapter 3

Elements, Isotopes and Radioactivity

3.1 PERIODIC TABLE OF ELEMENTS: A GEOCHEMICAL CLASSIFICATION (Figure 3.1)

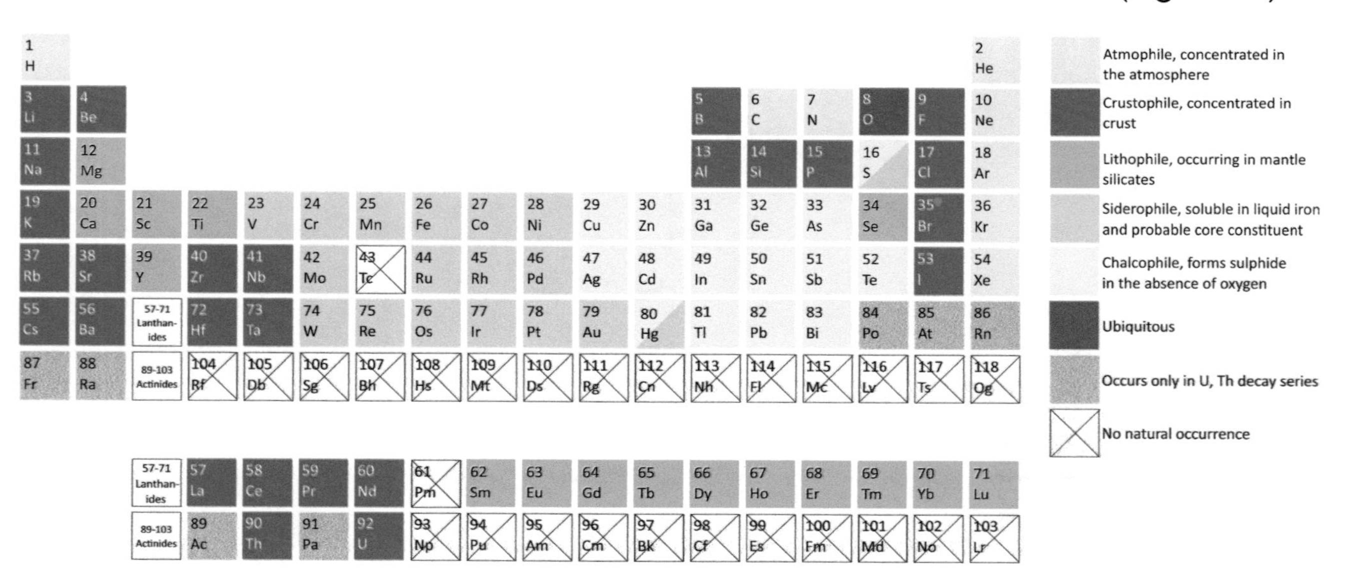

Figure 3.1 A geochemical classification of the elements.

3.2 PERIODIC TABLE OF ELEMENTS: A BIOLOGICAL CLASSIFICATION (Figure 3.2)

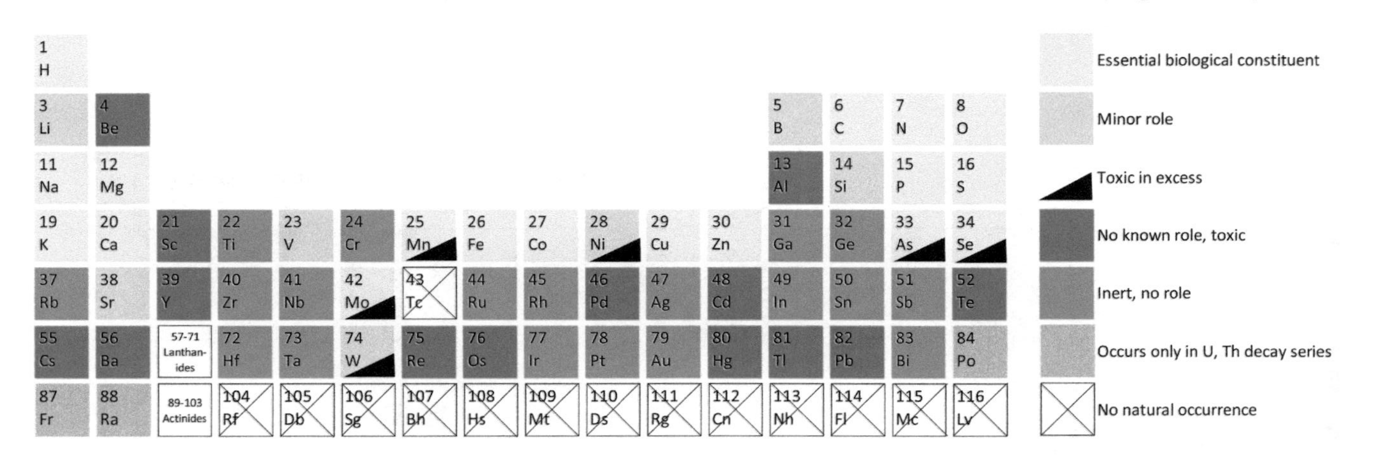

Figure 3.2 A biological classification of the elements.

3.3 ISOTOPES OF THE NATURALLY OCCURRING ELEMENTS

TABLE 3.1 ISOTOPIC ABUNDANCES AND MEAN ATOMIC WEIGHTS

Atomic No. z	Element	Symbol (mean atomic wt., in units of u = 1.66053878 × 10^{-27} kg)	Isotopic Masses, with Abundances in Atomic % in Parentheses
0	Neutron	n[a] (1.0886)	
1	Hydrogen	H (1.0079)	1 (99.985), 2 (0.015), 3[a] (atmospheric trace from cosmic ray bombardment)
2	Helium	He (4.00260)	3 (0.00013), 4 (99.99987)
3	Lithium	Li (6.940)	6 (7.59), 7 (92.41)
4	Beryllium	Be (9.01218)	9 (100), 10[a] (atmospheric trace from cosmic ray bombardment)
5	Boron	B (10.811)	10 (19.9), 11 (80.1)
6	Carbon	C (12.0107)	12 (98.93), 13 (1.07), 14[a] (1.6×10^{-10} in atmosphere)
7	Nitrogen	N (14.0067)	14 (99.635), 15 (0.365)
8	Oxygen	O (15.9994)	16 (99.757), 17 (0.038), 18 (0.205)
9	Fluorine	F (18.99480)	19 (100)
10	Neon	Ne (20.1797)	20 (90.48), 21 (0.27), 22 (9.25)
11	Sodium	Na (22.9898)	23 (100)
12	Magnesium	Mg (24.3050)	24 (78.99), 25 (10.00), 26 (11.01)
13	Aluminium	Al (26.98154)	27 (100)
14	Silicon	Si (28.0855)	28 (92.22), 29 (4.69), 30 (3.09)
15	Phosphorus	P (30.97376)	31 (100)
16	Sulphur	S (32.065)	32 (94.93), 33 (0.76), 34 (4.29), 36 (0.02)
17	Chlorine	Cl (35.453)	35 (75.76), 37 (24.24)
18	Argon		
	Atmosphere	Ar (39.948)	36 (0.337), 38 (0.063), 40 (99.600)
	Solar wind	Ar (36.67)	36 (75.3), 38 (14.2), 40 (10.5)

(Continued)

TABLE 3.1 *(Continued)* ISOTOPIC ABUNDANCES AND MEAN ATOMIC WEIGHTS

Atomic No. z	Element	Symbol (mean atomic wt., in units of u = 1.66053878 × 10^{-27} kg)	Isotopic Masses, with Abundances in Atomic % in Parentheses
19	Potassium	K (39.0983)	39 (93.258), 40[a] (0.01167), 41 (6.730)
20	Calcium	Ca (40.078)	40 (96.94), 42 (0.65), 43 (0.13), 44 (2.09), 46 (0.0041), 48[b] (0.19)
21	Scandium	Sc (44.95591)	45 (100)
22	Titanium	Ti (47.867)	46 (8.25), 47 (7.44), 48 (73.72), 49 (5.41), 50 (5.18)
23	Vanadium	V (50.9415)	50[b] (0.25), 51 (99.75)
24	Chromium	Cr (51.996)	50 (4.35), 52 (83.79), 53 (9.50), 54 (2.36)
25	Manganese	Mn (54.93804)	55 (100)
26	Iron	Fe (55.845)	54 (5.84), 56 (91.75), 57 (2.12), 58 (0.28)
27	Cobalt	Co (58.93319)	59 (100)
28	Nickel	Ni (58.6934)	58 (68.077), 60 (26.223), 61 (1.140), 62 (3.634), 64 (0.926)
29	Copper	Cu (63.546)	63 (69.2), 65 (30.8)
30	Zinc	Zn (65.409)	64 (48.27), 66 (27.98), 67 (4.10), 68 (19.02), 70 (0.63)
31	Gallium	Ga (69.723)	69 (60.108), 71 (39.892)
32	Germanium	Ge (72.63)	70 (20.38), 72 (27.31), 73 (7.76), 74 (36.72), 76[b] (7.83)
33	Arsenic	As (74.9216)	75 (100)
34	Selenium	Se (78.96)	74 (0.89), 76 (9.37), 77 (7.63), 78 (23.77), 80 (49.61), 82[b] (8.73)
35	Bromine	Br (79.904)	79 (50.69), 81 (49.31)
36	Krypton	Kr (83.798)	78 (0.35), 80 (2.29), 82 (11.59), 83 (11.50), 84 (56.99), 86 (17.28)
37	Rubidium	Rb (85.4678)	85 (72.165), 87[a] (27.835)
38	Strontium	Sr (87.62)	84 (0.56), 86 (9.86), 87 (7.00), 88 (82.58)

(Continued)

TABLE 3.1 *(Continued)* **ISOTOPIC ABUNDANCES AND MEAN ATOMIC WEIGHTS**

Atomic No. z	Element	Symbol (mean atomic wt., in units of u = 1.66053878 × 10^{-27} kg)	Isotopic Masses, with Abundances in Atomic % in Parentheses
39	Yttrium	Y (88.90585)	89 (100)
40	Zirconium	Zr (91.224)	90 (51.45), 91 (11.22), 92 (17.15), 94 (17.38), 96[b] (2.80)
41	Niobium	Nb (92.90638)	93 (100)
42	Molybdenum	Mo (95.96)	92 (14.77), 94 (9.23), 95 (15.90), 96 (16.68), 97 (9.56), 98 (24.19), 100[a] (9.67)
43	Technetium	Tc	No naturally occurring isotope
44	Ruthenium	Ru (101.07)	96 (5.44), 98 (1.87), 99 (12.76), 100 (12.60), 101 (17.06), 102 (31.55), 104 (18.62)
45	Rhodium	Rh (102.90550)	103 (100)
46	Palladium	Pd (106.42)	102 (1.02), 104 (11.14), 105 (22.33), 106 (27.33), 108 (26.46), 110 (11.72)
47	Silver	Ag (106.8682)	107 (51.839), 109 (48.161)
48	Cadmium	Cd (112.411)	106 (1.25), 108 (0.89), 110 (12.49), 111 (12.80), 112 (24.13), 113[b] (12.22), 114[b] (28.72), 116[b] (7.49)
49	Indium	In (114.818)	113 (4.29), 115[b] (95.71)
50	Tin	Sn (118.71)	112 (0.97), 114 (0.66), 115 (0.34), 116 (14.54), 117 (7.68), 118 (24.22), 119 (8.59), 120 (32.58), 122 (4.63), 124 (5.79)
51	Antimony	Sb (121.60)	121 (57.21), 123 (47.79)
52	Tellurium	Te (127.60)	120 (0.09), 122 (2.55), 123 (0.89), 124 (4.74), 125 (7.07), 126 (18.84), 128[b] (31.74), 130[b] (30.08)
53	Iodine	I (126.90448)	127 (100)
54	Xenon	Xe (131.293)	124 (0.095), 126 (0.089), 128 (1.910), 129 (26.401), 130 (4.071), 131 (21.232), 132 (26.909), 134 (10.436), 136[b] (8.857)

(Continued)

TABLE 3.1 *(Continued)* **ISOTOPIC ABUNDANCES AND MEAN ATOMIC WEIGHTS**

Atomic No. z	Element	Symbol (mean atomic wt., in units of u = 1.66053878 × 10^{-27} kg)	Isotopic Masses, with Abundances in Atomic % in Parentheses
55	Caesium	Cs (132.90552)	133 (100)
56	Barium	Ba (137.327)	130[b] (0.106), 132 (0.101), 134 (2.417), 135 (6.592), 136 (7.854), 137 (11.232), 138 (71.698)
57	Lanthanum	La (138.90547)	138[a] (0.090), 139 (99.910)
58	Cerium	Ce (140.116)	136 (0.190), 138 (0.251), 140 (88.450), 142 (11.114)
59	Praseodymium	Pr (140.90765)	141 (100)
60	Neodymium	Nd (144.242)	142 (27.2), 143 (12.2), 144[b] (23.8), 145 (8.23), 146 (17.2), 148 (5.72), 150[b] (5.60)
61	Promethium	Pm	No naturally occurring isotope
62	Samarium	Sm (150.36)	144 (3.07), 146[a] (trace), 147[a] (14.99), 148[b] (11.24), 149 (13.82), 150 (7.38), 152 (26.75), 154 (22.75)
63	Europium	Eu (151.964)	151[b] (47.81), 153 (52.19)
64	Gadolinium	Gd (157.25)	152[b] (0.20), 154 (2.18), 155 (14.80), 156 (20.47), 157 (15.65), 158 (24.84), 160 (21.86)
65	Terbium	Tb (158.92535)	159 (100)
66	Dysprosium	Dy (162.500)	156 (0.056), 158 (0.095), 160 (2.329), 161 (18.889), 162 (25.475), 163 (24.896), 164 (28.260)
67	Holmium	Ho (164 93032)	165 (100)
68	Erbium	Er (167.259)	162 (0.139), 164 (1.601), 166 (33.503), 167 (22.869), 168 (26.978), 170 (14.910)
69	Thulium	Tm (168.9342)	169 (100)
70	Ytterbium	Yb (173.04)	168 (0.13), 170 (3.04), 171 (14.28), 172 (21.83), 173 (16.13), 174 (31.83), 176 (12.76)
71	Lutetium	Lu (174.967)	175 (97.41), 176[a] (2.59)
72	Hafnium	Hf (178.49)	174[b] (0.162), 176 (5.26), 177 (18.60), 178 (27.28), 179 (13.63), 180 (35.08)

(Continued)

TABLE 3.1 *(Continued)* ISOTOPIC ABUNDANCES AND MEAN ATOMIC WEIGHTS

Atomic No. z	Element	Symbol (mean atomic wt., in units of u = 1.66053878 × 10^{-27} kg)	Isotopic Masses, with Abundances in Atomic % in Parentheses
73	Tantalum	Ta (180.9479)	180 (0.012), 181 (99.988)
74	Tungsten	W (183.84)	180[b] (0.12), 182 (26.55), 183 (14.31), 184 (30.64), 186 (28.45)
75	Rhenium	Re (186.207)	185 (37.40), 187[a] (62.60)
76	Osmium	Os (190.23)	184 (0.02), 186[b] (1.59), 187 (1.96), 188 (13.24), 189 (16.15), 190 (26.26), 192 (40.78)
77	Iridium	Ir (192.217)	191 (37.3), 193 (62,7)
78	Platinum	Pt (195.089)	190[a] (0.014), 192 (0.782), 194 (32.967), 195 (33.832), 196 (25.242), 198 (7.163)
79	Gold	Au (196.966569)	197 (100)
80	Mercury	Hg (200.592)	196 (0.15), 198 (9.97), 199 (16.87), 200 (23.10), 201 (13.18), 202 (29.86), 204 (6.87)
81	Thallium	Tl (204.3833)	203 (29.52), 205 (70.48)
82	Lead	Pb (207.21) (variable)	204[b] (1.347), 206 (25.03). 207 (21.25), 208 (52.37) (averages in marine sediments)
83	Bismuth	Bi (208.9804)	209[b] (100)
84	Polonium	Po	Intermediate daughters in uranium and thorium decay series
85	Astatine	At	
86	Radon	Rn	
87	Francium	Fr	
88	Radium	Ra	
89	Actinium	Ac	
90	Thorium	Th (232.0381)	232[a] (100)
91	Protactinium	Pa	Intermediate daughter in uranium decay
92	Uranium	U (238.0289)	234[a] (0.0055), 235[a] (0.7200), 238[a] (99.2745)

[a] Radioactive isotopes.
[b] Isotopes with half-lives exceeding the age of the universe.

3.4 NATURALLY OCCURRING LONG-LIVED RADIOACTIVE ISOTOPES

This list recognises half-lives exceeding the age of the universe for isotopes that have generally been regarded as stable. The possibility that β decay rates are affected by neutrino flux is under consideration. It is not clear whether any of the following numbers are seriously affected.

TABLE 3.2 DECAY MECHANISMS AND HALF-LIVES

Isotope	% of Element	Decay Mechanism	Half-Life (years)	Decay Product
^{40}K	0.01167	89.28% β 10.72% K 0.001% $β^{+}$	1.248×10^{9}	^{40}Ca ^{40}Ar ^{40}Ar
^{48}Ca	0.19	2β	4×10^{19}	^{48}Ti
^{50}V	0.25	β	1.4×10^{17}	^{50}Cr
^{76}Ge	7.83	2β	1.8×10^{21}	^{26}Se
^{82}Se	8.73	2β	9.7×10^{19}	^{82}Kr
^{87}Rb	27.835	β	4.92×10^{10}	^{87}Sr
^{96}Zr	2.80	2β	2×10^{19}	^{96}Mo
^{100}Mo	9.67	2β	8.5×10^{18}	^{100}Ru
^{113}Cd	12.22	β	8.04×10^{15}	^{113}I
^{116}Cd	7.49	2β	2.8×10^{19}	^{116}Sn
^{115}In	95.71	β	4.4×10^{14}	^{115}Sn
^{128}Te	31.74	2β	2.2×10^{24}	^{128}Xe
^{130}Te	30.08	2β	7.9×10^{20}	^{130}Xe
^{130}Ba	0.106	2K	$\sim 10^{21}$	^{130}Xe
^{136}Xe	8.857	2β	2.38×10^{21}	^{136}Ba
^{138}La	0.090	33.6% β 66.4% $β^{+}$	1.02×10^{11}	^{138}Ce ^{138}Ba
^{142}Ce	11.05	α	5.0×10^{15}	^{138}Ba
^{144}Nd	23.8	α	2.3×10^{15}	^{140}Ce
^{146}Sm	trace	α	6.8×10^{7}	^{142}Nd
^{147}Sm	14.99	α	1.06×10^{11}	^{143}Nd
^{148}Sm	11.24	α	7×10^{15}	$^{144}Nd \rightarrow ^{140}Ce$

(Continued)

TABLE 3.2 *(Continued)* DECAY MECHANISMS AND HALF-LIVES

Isotope	% of Element	Decay Mechanism	Half-Life (years)	Decay Product
^{150}Nd	5.6	2β	6.7×10^{18}	^{150}Sm
^{151}Eu	47.81	α, β	5×10^{18}	$^{147}Pm \rightarrow ^{147}Sm$
^{152}Gd	0.2	2α	1.08×10^{14}	$^{148}Sm \rightarrow ^{144}Nd$
^{156}Dy	0.0524	α	2×10^{14}	^{152}Gd
^{174}Hf	0.162	α	2.0×10^{15}	^{170}Yb
^{176}Lu	2.59	β	3.85×10^{10}	^{176}Hf
^{180}W	0.12	α	1.8×10^{18}	^{176}Hf
^{186}Os	1.89	α	2×10^{15}	^{182}W
^{187}Re	62.6	99.99% β 0.01% α	4.12×10^{10}	^{187}Os ^{183}Ta
^{190}Pt	0.014	α	6.5×10^{11}	^{186}Os
^{204}Pb	1.35	α	1.4×10^{17}	^{200}Hg
^{209}Bi	100	α	1.9×10^{19}	^{205}Tl
^{232}Th	100	6α + 4β	1.4010×10^{10}	^{208}Pb (final)
^{235}U	0.7201	7α + 4β	7.0381×10^{8}	^{207}Pb (final)
^{238}U	99.2743	8α + 6 β 5.4×10^{-5}% fission	4.4683×10^{9}	^{206}Pb (final)

Note: α, alpha particle emission; β, electron emission; $β^+$, positron emission; K, electron capture [normally from the innermost (K) shell of orbital electrons].

3.5 SOME EXTINCT ISOTOPES

TABLE 3.3 EXTINCT ISOTOPES WITH DECAY PRODUCTS THAT ARE IDENTIFIABLE IN METEORITES OR PROVIDE ISOTOPIC CLUES TO EARLY SOLAR SYSTEM PROCESSES

Isotope	Decay Mechanism	Half-Life (years)	Decay Product
^{22}Na	$β^+$	2.603	^{22}Ne
^{26}Al	85% $β^+$ 15% K	7.17×10^{5}	^{26}Mg
^{60}Fe	2β	3×10^{5}	^{60}Ni via ^{60}Co

(Continued)

TABLE 3.3 *(Continued)* EXTINCT ISOTOPES WITH DECAY PRODUCTS THAT ARE IDENTIFIABLE IN METEORITES OR PROVIDE ISOTOPIC CLUES TO EARLY SOLAR SYSTEM PROCESSES

Isotope	Decay Mechanism	Half-Life (years)	Decay Product
^{107}Pd	β	6.5×10^5	^{107}Ag
^{129}I	β	1.6×10^7	^{129}Xe
^{146}Sm	α	1.03×10^8	^{142}Nd
^{182}Hf	2β	8.9×10^6	^{182}W via ^{182}Ta
^{236}U	α	2.4×10^7	^{208}Pb via ^{232}Th
^{244}Pu	0.3% fission 99.7% α	8.3×10^7	0.3% fission products 99.7% ^{208}Pb via ^{232}Th

3.6 SHORT-LIVED ISOTOPES

TABLE 3.4 ISOTOPES PRODUCED BY COSMIC RAYS OR RADIOACTIVE DECAY

Isotope	Decay Mechanism	Half-Life	Decay Product
Neutron	β	611 s (10.18 minutes)	^{1}H
^{3}H, tritium	β	12.32 years	^{3}He
^{7}Be	K	53.22 days	^{7}Li
^{10}Be	β	1.39×10^6 years	^{10}B
^{14}C	β	5730 years	^{14}N
^{22}Na	β	2.603 years	^{22}Ne
^{32}Si	2β	153 years	^{32}S via ^{32}P
^{32}P	β	14.263 days	^{32}S
^{33}P	β	25 days	^{33}S
^{35}S	β	87.5 days	^{35}Cl
^{36}Cl	98.1% β 1.9% K	3.01×10^5 years	^{36}Ar ^{36}S
^{37}Ar	K	35 days	^{37}Cl

(Continued)

TABLE 3.4 *(Continued)* ISOTOPES PRODUCED BY COSMIC RAYS OR RADIOACTIVE DECAY

Isotope	Decay Mechanism	Half-Life	Decay Product
^{39}Ar	β	270 years	^{39}K
^{41}Ca	K	1.02×10^5 years	^{41}K
^{53}Mn	K	3.7×10^6 years	^{53}Cr
^{234}U	α	2.47×10^5 years	$^{230}Th \rightarrow ^{206}Pb$

3.7 FISSION PRODUCTS

Nuclear fission, principally of ^{235}U and ^{239}Pu, results in fragments with unequal atomic masses in ranges 85–110 and 125–155. Statistics of the fragments depend somewhat on the fissioning isotopes and on conditions such as energies of incident neutrons. Each fission event also produces two or three neutrons and, in a few cases, other small fragments such as tritium (^3H). The products are all neutron-rich, making them radioactive emitters of β particles (energetic electrons). There is a very wide range of radioactive fission products. Table 3.5 lists the ones of particular environmental concern on account of their abundances and likelihood of ingestion.

Releases from nuclear accidents also include isotopes of the actinides (elements close to thorium, uranium and plutonium in the periodic table), as well as unfissioned ^{235}U and ^{239}Pu (half-life 24,100 years).

The total intensity of radiation from fission products, R, decreases with time as a sum of numerous exponential decays, with complications arising from secondary decays. A rough empirical representation for fallout radiation at an open site, derived largely from Chernobyl data, is

$$R = R_1 t^{-1/2} \tag{3.1}$$

where t is time in days, starting at day 1, when the intensity was R_1. This simple equation is a useful approximation for $t = 1$ day to 10,000 days (27.4 years). In the

TABLE 3.5 ENVIRONMENTALLY PROBLEMATIC FISSION PRODUCTS

Isotope	^{90}Sr	^{137}Cs	^{89}Sr	^{3}H	^{140}Ba	^{131}I
Half-life	28 years	30 years	15 days	13 years	128 days	8.05 days
Ingestion	30%	~100%	30%	High[a]	5%	100%

[a] Readily absorbed but also generally quickly excreted.

much longer term, this relationship will fail because fission products all have half-lives of either less than 100 years or more than 200,000 years and the half-life distribution of still active isotopes will change dramatically.

3.8 RADIOGENIC HEAT

TABLE 3.6 THERMALLY IMPORTANT RADIOACTIVE ISOTOPES

Isotope	μW/kg	μW/kg of Element	Total Earth Content (kg)	Heat (10^{12} W)		
				Now	4.5×10^9 years ago	In 10^9 Years' Time
^{238}U	95.0	94.35	15.02×10^{16}	14.25	28.6	12.2
^{235}U	562.0	4.05	0.11×10^{16}	0.60	50.1	0.22
^{232}Th	26.6	26.6	55.98×10^{16}	14.87	18.6	14.1
^{40}K	30.0	0.00350	8.06×10^{20} (total K)	2.82	34.2	1.6
	Total heat			31.2	132	28.1

TABLE 3.7 AVERAGE RADIOGENIC HEAT IN EARTH MATERIALS

Material		Concentration (ppm by mass)			K/U	Heat (10^{-12} W/kg)
		U	Th	K		
Igneous rocks	Granites	4.6	18	33,000	7000	1050
	Alkali basalts	0.75	2.5	12,000	16,000	180
	Tholeiitic basalts	0.11	0.04	1500	13,600	27
	Eclogites	0.035	0.15	500	14,000	9.2
	Peridotites	0.006	0.02	100	17,000	1.5
Meteorites	Carbonaceous Chondrites	0.020	0.070	400	20,000	5.2
	Ordinary Chondrites	0.015	0.046	900	60,000	5.8
	Iron Meteorites	Nil	Nil	Nil	–	$<3 \times 10^{-4}$
Moon	Apollo samples	0.23	0.85	590	2500	47
Global averages	Crust	1.2	4.5	15,500	13,000	293
	Mantle	0.029	0.109	81	2800	5.7
	Core	0	0	29	–	0.1
	Whole Earth	0.025	0.093	135	5400	5.2

Chapter 4

The Solar System

4.1 PHYSICS (TABLES 4.1 THROUGH 4.3)

TABLE 4.1 MECHANICAL PROPERTIES

Body	Orbit Radius, R (AU)[a]	Orbital Period (tropical years)	Orbital Eccentricity[b] (e)	Rotation Period (days)	Axis Tilt (degrees)	Mean Radius (earth radii)[c]	Mass (earth masses)[d]	Known Satellites
Sun				27 (av.)[e]		109.2	332 946.8	
Mercury	0.387098	0.240851	0.205630	58.646	2.1	0.3829	0.055284	0
Venus	0.723327	0.615198	0.00677	243.02	177.3	0.9499	0.81500	0
Earth	1	1.0000388 (1 sidereal year)	0.0167102	0.9972697	23.45	1	1	1
Moon	0.002570 (about Earth)[f]	Rotational period	0.0549	27.32166		0.2728	0.0123000	
Mars	1.523679	1.8808	0.093315	1.025957	25.2	0.5320	0.10745	2
Asteroids	2.8 (av.)	3.9 (av.)		Various			<0.0002	
Jupiter	5.2033630	11.85678	0.0483927	0.41354	3.12	10.981	317.82	67
Saturn	9.5370703	29.42415	0.0541506	0.43777	26.7	9.168	95.161	62
Uranus	19.191264	83.74920	0.0471677	0.71833	97.9	3.977	14.371	27
Neptune	30.068963	163.7267	0.0085859	0.67125	29.6	3.861	17.147	14
Pluto	39.482	248.03	0.2488	6.3872	120	0.186	0.00218	5
Kuiper belt	~40–1000							
Oort cloud	~75,000–150,000							

[a] Semi-major axis.

[b] $e = (1-b^2/a^2)^{1/2}$ where b, a are the semi-minor and semi-major axes, respectively.

[c] $R_{Earth} = 6.3710\times10^6$ m.

[d] $M_{Earth} = 5.9724\times10^{24}$ kg.

[e] latitude dependent, 25–32 days, slowest at poles.

[f] mean distance 3.8440×10^8 m, increasing by 3.7 cm/year (see Figure 9.6).

TABLE 4.2 PHYSICAL PROPERTIES

Body	Mean Density (kg m^{-3})	Decompressed Density (kg m^{-3})	Surface Gravity[a] (m s^{-2})	Escape Velocity[a] (km s^{-1})	Surface Magnetic Field (nT)	Net Energy Output (W)	Albedo[b]
Sun	1408	Gaseous	274	618	10^5–10^7	3.829×10^{26}	
Mercury	5429	5017	3.703	4.25	430	$\sim1.5\times10^{12}$	0.119
Venus	5243	3868	8.870	10.36	0	$\sim1.7\times10^{13}$	0.90
Earth	5513.7	3995	9.780	11.19	41,000	4.7×10^{13}	0.306
Moon	3340.5	3269	1.617	2.38	~1	$\sim7\times10^{11}$	0.123
Mars	3933.5	3697	3.712	5.02	~3.5	$\sim3\times10^{12}$	0.25
Asteroids[c]	~3700	~3700					Various
Jupiter	1326	Gaseous	22.9	59.54	595,000	8×10^{17}	0.343
Saturn	687	Gaseous	11.15	35.49	29,000	8×10^{16}	0.342
Uranus	1270	Gaseous	8.69	21.29	32,000	4×10^{14}	0.300
Neptune	1638	Gaseous	11.15	23.71	20,000	4×10^{15}	0.290
Pluto	~2090	~2000	0.66	1.27			~0.5

[a] Equatorial values.
[b] Bond albedo, the total fraction of incident radiation that is reflected in all directions.
[c] Average of observed meteorite falls.

TABLE 4.3 SUN, MOON AND TERRESTRIAL PLANETS: ADDITIONAL DETAILS

	Central P (GPa)	Moment of Inertia I/MR^2	M_{core}/M_{planet}	R_{core}/R_{planet}	Mean Surface Temp. (K)
Sun	2.34×10^7	0.059			5778
Mercury	39	0.346	0.677	0.828	440
Venus	286	0.336	0.286	0.522	737
Earth	364	0.3307	0.326	0.546	288
Moon	5.9	0.3935	~0.02	~0.2	278
Mars	43	0.366	0.156	0.422	210

4.2 CHEMISTRY (TABLES 4.4 THROUGH 4.7)

TABLE 4.4 ABUNDANCES OF ATMOSPHERIC CONSTITUENTS OF TERRESTRIAL PLANETS AS FRACTIONS OF PLANETARY MASSES

Components	Venus (ppb)	Earth (ppb)	Mars (ppb)
N_2	2278	664	6.8
O_2	~0.3	203	0.37
H_2O	~3	Up to 22	Up to 0.035
Ar	6.2	11.3	5.8
CO_2	98,720	0.55	377
Ne	0.33	0.011	4.5×10^{-4}
He	0.11	6.4×10^{-4}	~0
Kr	0.005	0.0028	2.3×10^{-4}
H_2	Trace	3.4×10^{-5}	Trace
CO	2.0	1.1×10^{-4}	0.20
Xe	$\sim6\times10^{-4}$	4.8×10^{-4}	1.3×10^{-4}
SO_2	23	$\sim1\times10^{-7}$	~0
CH_4	~0	0.015	~0
Atmosphere mass/planet mass (ppb)	1.01×10^5	879	390
Mean mol. wt.	43.45	28.97	43.34
Properties:			
Surface pressure (10^5 Pa)	95	1.01	0.064
Surface gravity (m s^{-2})	8.87	9.78	3.69
Gravitational potential (10^7 m^2 s^{-2})	-5.369	-6.258	-1.264

Note: For the Earth, values are listed as volume fractions of the atmosphere in Table 12.2. Mercury and the Moon have only transient traces.

TABLE 4.5 VOLUME PERCENTAGES OF CONSTITUENTS IN ATMOSPHERES OF GIANT PLANETS

Constituent	Jupiter	Saturn	Uranus	Neptune
H_2	89.8	96.3	82.5	80
He	10.2	3.2	15.2	19
CH_4	0.3	0.4	2.3	1.5
CO		~1 ppb		0.6 ppm
NH_3	0.026	50 ppm		
H_2O	30 ppm			

TABLE 4.6 ISOTOPIC RATIOS (BY MASS) IN PLANETARY ATMOSPHERES

	Venus	Earth	Mars	Jupiter	Saturn	Uranus	Neptune
D/H	0.032	3.115×10^{-4}	15.6×10^{-4}	5.2×10^{-5}	3.4×10^{-5}	$\sim1.8\times10^{-4}$	$\sim2.4\times10^{-4}$
$^{13}C/^{12}C$	0.012	0.0122	0.012	0.012	~0.012		
$^{18}O/^{16}O$	0.0022	0.002256	0.0023				
$^{36}Ar/^{40}Ar$	0.93	0.00304	0.00030				
$^{15}N/^{14}N$	0.0039	0.00394	0.0063	0.0086			

Note: For volume ratios in the terrestrial atmosphere, see Table 12.4.

TABLE 4.7 ELEMENT ABUNDANCE COMPARISONS: SOLAR ATMOSPHERE, METEORITES AND EARTH NORMALISED TO THE ABUNDANCE OF SILICON

Element	Sun	Meteorites	Earth
H	1003	0.003	0.002
He	392	8×10^{-9}	5×10^{-9}
O	13.6	2.2	2.22
C	4.4	0.33	0.012
Ne	3.5	trace	1.1×10^{-10}
Fe	2.6	1.81	2.14
N	1.6	0.001	20 ppm
Si	1	1	1
Mg	0.91	0.91	1.09

(Continued)

TABLE 4.7 *(Continued)* ELEMENT ABUNDANCE COMPARISONS: SOLAR ATMOSPHERE, METEORITES AND EARTH NORMALISED TO THE ABUNDANCE OF SILICON

Element	Sun	Meteorites	Earth
S	0.52	0.60	0.20
Ar	0.13	~20 ppb	~20 ppb
Ni	0.105	0.105	0.16
Ca	0.092	0.088	0.12
Al	0.080	0.082	0.11
Na	0.049	0.047	0.013

Section III

Global Geophysics

Chapter 5

Whole Earth Properties

5.1 PLANETARY PARAMETERS

(For properties of parts of the Earth, see Chapters 6 through 12).

Gravitational constant × mass: $GM = 3.986004418 \times 10^{14}$ m^3 s^2 (including atmosphere).

Mass: $M = 5.9724 \times 10^{24}$ kg

Equatorial radius: $a = 6378.137$ km

Polar radius: $c = 6356.751$ km

Mean radius: $R = (2a + c)/3 = 6371.000$ km

Polar flattening: $f = (a - c)/a = 3.353 \times 10^{-3}$

Volume: $V = (4\pi/3)\, a^2c = (4\pi/3)\, R^3 = 1.08320 \times 10^{21}$ m^3

Mean density: $\rho = M/V = 5513.7$ kg m^{-3}

Geoid potential: $W_0 = -6.26 \times 10^7$ m^2 s^{-2}

Moments of inertia

about polar axis $C = 8.0359 \times 10^{37}$ kg m^2

about equatorial axis $A = 8.0096 \times 10^{37}$ kg m^2

Dynamical ellipticity: $H = (C - A)/C = 3.273795 \times 10^{-3} = 1/305.456$

Ellipticity coefficient: $J_2 = (C - A)/Ma^2 = 1.08262982 \times 10^{-3}$

Moment of inertia coefficient: $C/Ma^2 = J_2/H = 0.330698$

Surface areas

Total 5.1006×10^{14} m^2

Land 1.48×10^{14} m^2

Sea 3.62×10^{14} m^2

Continents 2.0×10^{14} m^2 (including margins)

Ocean basins 3.1×10^{14} m^2

Obliquity of the ecliptic: 23.4523°

Orbital eccentricity, $e = (1-b^2/a^2)^{1/2}$ where b, a are the semi-minor and semi-major axes, respectively: 0.01673

Orbital angular velocity: 1.990987×10^{-7} rad s^{-1}

Orbital angular momentum: 2.662×10^{40} kg m^2 s^{-1}

Date of perihelion (closest approach to Sun): ~3 January
Date of aphelion (most remote point): ~4 July
Ratio of most distant and closest approaches: 1.03403
Ratio, max/min insolation: 1.0692
Rate of precession of the equinox: $\omega_p = 50.291''/\text{year} = 7.7260 \times 10^{-12}$ rad s^{-1}
Period of precession: 8.132×10^{11} s = 25,770 years
Rotational angular velocity: $\Omega = 7.292116 \times 10^{-5}$ rad s^{-1}
Sidereal day = $2\pi/\Omega$ = 86164.10 s, increasing by 2.4 μs per year by tidal friction
Surface gravity at latitude φ (International reference gravity formula):

$$g = 9.780327\,(1 + 0.0053024 \sin^2\varphi - 5.8 \times 10^{-6} \sin^2 2\varphi)\ \text{m s}^{-2}$$

(see Equation 5.1 for an alternative, equivalent form)
Ratio, centrifugal force/gravity on equator: $m = \omega^2 a/g = 3.46775 \times 10^{-3}$
Magnetic dipole moment (2015): 7.724×10^{22} A m^2,
  decreasing by 0.05% per year.
Global heat flux: 47×10^{12} W (see also Table 5.2)
Continental heat flux: 13×10^{12} W
Ocean floor heat flux: 34×10^{12} W
Mean continental temperature gradient: 0.025 K m^{-1}
Tidal dissipation,
  solid Earth: ~2.5×10^{11} W
  oceans: 3.5×10^{12} W
Ratio, mass of Sun/mass of Earth = 332946.8
Total solar radiation,
  top of atmosphere: 174 000 TW
  at surface: 98 000 TW
  retained as stored heat: ~200 TW
Solar constant (mean solar radiation at 1 AU): 1361.6 W m^{-2}
Ratio, mass of Earth/mass of Moon = 81.30059
Mean distance of Moon: 3.8440×10^8 m, increasing by 3.7 cm/year (see Figure 9.6)
Mean lunar orbital angular velocity: $\omega_L = 2.661698 \times 10^{-8}$ rad s^{-1}

5.2 GLOBAL ENERGY (TABLES 5.1 AND 5.2)

TABLE 5.1 COMPONENTS OF THE ENERGY OF EARTH FORMATION

Component	Energy (10^{30} J)
Accretion of a homogeneous mass	219.0
Core separation less strain energy	13.9
Inner core formation	0.09
Mantle differentiation	0.03
Stored elastic strain energy	15.8
Radiogenic heat in 4.5×10^9 years	8.4
Radiogenic heat now to $t = \infty$	12.8
Residual stored heat	13.3
Heat loss in 4.5×10^9 years	14.2
Present rotational energy	0.2
Tidal dissipation in 4.5×10^9 years	~1.1
Solar energy received at surface in 4.5×10^9 years	14,000

TABLE 5.2 THE CURRENT GLOBAL ENERGY BUDGET

Energy Components Converted to Heat	10^{12} W
Crustal radioactivity	8.2
Mantle radioactivity	22.8
Core radioactivity	0.2
Energy of core evolution[a]	1.0
Mantle differentiation[a]	0.1
Thermal contraction[a]	3.1
Tidal dissipation	0.1
Total	*35.5*
Heat loss	
Crustal heat	8.2
Mantle heat	34.8
Core heat	4.0
Total	*47.0*
Net loss	*11.5*

[a] Gravitational energies.

5.3 PRECESSION, WOBBLE AND THE MILANKOVITCH CYCLES

If the Earth had perfect spherical symmetry, gravitational interactions with all other bodies would be central, with no angular dependence; the Earth would rotate about an axis fixed in space. The major departure from sphericity is alternatively referred to as the equatorial bulge or polar flattening, the result of a balance between the centrifugal effect of rotation and self-gravitation. Gravitational interactions of the Sun and Moon with the bulge cause precession, which is the biggest and most obvious departure from fixed-axis rotation. The present axis of rotation is inclined to the ecliptic pole (normal to the plane of the orbit) by 23.45°, an angle termed obliquity, and precesses about it with a period of 25,770 years. Referred to the distant stars, the orientation of the pole shifts by almost 47° in 13,000 years. Because the orbit about the Sun is elliptical, there is a consequent climatic effect. The Earth is closest to the Sun around January 3, the peak of the southern hemisphere summer (but is then orbiting fastest, making the summer shorter), but this date is progressively shifted by the precession. The consequent climatic precessional cycle is shorter than the astronomical cycle because of a change in direction of the orbital axis (apsidal precession) and is nearer to 21,600 years. This is one of the astronomically driven Milankovitch climatic cycles. The dynamical ellipticity of the Earth (H in Section 5.1) is determined from the period of the 25,770-year astronomical cycle.

The obliquity (angle of equator to orbital plane) also varies and has a 41,000-year period. The effect of increasing obliquity is to intensify seasonal variations in insolation. Ice core data indicate that in the period between 1 million and 500,000 years ago, this was the dominant control on ice age advances and retreats.

There are two effects that cause climatic variations with periods of about 100,000 years, and this period is most pronounced in ice core data for the last 500,000 years. Theory indicates variations in orbital eccentricity with periods of 413,000, 125,000 and 95,000 years, the last two of these having a beat period of 396,000 years, which is doubtfully distinguishable from 413,000 years. There is also a 100,000-year variation in the inclination of the Earth's orbit relative to the plane of the solar system as a whole. It is not clear which mechanism is responsible for the 100,000-year ice age cycle. Coincidence of Milankovitch periods with ice age cycles is convincing, but the Earth's response is complicated and detailed, quantitative explanations are elusive.

An effect of the Earth's ellipticity that arises internally is the wobble or, in the name of its discoverer, Chandler wobble. A departure of the rotation axis from the axis of maximum moment of inertia causes a precessional torque, with the resulting motion apparent as a cyclic variation in the latitude of any observatory. It is excited by shifts in the mass distribution of the Earth, now identified as arising from variations in deep ocean currents. Its period (about 430 days) is, like precession, controlled by the dynamical ellipticity, H, and its amplitude averages 0.15 arc seconds. There is a superimposed annual cycle with 0.1 arc second amplitude, driven by the seasonal redistribution of snow and ice. The observed variation in latitude, which is a combination of these effects, has a 6-year beat period.

Independently of these effects, there are variations in the rate of the Earth's axial rotation, generally referred to as the length-of-day (LOD) variations. There is a slow, steady increase in the LOD amounting to 2.4 μs per year caused by tidal friction, the subject of Section 9.3.2. More rapid fluctuations arise from coupling of the mantle to irregular motions in the atmosphere and core, with an annual variation averaging about 1 ms, attributed to deposition of snow and ice in the northern hemisphere. Coupling to motion in the fluid core causes variations of a few milliseconds on a decadal time scale.

5.4 GRAVITY

Rotation and ellipticity cause gravity on the surface of the Earth to differ from the Newtonian formula for a uniform sphere, $g = GM/r^2$, and vary with latitude φ in a manner represented by the International Reference Gravity Formula:

$$g = 9.780327\,(1 + 0.0052792\,\sin^2\varphi + 2.32 \times 10^{-5}\,\sin^4\varphi)\ \mathrm{m\,s^{-2}} \tag{5.1}$$

$$= 9.780327\,(1 + 0.0053024\,\sin^2\varphi - 5.8 \times 10^{-6}\,\sin^2 2\varphi)\ \mathrm{m\,s^{-2}}$$

Departures of local values from this formula, arising from heterogeneities in the Earth and referred to as gravity anomalies, are smaller than this latitude effect by factors exceeding 10 and generally 100 (see Section 5.4.3). The broad-scale features of these departures are calculated from satellite orbits in terms of spherical harmonics. The total gravitational potential at latitude φ, longitude λ and radius r (referred to a standard radius a) is expressed as

$$V = -\frac{GM}{r}\left\{1 + \sum_{l=2}^{\infty}\left(\frac{a}{r}\right)^{1}\sum_{m=0}^{l} \mathrm{p}_l^m(\sin\varphi)\left[\bar{C}_l^m \cos m\lambda + \bar{S}_l^m \sin m\lambda\right]\right\} \tag{5.2}$$

where l, termed the degree, is the number of world-encircling nodes in the surface pattern and m, termed the order, is the number of meridional nodes, so that $(l - m)$ is the number of latitudinal nodes, as in Figure 5.1. The spherical harmonics are conventionally written in terms of colatitude, $\theta = 90° - \varphi$ taken

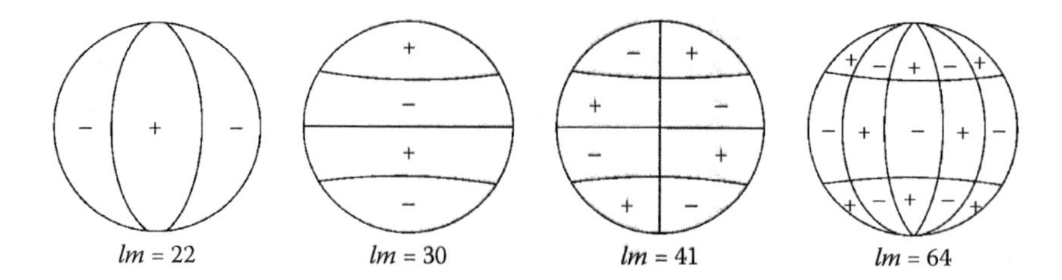

Figure 5.1 Examples of spherical harmonics. $m = 0$ gives zonal harmonics, $m = l$ gives sectoral harmonics and the general case $0 < m < l$ are known as tesseral harmonics. All of the nodal lines are world-encircling.

as zero at the north pole. p_l^m is the fully normalised spherical harmonic given in terms of the corresponding standard harmonic, P_{lm}, by

$$p_l^m(\cos\theta) = \left[(2-\delta_{m,0})(2l+1)\frac{(l-m)!}{(l+m)!}\right]^{1/2} P_{lm}(\cos\theta) \tag{5.3}$$

where the factor $(2 - \delta_{m.0})$ is 2 for $m = 0$ but 1 otherwise, and

$$P_{lm}(\cos\theta) = \frac{\sin^m\theta}{2^l}\sum_{t=0}^{Int[(l-m)/2]}\frac{(-1)^t(2l-2t)!}{t!(l-t)!(l-m-2t)!}\cos^{l-m-2t}\theta \tag{5.4}$$

$\overline{C}_l^m, \overline{S}_l^m$ in Equation 5.2 are coefficients representing different harmonic components of the overall pattern. The significance of the normalisation, using p_l^m rather than P_{lm}, is that the p_l^m are so adjusted that their mean square values over the surface of a sphere are unity and the coefficients $\overline{C}_l^m, \overline{S}_l^m$ give the true magnitudes of the effects they represent.

5.4.1 Algebraic Forms of Spherical Harmonics

Simple algebraic expressions for the first few P_{lm} with numerical factors [in square brackets] that convert to p_{lm} are given in Table 5.3. Note: θ is colatitude, so that $\cos\theta = \sin\varphi$ and $\sin\theta = \cos\varphi$.

5.4.2 Harmonic Coefficients of the Earth's Gravitational Potential

Spherical harmonic coefficients of the Earth's gravitational potential to degree (l) and order (m) 8,8, are listed in Table 5.4. For each (l,m), the coefficients

TABLE 5.3 LOWEST DEGREE HARMONICS

		Order (m)				
		0	**1**	**2**	**3**	**4**
Degree (l)	0	1 [1]				
	1	$\cos\theta$ [$\sqrt{3}$]	$\sin\theta$ [$\sqrt{3}$]			
	2	$(3\cos^2\theta - 1)/2$ [$\sqrt{5}$]	$3\cos\theta \sin\theta$ [$\sqrt{(5/3)}$]	$3\sin^2\theta$ [$\sqrt{(5/12)}$]		
	3	$(5\cos^3\theta - 3\cos\theta)/2$ [$\sqrt{7}$]	$(3/2)(5\cos^2\theta - 1)\sin\theta$ [$\sqrt{(7/6)}$]	$15\cos\theta \sin^2\theta$ [$\sqrt{(7/60)}$]	$15\sin^3\theta$ [$\sqrt{(7/360)}$]	
	4	$(35\cos^4\theta - 30\cos^2\theta + 3)/8$ [3]	$(5/2)(7\cos^3\theta - 3\cos\theta)\sin\theta$ [$\sqrt{(9/10)}$]	$(15/2)(7\cos^2\theta - 1)\sin^2\theta$ [$\sqrt{(1/20)}$]	$105\cos\theta \sin^3\theta$ [$\sqrt{(1/280)}$]	$105\sin^4\theta$ [$\sqrt{(1/2240)}$]

TABLE 5.4 GRAVITATIONAL POTENTIAL COEFFICIENTS TO DEGREE (l) AND ORDER (m) 8,8

l/m	0	1	2	3	4	5	6	7	8
2	−484.165	-	2.439	-	-	-	-	-	-
	-	-	−1.400	-	-	-	-	-	-
3	0.957	2.030	0.905	0.721	-	-	-	-	-
	-	0.249	−0.619	1.414	-	-	-	-	-
4	0.540	−0.536	0.351	0.991	−0.189	-	-	-	-
	-	−0.473	0.663	−0.201	0.309	-	-	-	-
5	0.069	−0.062	0.652	−0.452	0.295	0.175	-	-	-
	-	−0.094	−0.323	−0.215	0.050	−0.669	-	-	
6	−0.150	−0.076	0.048	0.057	−0.086	−0.267	0.010	-	-
	-	0.026	−0.374	0.009	−0.471	−0.536	−0.237		
7	0.091	0.280	0.323	0.250	−0.275	0.002	−0.359	0.001	-
	-	0.095	0.093	−0.217	−0.124	0.018	0.152	0.024	-
8	0.050	0.023	0.080	−0.019	−0.245	−0.026	−0.066	0.067	0.124
	-	0.059	0.065	−0.086	0.070	0.089	0.309	0.075	0.121

Note: All values in units 10^{-6}.

are $\bar{C}_l^m$ followed by $\bar{S}_l^m$, in units of 10^{-6}. There are no $\bar{S}_l^m$ terms for $m = 0$. Note that the first term ($l = 2$, $m = 0$) represents the ellipticity, which is given by $J_2 = 1.082626 \times 10^{-3}$ in Section 5.1, and that this differs from C_2^0 by the factor $-\sqrt{5}$.

5.4.3 Gravity Anomalies

Local departures of observed gravity from the international formula for the latitude variation (Equation 5.1) are referred to as gravity anomalies. There are three principal ways of representing them, each giving somewhat different information about the density variations within the Earth. All of them refer to the geoid, a hypothetical surface of constant gravitational potential roughly corresponding to sea level. In continental areas, this can be envisaged as the water level in imaginary narrow canals connected to the oceans. From a global perspective, the most fundamental representation is the geoid itself, but this was not well observed before satellite measurements. Figure 5.2 is a contour plot of the undulations of the geoidal surface. The highs and lows of this broad-scale plot reflect the deep distribution of density variations. They do not correspond to large surface features of the Earth, such as the distribution of continents and ocean basins, which do not show because they are gravitationally balanced. This is the principle of isostasy, illustrated for the crust in Figure 8.1.

The two methods of plotting gravity anomalies to infer geological structure extrapolate to the geoid from observations on an elevated land surface, using different measures of the gravity gradient. One uses the free-air gradient, that is, the vertical variation of gravity above ground, assuming this gradient to apply down to the geoid. The standard value, which is generally assumed, is 0.3086 mGal/m (3.086 s^{-2}), and variations are much smaller than this. Observations adjusted in this way yield what are termed free-air anomalies. They effectively assume that all of the material below a point of observation is collapsed down to the geoid, so that it is included in the local measure of observed gravity on the reference surface. The other method, attributed to M. Bouguer, uses a smaller gradient, by discounting an assumed uniform slab of rock between the surface and the geoid. Usually a standard rock density, 2670 kg m^{-3}, is assumed and subtraction of its gravitational effect from the free-air gradient gives the Bouguer gradient, 0.1967 mGal/m. The Bouguer anomalies obtained by adding this gradient to the observations are generally favoured in local exploration for minerals because they most closely reflect the geology of the immediate area, but free-air anomalies better represent broader scale features such as continental coastlines. For interpretation of gravity observations at sea (obtained using survey vessels with stabilised platforms),

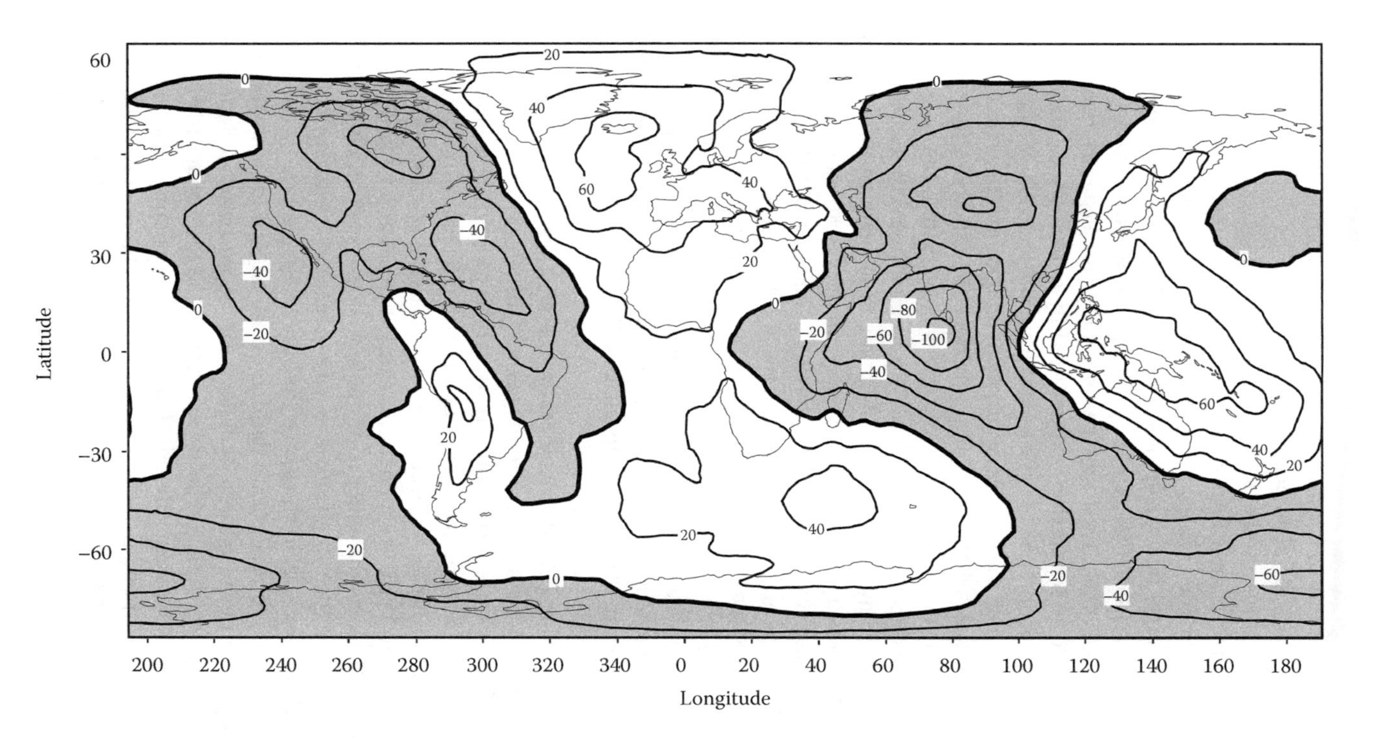

Figure 5.2 Departures of the geoid, the reference surface of constant gravitational potential, from that of a uniform elipsoid. Numbers on contours are heights in metres with negative areas shaded.

a standard procedure is to use an inverted Bouguer method by computationally replacing the sea with sea-floor sediment.

Broad-scale features of the Earth's gravity, as displayed in Figure 5.2, have been obtained primarily from satellite observations, with reliance on surface data for smaller scale anomalies. Satellite methods were extended to the smaller features by development of the GRACE (Gravity Recovery and Climate Experiment) satellites. These are 'identical twins' following the same polar orbit at about 500 km altitude with an along-path separation averaging 220 km. Their precise separation is monitored by a microwave link between them and adjusted from time to time to maintain suitable relative positions. It increases slightly when the first one approaches a region of higher than average gravity and decreases when the second one does so, although these changes are smeared out by the fact that the altitude is greater than their separation. The orientation of the orbit remains fixed in space as the Earth rotates and the second twin passes any point in the orbit about 28 seconds after the first one. Since the Earth turns by 7 arcminutes in that time, it corresponds to a displacement of 13 km between the orbital paths at the equator, but that is a minor effect. The orbital period is about 90 minutes and successive passes are separated by about 2500 km (22.5°) on the ground, so that there are 480 passes each way (N-S and S-N) in a 30-day period, or 960 passes counting both directions. Their average separation is 42 km at the equator and less at the poles, giving sufficient detail for monthly models of the Earth's gravity to be developed by unravelling the record of variations in the satellite separation. These numbers indicate the resolution that is possible.

Averages over long periods give details of static gravity anomalies, confirming, as examples, that the Himalayas and Indonesia are gravity highs and the Indian Ocean and Hudson Bay are gravity lows. However, the time variations, both trends and transients, apparent in the monthly models are of greater interest. Ice losses by Greenland, Antarctica and major glaciers and consequent sea level rise are a particular target. Since the GRACE observations measure these as movements of mass, they are partially offset by isostatic adjustments. The rate of sea level rise measured by satellite altimetry (Section 9.1.1) is absolute, in the sense that the Earth's centre of mass is the ultimate reference point. The rate deduced from GRACE data is systematically slightly lower because there is an isostatic response to sea-floor depression and, similarly, the losses of land-based ice are slightly underestimated by GRACE. Without allowing for this effect, ice mass losses, commonly expressed in gigatonnes, correspond to global sea level rises of 0.276 mm/100 Gt, and it is convenient to represent all major water movements as equivalent millimetres of sea level. With this convention, Greenland ice melting contributes 0.83 mm/year to sea level, but Antarctica contributes only ~0.4 mm/year and major glaciers 0.3 mm/year. Water is held on land in various ways, some being seasonal, particularly the winter build-up of snow and ice in the northern hemisphere, which accounts for ~1 mm sea level equivalent, but there are significant longer term effects. The pumping of groundwater for domestic, industrial and agricultural use is widespread in arid areas, with progressive depletion of major aquifers that have limited prospects of naturally recharging. GRACE observations identify Arabia, the Indus basin, North Africa and California as areas of concern because local societies depend on continued availability of groundwater. In spite of this, over the last decade, the global total land storage of water has been increasing due to increased rainfall over wetter areas, with a noticeable rise in the levels of the North American great lakes. The net store of water on (and in) land has increased by ~0.7 mm/year sea level equivalent for several years, and this must be seen as a transient reduction in the rate of sea level rise.

An observation of fundamental interest is the isostatic rebound of formerly glaciated areas, especially Fennoscandia, centred on the Gulf of Bothnia, and Laurentia, centred on Hudson Bay. This is a long-term, ongoing process studied by the classical method of measuring the rise of former shorelines, but now subject to direct observation with GRACE data as well as GPS measurements. GPS records from the Hudson Bay area (Sella et al. 2007) show ground-level rises of order 1 cm/year in a wide area around the bay, but GRACE data indicate mass increases east and west of the bay clearly greater than over the area of the bay itself. This is not an indication of separated centres of rebound but is due to expulsion of water from the bay by the rising sea floor. Correction for it shows that Hudson Bay is indeed the centre of Laurentide rebound, an indication of the care required in detailed interpretation of GRACE observations.

5.5 THE GEOMAGNETIC FIELD

5.5.1 The Main Field

Long before there was any understanding of magnetism or magnetic fields, compasses made of lodestone (magnetite) were used for navigation, with the supposition that they were oriented by an influence of the celestial poles. The lodestone magnets were also affected by one another and so were understood to have poles themselves, designated either north or south according to the orientations adopted when they were freely suspended. This remains a convenient way of referring to the polarities of magnets, and the original concept of poles has led to wider use of the words 'polarity' and 'polarise'. Recognition that the Earth itself is a great magnet, and that the celestial poles are not relevant to the alignment of a compass, originated in the late 16th century with the observation by William Gilbert and Robert Norman that the pattern of the magnetic field of a lodestone sphere was similar to that of the Earth. By then, navigators had maps showing magnetic declination, that is, the angle between magnetic and geographic north, and a progressive change of the declination in London had been noticed. These observations came together in a systematic way with the work of Edmund Halley (of comet fame) almost 100 years later. From surveys of magnetic declination over wide ocean areas, Halley found a field pattern departing from that of a simple (dipole) magnet, with features that slowly drifted westwards. His interpretation was that, within the Earth, there were layers rotating slightly more slowly than the outer part and carrying with them embedded magnets. This was remarkably prescient of the theory, developed more than 300 years later, involving differential rotation within the fluid core, where the field is generated. Rigorous confirmation that the field originates within the Earth was presented in 1838 by C. F. Gauss, using a mathematical development in the form of the equations in Section 5.5.2. But, although Faraday's discovery of electromagnetic induction was published in 1832, the idea that the magnetic field was driven by deep electric currents only emerged in about 1900 from the insight of L. A. Bauer at the Carnegie Institution in Washington, DC. Even then, an understanding of how the currents could be generated began to develop only in the 1940s with the work of W. M. Elsasser.

As now understood, the geomagnetic field is driven by complex motion in a fluid electrical conductor (the liquid iron alloy in the outer core, below 2900 km depth). Motion of the conductor through a field generates a current, which causes an additional field. Motion with respect to that causes a further current, which, with suitable geometry, reinforces the original field. But no initial field needs to be assumed because the state of zero field is unstable and the process is spontaneously self-exciting. The ultimate driving force is convective motion in the core, but there is a mutual control of the motion and the field which it causes. The principles are outlined by Merrill et al. (1996), but detailed modelling is necessarily numerical and, although several models represent the behaviour of the field reasonably well, a complete unambiguous theory is out of reach. Such homogeneous dynamo processes are ubiquitous in the universe; the Sun and most of the planets have them (Table 5.5), and there are galactic and intergalactic fields in the very tenuous conducting plasma of space.

5.5.2 Spherical Harmonic Representation

The Earth's magnetic field is represented by a magnetic potential, V_m, that is expressed in spherical harmonics, almost the same as those used for

TABLE 5.5 MAGNETIC FIELDS OF PLANETS		
Planet	**Mean Surface Field Strength (nT)**	**Cause**
Mercury	430	Weak dynamo
Venus	None detected	Probably none at all
Earth	41,000	Dynamo
Moon	~1	Magnetised rock
Mars	~3.5	Magnetised rock
Jupiter	595,000	Dynamo
Saturn	29,000	Dynamo
Uranus	32,000	Dynamo
Neptune	20,000	Dynamo

gravitational potential in Section 5.4. The only difference between gravitational and magnetic harmonics is in their normalisations. Gravitational harmonics are fully normalised. In geomagnetism, no normalisation is applied to zonal harmonics of zero order ($m = 0$), but for tesseral and sectoral harmonics ($m \neq 0$) the polynomials P_l^m are normalised by applying the factor $\left[(2-\delta_{m,0})(l-m)!/(l+m)!\right]^{1/2}$, which gives them the same rms values over a spherical surface as the zonal harmonics of the same degree, l. The total potential is

$$V_m = \frac{a}{\mu_0}\sum_{l=1}^{\infty}\left(\frac{a}{r}\right)^{l+1}\sum_{m=0}^{l}\left(g_l^m \cos m\lambda + h_l^m \sin m\lambda\right)P_l^m(\cos\theta) \tag{5.5}$$

where g_l^m and h_l^m are coefficients that give the magnitudes of the harmonic terms, conventionally expressed in nanotesla, as in Table 5.6. The magnetic potential, V_m, is not observable but its derivative in any direction gives the field component in that direction. Components may be measured and recorded northwards (X), eastwards (Y) and upwards (Z), but more usually the total horizontal component, $H = \sqrt{(X^2 + Y^2)}$ and its departure from geographic north (declination), $D = \arctan(Y/X)$, are recorded. For full normalisation, the rms field strength of any harmonic term is given by $\left(B_l^m\right)_{\mathrm{rms}} = (l+1)^{1/2}\left(g_l^m, h_l^m\right)$. For the axial dipole component, $g_l^0 = 29422.0$ nT is the field strength on the magnetic equator. The total dipole field strength is $\left[\left(g_1^0\right)^2 + \left(g_1^1\right)^2 + \left(h_1^1\right)^2\right]^{1/2} = 29868\,\mathrm{nT}$. These values correspond to dipole moments of 7.6091×10^{22} A m^2 and 7.7245×10^{22} A m^2, respectively.

TABLE 5.6 HARMONIC COEFFICIENTS OF THE IGRF 2015

	$m = 0$	1	2	3	4	5	6	7	8
$l = 1$	-29442.0	-1501.0							
	10.3	*18.1*							
		4797.1							
		-26.6							
2	-2445.1	3012.9	1676.7						
	-8.7	*-3.3*	*2.1*						
		-2845.6	-641.9						
		-27.4	*-14.4*						
3	1350.7	-2352.3	1225.6	582.0					
	3.4	*-5.5*	*-0.7*	*-10.1*					
		-115.3	244.9	-538.4					
		8.2	*-0.4*	*1.8*					
4	907.6	813.7	120.4	-334.9	70.4				
	-0.7	*0.2*	*-9.1*	*4.1*	*-4.3*				
		283.3	-188.7	180.9	-329.5				
		-1.3	*5.3*	*2.9*	*-5.2*				
5	-232.6	360.1	192.4	-140.9	-157.5	4.1			
	-0.2	*0.5*	*-1.3*	*-0.1*	*1.4*	*3.9*			
		47.3	197.0	-119.3	16.0	100.2			
		0.6	*1.7*	*-1.2*	*3.4*	*0.0*			
6	70.0	67.7	72.7	-129.9	-28.9	13.2	-70.9		
	-0.3	*-0.1*	*-0.7*	*2.1*	*-1.2*	*0.3*	*1.6*		
		-20.8	33.2	58.9	-66.7	7.3	62.6		
		0.0	*-2.1*	*-0.7*	*0.2*	*0.9*	*1.0*		
7	81.6	-76.1	-6.8	51.8	15.0	9.4	-2.8	6.8	
	0.3	*-0.2*	*-0.5*	*1.3*	*0.1*	*-0.6*	*-0.8*	*0.2*	
		-54.1	-19.5	5.7	24.4	3.4	-27.4	-2.2	
		0.8	*0.4*	*-0.2*	*-0.3*	*-0.6*	*0.1*	*-0.2*	
8	24.2	8.8	-16.9	-3.2	-20.6	13.4	11.7	-15.9	-2.0
	0.2	*0.0*	*-0.6*	*0.5*	*-0.2*	*0.4*	*0.1*	*-0.4*	*0.3*
		10.1	-18.3	13.3	-14.6	16.2	5.7	-9.1	2.1
		-0.3	*0.3*	*0.1*	*0.5*	*-0.2*	*-0.3*	*0.3*	*0.0*

5.5.3 Geomagnetic Reference Field

Table 5.6 gives harmonic coefficients of the International Geomagnetic Reference Field (IGRF). For each degree (l) and order (m), the values are g_l^m in nanotesla (nT), followed *in italics* by its rate of variation in nT/year and then h_l^m with its rate of variation. There are no h terms for $m = 0$ because $\sin m\lambda = 0$.

5.5.4 Secular Variation

In the context of geomagnetism, the expression 'secular variation' means the progressive change in the pattern of the internally generated field, represented by the coefficients with italic numbers in Table 5.6. The parameters g_l^m, h_l^m and their italicised time derivatives are not fully normalised in the manner of harmonics used in representing the gravitational field in Table 5.4. The rms value of the surface magnetic field strength of any harmonic component is $(l + 1)^{1/2}$ times the value of its coefficient in Table 5.6. Combining all features of the same scale (with the same value of l), the mean square surface field strengths of the harmonic degree fields, that is, the sums of all harmonics with the same degree, l, are given by

$$R_l = (l+1)\sum_{m=0}^{l}\left[\left(g_l^m\right)^2 + \left(h_l^m\right)^2\right] \tag{5.6}$$

It is convenient to refer to an analogous quantity for the time variations, $\dot{g}$ and $\dot{h}$

$$Q_l = (l+1)\sum_{m=0}^{l}\left[\left(\dot{g}_l^m\right)^2 + \left(\dot{h}_l^m\right)^2\right] \tag{5.7}$$

Then, the ratio of these quantities gives reorganisation times for the harmonic degree fields

$$\tau_l = \left(R_l / Q_l\right)^{1/2} \tag{5.8}$$

R_l, Q_l and τ_l are listed in Table 5.7 for the 2015 field coefficients in Table 5.6.

The reorganisation times vary in the manner expected for the sizes of core current loops required to produce the harmonics of the field. They are of the order 10 times shorter than the magnetic diffusion times for the corresponding current loop dimensions, demonstrating that the secular variation is controlled primarily by variations in core motion (the frozen flux principle) and not by field diffusion. It appears significant that the trend in reorganisation times is systematically shorter for the terms in the series with even values of l than for those with odd values (which include the dipole field, $l = 1$). The dipole field is clearly above the trends of the other harmonics, in both strength and slowness of its change, justifying the separate identifications of the axial dipole field (the g_1^0 term) and the non-dipole field (all the other harmonics together) in discussing the fundamental physics. There is some physical evidence that the distinction between odd and even values of $(l - m)$, that is, field components that are antisymmetric (dipole-type) and symmetric (quadrupole-type),

TABLE 5.7 PROPERTIES OF HARMONIC COMPONENTS OF THE GEOMAGNETIC FIELD FOR 2015

Harmonic Degree, l	rms Strength, $\sqrt{R_l}$ (Equation 5.6) (nT)	Rate of Change, $\sqrt{Q_l}$ (Equation 5.7) (nT/year)	Reorganisation Time, τ_l (Equation 5.8) (years)
1	42,240	47.8	884
2	8896	56.1	159
3	6184	29.3	211
4	3062	27.1	101
5	1287	13.3	97
6	554	9.75	57
7	376	5.58	67
8	159	3.86	41

respectively, about the equator, is more significant than odd and even values of l, but no obvious trends are seen when the data are separated in that way.

Prompted by a report by Edmund Halley (of comet fame), Bullard et al. (1950) compared the global patterns of the non-dipole field for 1907 and 1945 and observed that the entire field drifted westwards at 0.18 degree/year. This is consistent with the secular variation seen at individual sites, notably London, where there is a 400-year record, showing a rotation of the local magnetic field direction about the direction of the inclined dipole field. This is what would be expected from drift of the non-dipole pattern past a fixed point, with a much more slowly varying dipole field, as is indicated by the numbers in Table 5.7. The inclined dipole is the total dipole field, including the equatorial components represented by the g_1^1 and h_1^1 coefficients in Table 5.6 as well as the g_1^0 coefficient for the axial dipole (the component parallel to the rotational axis). However, although it became central to interpretation of geomagnetic observations (and requires explanation, e.g., Stacey and Davis 2008, Section 24.6), the westward drift of the most recent few hundred years is not representative of the longer-term secular variation revealed by palaeomagnetism. This shows the field direction wandering about the axial dipole, not the inclined dipole, and that there is no regular cyclic pattern. Over a few thousand years or more, the average field is that of a dipole, coinciding with a rotational axis. This is the axial dipole principle, the basis of palaeomagnetic measurements of polar wander and continental drift, because, with suitable averaging, the measurements record the motion of any site relative to the geographic pole and not just the magnetic pole.

Palaeomagnetic measurements not specifically concerned with the secular variation average it out by taking samples with a range of ages spanning a few thousand years. Then, for rocks with ages up to 10 million years or so, within the uncertainty of the observations, the average magnetic pole coincides with the present geographic pole, confirming the axial dipole principle, but with a sign ambiguity arising from reversals of the dipole field, the subject of Section 5.5.5. For greater ages, there is an increasing divergence of the magnetic directions indicative of polar wander, typically amounting to about 0.3 degree/million years, and differences between the polar wander paths for different continents present a quantitative basis for studying continental drift, now attributed to plate tectonics.

Since the main field is generated in the core, nearly 3000 km from the surface observations, there is some loss of detail, partly because of the distant perspective but more importantly because the electrical conductivity of the intervening mantle attenuates the time variations. These are related effects because the smallest scale features are the most rapidly varying. An upper bound on the mean mantle conductivity is estimated from the highest observed frequencies of the secular variation. The sharpest change seen with modern observations and analysis methods occurred in 1969, when there was a global change in the rate of secular variation. Compared with other features of the secular variation it was sudden, occurring within about a year, and is referred to as the 1969 geomagnetic jerk. It imposes an upper limit of about 3 S m^{-1} on the mean lower mantle conductivity. This is much higher than the upper mantle conductivity deduced from currents induced by extra-terrestrial disturbances (see Section 5.6), which cannot give information on the more conducting deeper mantle. However, the lower mantle conductivity is characteristic of hot compressed silicate and is lower than that of the metallic core by a factor 10^5.

5.5.5 Geomagnetic Reversals

Reversals of the polarity of the dipole field, seen in the palaeomagnetic record, are distinct in character from the secular variation. Although the dipole field participates in the secular variation, in 'normal' times it remains oriented within about 15° of the rotational axis. The equatorial component of the dipole is comparable in strength to features of the non-dipole field, sharing its transience and reasonably regarded as part of it. When a reversal is about to occur, the dipole is weakened by a factor, generally estimated to be 5 to 10, but an alternative view is that it disappears completely and redevelops in the opposite direction, leaving only the weaker equatorial dipole during a reversal. Several authors have drawn attention to evidence that this is not just a feature of the non-dipole field, but a remnant dipole, which tends to follow certain preferred paths across the equator (close to 75°W and 120°E) during reversals. However, that appears to be a consequence of mantle control of features of the field, with areas of low mantle temperature causing downwelling of core fluid and, by the frozen flux principle, concentrating the vertical field components in those areas, giving the appearance of preferred pole paths (Gubbins 2003).

Detailed records of individual reversals indicate that they take a few thousand years to occur, which is consistent with the electromagnetic relaxation time for the core, and that the weakened field lasts longer than the time required for a reversal in direction.

The statistics of reversals have attracted attention as a possible indicator of the mechanism. If reversals are regarded as independent, random events, then the duration of intervals of constant polarity can be treated as following a Poisson distribution, with the probability decreasing exponentially with duration. Suggestions that immediately after a reversal another one becomes either less likely or more likely lead to the inference that the available data would be fitted better by a gamma distribution. Since this introduces another fitting parameter, it allows a closer fit to any data set without necessarily being significant, and it does not yield consistent results. This indicates that neither the Poisson nor gamma distribution really captures the essence of what provokes or controls reversals. Serious doubt about the relevance of these distributions arises from the fact that they are scaled by the mean duration of intervals of fixed polarity, but this has varied dramatically, from periods with four or five reversals per million years to constant polarity for tens of millions of years. This variability points to mantle control and the interpretation of the preferred reversal pole paths suggests an explanation. In the course of mantle convection, areas of higher or lower temperature at the core–mantle boundary change, on a time scale of many million years, cause varying patches of magnetic flux by changing thermally driven core motion. Although there is no theory of the effect, as a dynamo model by Takahashi et al. (2005) demonstrates, instability of the dynamo could be provoked by such mantle-imposed influences on core motion.

Dating of reversals using igneous rocks is secure enough to be confident of the detailed record for the most recent 6 million years, as in Figure 5.3. For the preceding 100 million years or so, the ocean floors provide a record that confirms the broad features of the rock-derived polarity sequence. Fresh, basaltic oceanic crust at spreading centres is magnetised by the ambient field and carries its remanence towards eventual subduction, leaving the ocean floors with magnetised stripes, so that magnetic surveys record the reversal sequence. For the magnetisation process, and the special conditions for self-reversing remanence, see Section 21.4.3. The ocean floor record extends the reversal record almost to the age of the oldest ocean floor, with dating based on

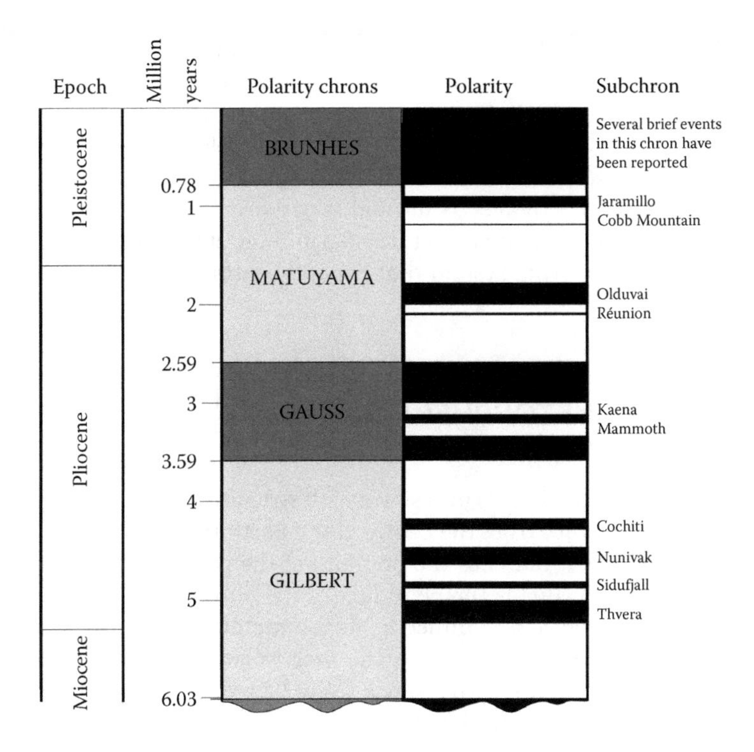

Figure 5.3 The sequence of geomagnetic reversals for the most recent 6 million years, the period for which dating is most reliable.

an assumption that the rate of sea floor spreading has been constant, although not the same for each ocean, and is calibrated by comparison with the data in Figure 5.3. Being the same everywhere, but irregular, the reversal sequence at any time is characteristic of that time. It allows identification of correlations that establish coincidences in time between widely separated geological events, for which sufficiently accurate radiometric dating is not possible. This is magnetic stratigraphy.

There appears to be no correlation between the reversal record and the established geological periods. Reversals are not responsible for or accompanied by biological changes, in particular species extinctions or dramatic climatic effects that could cause them.

In Figure 5.3, intervals with polarity the same as at present (referred to as 'normal') are shown in black and periods of reversed polarity in white. A more extended record shows that the two polarities occur equally often. Long periods of one dominant polarity are termed chrons, with the most recent four, shown in Figure 5.3, named after pioneers in geomagnetism. Within the chrons are shorter subchrons opposite to them, identified by the locations of rocks in which they were first recognised, and there are also brief 'events' and 'excursions' that may be either double reversals or unsuccessful 'attempts'. There are two well-documented superchrons, lasting more than 10 million years, the Cretaceous normal superchron (120–83 million years) and the Kiama reverse superchron (312–262 million years). Reversals are now identified by a numbering system that permits additions and corrections to the record.

5.6 THE MAGNETOSPHERE AND RAPID MAGNETIC VARIATIONS

The geomagnetic field extends into space with strength varying roughly as the inverse cube of distance from the centre of the Earth to the point at which its strength, about 50 nT, matches the pressure of the plasma flowing from the Sun (the *solar wind*). This is the boundary of the *magnetosphere*, the volume to which the Earth's field is confined by its interaction with the solar plasma. On the sunward side, the boundary, the *magnetopause*, is at a distance of ~10 Earth radii, 60,000 km, with a shock front 2 Earth radii further away, but on the downwind side, the field is drawn out into a long tail. Most of the particles from the Sun, the solar wind and low-energy cosmic rays are deflected around the magnetosphere and are prevented from reaching the Earth or lower atmosphere directly, but some sneak into the magnetosphere via its tail and others cross the barrier during disturbances in the solar wind. Fluctuations in the magnetic field, more rapid than the secular variation of the internally generated field (Section 5.5.4), are triggered by activity of the Sun and interpreted in terms of the response of the magnetosphere to impulses by the solar wind. Following is a summary of several effects that can be identified with particular causes, although they are not always independent. Words with technical meanings requiring explanation are italicised.

Diurnal variation. Observed during times of little solar disturbance, this is designated S_q (Solar, quiet) in geomagnetic literature. The strongest variations are in the horizontal component of the field in daylight hours, with peak-to-peak amplitude at mid-latitudes ~75 nT. Compression of the magnetosphere on the daylight side is an obvious cause but greater emphasis is given to a dynamo mechanism driven by ionospheric winds, which generate currents, primarily in the E layer of the ionosphere, by moving the conducting plasma through the main field, which is effectively steady.

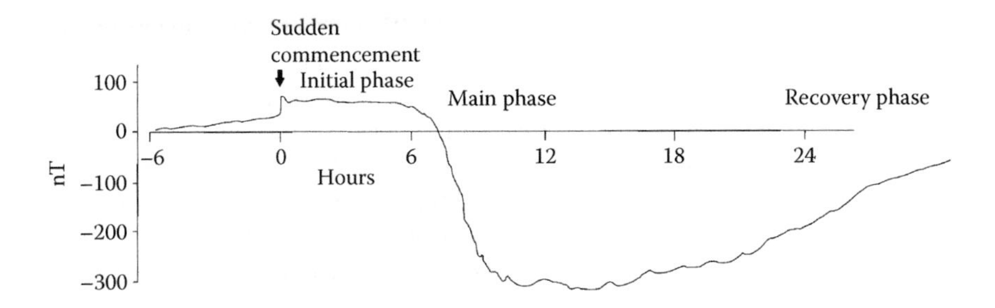

Figure 5.4 Characteristic sequence of field fluctuations during a magnetic storm.

Magnetic storms. These are the major disturbances, generally observed to have four phases, as in Figure 5.4: (i) A *sudden commencement,* coincident over both hemispheres, caused by compression of the magnetosphere, resulting from a sharply increased particle flux from the Sun. The strongest effect on the field is an increase in the horizontal component, typically 100 nT. (ii) An *initial phase,* lasting ~4 hours, driven by a continued strong, but erratic, solar particle flux. (iii) A *main phase* with a reversed field disturbance lasting ~18 hours, typically peaking at 500 nT. This is caused by a *ring current* in the outer magnetosphere, arising from an injection of energetic particles into the *radiation belts* (see below). (iv) A *recovery phase* with return to pre-storm conditions over a few days but commonly interrupted by briefer *substorms.*

Magnetic storms have a wide range of intensities and the stronger ones affect the Earth in several ways that have prompted a continuous watch of the solar surface to give the maximum warning. Some details are given below under *Space weather.* The most dramatic storm on record occurred in 1859 when electrical communications and power grids were seriously affected, although primitive by today's standards, and an understanding of magnetic storms was restricted to a few scientific specialists. The most notable of these was the English astronomer, Richard Carrington, who had kept detailed records of sunspots and solar behaviour generally. Carrington recognised that surges on power lines, sparking from telegraph wires and aurorae extending to low latitudes were all attributable to effects on the Sun and the 1859 storm is referred to as the Carrington event. A comparable event now would cause extensive damage and disruption.

Substorms. These are disturbances caused by an enhanced flux of solar particles, generally weaker than full-blown storms and lacking some of their features, but often accompanying their recovery phases. Increased radiation belt populations and dumping into the auroral zone ionosphere causes aurorae and increased conductivity. The resulting current in the auroral electrojet produces a localised magnetic disturbance.

Equatorial electrojet. Within a few hundred kilometres of the magnetic equator (where the field has no vertical component), disturbances to the horizontal component are enhanced in the daylight hemisphere. This includes the S_q (quiet Sun) variation, with a noticeable lunar component. It arises from the fact that, in the presence of the magnetic field, the conductivity of the ionosphere is anisotropic, with a high conductivity in the horizontal direction on the magnetic equator.

Radiation belts and the *ring current.* In the outer magnetosphere, the density is low enough to make the motions of energetic particles almost collisionless, so that they are controlled by the field. They move freely in the direction of the field and circulate about it, describing spiral paths that effectively enclose 'tubes' of magnetic flux. But the field lines converge towards the poles 'reflecting' the particles, so that they are trapped and 'bounce' to and fro, within their tubes, between the high latitudes. The zones of trapped energetic particles, mostly protons and electrons, constitute the *radiation belts*, often called Van Allen belts after the leader of the group that discovered them. If the field were uniform, the particle motion perpendicular to it would be circular, with curvature proportional to the field strength, that is, radii of the circles increasing as the field strength decreases. But the Earth's field is not uniform. It decreases with distance from the Earth, so that the spiral paths of the particles are tighter near the Earth than further out, causing a drift in longitude, westwards for protons and eastwards for electrons (which circulate in opposite directions), both contributing to a westward current. This is the *ring current*, invoked to explain the main phases of magnetic storms many years before recognition of the radiation belts. It is always present, but is strongly enhanced during storms by injection of additional particles into the radiation belts. On the sunward side, the main radiation belts extend from 1.3 to 2.4 Earth radii and 3.5 to 10 Earth radii.

Ionosphere. This is the lower boundary of the magnetosphere, roughly extending from 60 km altitude to 500+ km, where solar ultraviolet radiation ionises the air. The atmospheric composition varies with altitude (see Table 12.5) and the radiation penetrates to different depths according to wavelength, so that there is a series of ionospheric layers of increasing electron density, designated by letters D (~60–90 km), E (~90–150 km) and F (above 150 km). Ionic recombination is rapid at the lower levels and D more or less disappears at night. In daytime, F may be split into F1 and F2. E_S (sporadic E) is observed irregularly as a strong enhancement of E ionisation that may arise either from particular wind patterns or from unusual ions of vaporised meteors. The sequence of layers is observed from the ground by the increasing frequencies of radio signals that are reflected from them, that is, E reflects frequencies that penetrate D and F reflects even higher frequencies. The F layer is the most important for the ground–ionosphere reflections that allow transmission of radio signals over long distances. The maximum reflected frequency (above which signals escape to space) depends on location and time of day as well as varying ionospheric conditions and the angle of incidence. It is typically about 12 Mz for vertical incidence. *Sudden ionospheric disturbances* (SIDs), caused by X-ray emission from *solar flares* (see below), are observed as strong radio wave absorption by the D layer. The atmosphere at ionospheric levels is partially ionised, with the proportion of neutral atoms decreasing with altitude to virtually zero in the magnetosphere, above about 1000 km. Nevertheless, the density of free electrons is greatest in the ionosphere, giving it higher conductivity, so that electric currents are concentrated in it. But, although the ionosphere is the most conducting part of the magnetosphere, the plasma is very tenuous and the total conductance of the ionospheric layer is less than that of a 1-micron-thick layer of copper.

Aurorae. Energetic electrons escaping from the radiation belts impact atoms at ionospheric heights and cause spectacular visual displays in the *auroral zones*, at latitudes around 70° in both hemispheres. They are a normal accompaniment to magnetic storms and substorms, when there are increases in the radiation belt populations. They can produce waving curtains of light as the electrons are guided by motion of disturbed magnetic field lines. The light is generally green or red and occasionally both. Green light is produced by atomic oxygen, which can also produce red light at high altitudes. Red light from lower altitudes comes from molecular nitrogen (N_2). The auroral display at any time is not a circle symmetrical about the pole but occurs in an *auroral oval*, reaching lower latitude at local midnight than at mid-day (when it can only be seen instrumentally or by satellite). The whole auroral zones move equatorwards during magnetic storms, so that, although aurorae are recognised as high-latitude phenomena, they may be seen at latitudes to 40° during particularly strong storms. The atomic excitations are caused primarily by electrons, which have higher energies than dumped protons because radiation belt protons tend to make a closer approach to the ionosphere than the electrons, establishing a charge separation and consequent strong electric field that accelerates the electrons.

Micropulsations. There is a variety of rapid periodic magnetic fluctuations that are seen on magnetic records, mostly during other disturbances that are attributed to cavity resonances in the magnetosphere or, for the higher frequencies, hydromagnetic waves propagating along field lines.

Solar flares. emit X-rays that cause short (~1 hour) periods of increased ionisation in the ionospheric D layer in the daylight hemisphere. These are termed *sudden ionospheric disturbances* (SIDs) and are identified on magnetic records by *bays*, deflections of the traces on photographic records that resemble coastline bays on maps. *Solar flare effects (SFEs)* are caused by electromagnetic radiation, not charged particles, affecting only the daylight D layer, with no effect on the auroral zones.

Currents induced in the ground. Although the mantle is a poor conductor by the standard of metals, its physical scale ensures that currents induced in it by geomagnetic disturbances have significant effects on observations of the disturbances, which combine the primary (magnetospheric) effects and secondary (induced) effects. Spherical harmonic analysis allows separation of these effects, providing a means of investigating the mantle conductivity. As a magnetic 'wave' penetrates the Earth, it undergoes both attenuation and a phase delay. The characteristic penetration depth, termed *skin depth*, at which the amplitude is reduced to $1/e = 0.37$ of the surface value, is $z_0 = (\mu_0\sigma\omega/2)^{1/2}$, where σ is the electrical conductivity and Ω is the angular frequency of a magnetic fluctuation. This is the measure of the depth to which mantle conductivity can be estimated from induction, but the problem is far from simple for several reasons. One is that it is not a static problem and the Earth rotates as the magnetic wave penetrates with a delay. Another is that, at least in the upper layers, there are lateral heterogeneities, which are of interest for their own sake but can confuse inferences about deeper layers. A few simple conclusions are secure: (i) the continental surface is more conducting ($\sim 3 \times 10^{-2}$ S m^{-1}) than the uppermost mantle ($\sim 1 \times 10^{-2}$ S m^{-1}), (ii) the ocean is more conducting than either of them (~ 4 S m^{-1}) and (iii) the mantle conductivity rises steeply in the phase transition zone to about 1 S m^{-1} at the top of the lower mantle. For greater depths the only evidence, obtained from penetration of the secular variation (Section 5.5.4), is that there is no further dramatic increase, as expressed by the values of electrical conductivity in Section 7.2.

Periodicities. The most obvious periodic effect in geomagnetic disturbances arises from the 27-day period of the Sun's rotation (relative to the orbiting Earth). The Sun is not rotating coherently and most of the activity that affects the Earth originates in a ±40° latitude range about the solar equator, where the 27-day rotation period is observed. The fact that this period is seen in indices of geomagnetic activity is evidence that active patches on the Sun (and the sunspots that mark them) often survive longer than one solar rotation. An 11-year geomagnetic cycle, accompanying the cycle of solar activity, is of particular interest because of its deep penetration of the mantle, but the signal is weak and analysis has not led to secure estimates of lower mantle conductivity.

Space weather. The analogy between meteorological conditions and solar-magnetospheric activity has been recognised for many years by the use of the term 'magnetic storm' and is emphasised by the more recent expression 'space weather'. This now means continuous monitoring of the state of the surface of the Sun, the solar wind and the magnetosphere to give warning of effects that are likely to interfere with communication systems and electrical power grids or damage satellites. A general indicator of solar activity and the likelihood of a disturbance is the abundance of sunspots, but a more explicit warning is obtained from observations of the outermost atmosphere of the Sun, the *corona*. This is a layer of very hot but tenuous plasma enveloping the *photosphere*, from which the visible light radiates, and extends for 1 to 3 million kilometres, but with no clear outer boundary. The solar wind originates in the *corona* and the disturbances in it are *coronal mass ejections*, produced by magnetic loops of energetic plasma which explode out of the *corona*. They travel several times as fast as the quiet time solar wind and can reach the Earth in less than a day.

The heliosphere. The Sun's magnetic field extends past all the planets to a distance of about 50 AU, where it balances the magnetic pressure of the interstellar plasma. The enclosed volume is the *heliosphere*, analogous to the Earth's magnetosphere, and is effectively the total atmosphere of the Sun. Its outer boundary is the *heliopause*, which has now been visited by space probes. Being much more extensive than the magnetosphere, the heliosphere is more effective in protecting the Earth from high-energy cosmic rays of galactic or intergalactic origins. A general increase in the Sun's magnetic field compresses the magnetosphere, reducing its protection, but its pressure on the *heliopause* expands the *heliosphere*, increasing the protection. However, there is a delay of several months before that is effective, this being the time taken by vortices of increased energy in the magnetically controlled solar wind plasma to reach the *heliopause*. There is a suggested climatic effect of such variations arising from the nucleation of clouds by energetic cosmic rays. If this is so, it invites a further inference. Compression of the heliosphere by increases in the density of the interstellar plasma could be a cause of much more prolonged global cooling than the ice ages of the last few million years.

Section IV

Major Subdivisions of the Earth

Structures and Properties

Chapter 6

The Core

6.1 CORE DETAILS FROM THE PRELIMINARY REFERENCE EARTH MODEL

Table 6.1 shows selected details from Preliminary Reference Earth Model (PREM; Dziewonski and Anderson 1981). Although later work indicated departures from the model in some details (as shown in Table 6.2) this model remains the most widely used reference for Earth structure.

6.2 CORE PROPERTIES

Table 6.4 recognises that the density difference between the inner and outer cores must be explained by the different solubilities of oxygen in solid and liquid iron and that this disallows the inclusion of silicon (Stacey and Davis 2008, Section 2.8). The thermal properties of the core (Table 6.3) are crucial to understanding its behaviour.

6.3 CORE ENERGETICS

Core–mantle heat flux: 4.0×10^{12} W

- Inner core–outer core heat flux: 1.2×10^{11} W
- Cooling rate:
  - Core–mantle boundary: 21.9 K/10^9 years
  - Inner core boundary: 28.6 K/10^9 years
- Inner core growth rate: 8.6×10^{-12} m s^{-1} = 0.27 km/million years
- Heat released by cooling: 1.97×10^{12} W
- Latent heat release: 0.71×10^{12} W

Gravitational energy:

Light solute redistribution: 0.54×10^{12} W

Contraction: 0.58×10^{12} W

Radiogenic heat: 2×10^{11} W

Ohmic dissipation: 1 to 3×10^{11} W

Precessional dissipation: 6.4×10^{9} W

TABLE 6.1 SELECTED CORE DETAILS OF PREM

	Radius r (km)	Pressure P (GPa)	Density ρ (kg m^{-3})	Bulk Modulus K_S (GPa)	Rigidity Modulus μ (GPa)	Gravity g (m s^{-2})
Inner core (Solid)	0	363.85	13088.48	1425.3	176.1	0
	200	362.90	13078.77	1423.1	175.5	0.7311
	400	360.03	13053.64	1416.4	173.9	1.4604
	600	355.28	13010.09	1405.3	171.3	2.1862
	800	348.67	12949.12	1389.8	167.6	2.9068
	1000	340.24	12870.73	1370.1	163.0	3.6203
	1200	330.05	12774.93	1346.2	157.4	4.3251
	1221.5	328.85	12763.60	1343.4	156.7	4.4002
Outer core (Liquid)	1221.5	328.85	12166.34	1304.7	0	4.4002
	1400	318.75	12069.24	1267.9	0	4.9413
	1600	306.15	11946.82	1224.2	0	5.5548
	1800	292.22	11809.00	1177.5	0	6.1669
	2000	277.04	11654.78	1127.3	0	6.7715
	2200	260.68	11483.11	1073.5	0	7.3645
	2400	243.25	11292.98	1015.8	0	7.9425
	2600	224.85	11083.35	954.2	0	8.5023
	2800	205.60	10853.21	888.9	0	9.0414
	3000	185.64	10601.52	820.2	0	9.5570
	3200	165.12	10327.26	748.4	0	10.0464
	3400	144.19	10029.40	674.3	0	10.5065
	3480	135.75	9903.49	644.1	0	10.6823

Note: By this model, the inner core mass is 9.8×10^{22} kg, and the outer core mass is 1.84×10^{24} kg.

TABLE 6.2 GENERAL PROPERTIES

Property	Outer Core at Core–Mantle Boundary	Outer Core at Inner Core Boundary	Inner Core at Inner Core Boundary
Density, ρ (kg m^{-3})	9902	12,163	12,983
Bulk modulus, K_S (GPa)	645.9	1301.3	1303.7
$(\partial K_S/\partial P)_S$	3.513	3.317	3.338
Rigidity modulus, μ (GPa)	0	0	169.4
$(\partial \mu/\partial P)_S$	0	0	0.207
Mean atomic wt.	44.43	44.43	50.10
Specific heat, C_P (J K^{-1} kg^{-1})	814	793	728
Temp. gradient, $-dT/dr$ (K/km)	0.884	0.286	0.049
Grüneisen parameter, γ	1.443	1.390	1.391
Volume expansion co-efficient, α (K^{-1})	18.0	10.3	9.7
Electrical resistivity, ρ_e ($\mu\Omega$ m)	3.62	4.65	3.7
Electrical conductivity, σ_e (10^5 S m^{-1})	2.76	2.15	2.7
Magnetic diffusivity, η_m (m^2 s^{-1})	2.88	3.70	2.9
Thermal conductivity, κ (W m^{-1} K^{-1})	28.3	29.3	36
Thermal diffusivity, η (10^{-6} m^2 s^{-1})	3.5	3.0	4.0
Viscosity (Pa s)	~10^{-2}	~10^{-2}	solid
Latent heat, solidification (10^5 J kg^{-1})		9.6	
Solidification density increment (kg m^{-3})		200	
Extrapolated zero pressure density, cooled to 300 K and solidified to ε phase (cf pure Fe 8352 kg m^{-3})		Outer core, 7488 kg m^{-3}	Inner core, 7993 kg m^{-3}

TABLE 6.3 CORE THERMAL PROPERTIES AT DEPTHS LISTED IN TABLE 6.1

	Radius (km)	Temperature, T (K)	Melting Point, T_M (K)	Volume Expansion Co-efficient, α (10^{-6} K^{-1})	Specific Heat, C_p (J K^{-1} kg^{-1})	Thermal Conductivity, κ (W m^{-1} K^{-1})
Inner core	0	5030	5330	9.015	693	36
	200	5029	5321	9.033	694	36
	400	5027	5294	9.088	694	36
	600	5023	5250	9.181	695	36
	800	5017	5188	9.314	697	36
	1000	5010	5107	9.490	700	36
	1200	5001	5012	9.713	703	36
	1221.5	5000	5000	9.740	703	36
Outer core	1221.5	5000	5000	10.314	794	29.3
	1400	4946	4890	10.525	794	29.3
	1600	4877	4772	10.805	796	29.2
	1800	4799	4629	11.135	797	29.1
	2000	4711	4545	11.525	799	29.1
	2200	4614	4452	11.985	800	29.0
	2400	4507	4261	12.528	802	28.9
	2600	4390	4057	13.172	804	28.8
	2800	4263	3840	13.940	806	28.7
	3000	4123	3609	14.865	808	28.6
	3200	3972	3367	15.991	811	28.5
	3400	3808	3112	17.386	814	28.4
	3480	3739	3007	18.040	815	28.3

TABLE 6.4 COMPOSITION OF THE CORE

Element	Outer Core		Inner Core	
	Mass %	Atomic %	Mass %	Atomic %
Fe	77.08	61.34	82.54	74.04
Ni	6.43	4.87	6.83	5.83
S	8.44	11.70	8.02	12.53
O	5.34	14.83	0.11	0.34
C	0.50	1.85	0.45	1.88
H	0.08	3.53	0.07	3.48
P	0.33	0.47	0.18	0.29
Cr	1.07	0.91	1.07	1.03
Mn	0.35	0.28	0.35	0.32
Co	0.28	0.21	0.28	0.24
K[a]	~30 ppm		~30 ppm	
Others[b]	~0.1 by mass			
Mean At. Wt.	44.43		50.10	
	Whole Core 44.71			

[a] Minor and uncertain, but important because of its radiogenic heat.

[b] Minor constituents include heavy siderophiles, Pd, W, Re, Os, Ir, Pt, Au and Hg, but they are not sufficiently abundant to have a significant effect on density.

Chapter 7

The Mantle

7.1 REFERENCE EARTH MODEL FROM SEISMOLOGY

TABLE 7.1 SELECTED MANTLE DETAILS FROM THE PRELIMINARY REFERENCE EARTH MODEL (PREM) BY DZIEWONSKI AND ANDERSON (1981)

	Radius (km)	Pressure (GPa)	Density (kg m^{-3})	Bulk Modulus (GPa)	Rigidity Modulus (GPa)	Gravity (m s^{-2})
D″	3480	135.7509	5566.45	655.6	293.8	10.6823
	3600	128.7067	5506.42	644.0	290.7	10.5204
	3630	126.9742	5491.45	641.2	289.9	10.4844
Lower mantle	3800	117.3465	5406.81	609.5	279.4	10.3095
	4000	106.3864	5307.24	574.4	267.5	10.1580
	4200	95.7641	5207.13	540.9	255.9	10.0535
	4400	85.4332	5105.90	508.5	244.5	9.9859
	4600	75.3598	5002.99	476.6	233.1	9.9474
	4800	65.5202	4897.83	444.8	221.5	9.9314
	5000	55.8991	4789.83	412.8	209.8	9.9326
	5200	46.4882	4678.44	380.3	197.9	9.9467
	5400	37.2852	4563.07	347.1	185.6	9.9698
	5600	28.2928	4443.17	313.3	173.0	9.9985
	5701	23.8342	4380.71	299.9	154.8	10.0143

(Continued)

TABLE 7.1 ***(Continued)*** **SELECTED MANTLE DETAILS FROM THE PRELIMINARY REFERENCE EARTH MODEL (PREM) BY DZIEWONSKI AND ANDERSON (1981)**

	Radius (km)	Pressure (GPa)	Density (kg m^{-3})	Bulk Modulus (GPa)	Rigidity Modulus (GPa)	Gravity (m s^{-2})
Upper mantle	5701	23.8342	3992.14	255.6	123.9	10.0143
	5821	19.0703	3912.82	233.2	112.8	9.9965
	5971	13.3527	3723.78	189.9	90.6	9.9686
	5971	13.3527	3543.25	173.5	80.6	9.0686
	6061	10.2027	3489.51	163.0	77.3	9.9361
	6151	7.1111	3435.78	152.9	74.1	9.9048
	6151	7.1111	3359.50	127.0	65.6	9.9048
	6251	3.7846	3370.37	129.4	66.8	9.8681
Cont. Moho	6331	1.1239	3379.06	131.1	68.0	9.8437
Oceanic Moho	6364	0.1323	3378.00	132.0	68.4	9.8258

Note: The upper boundary of the mantle (Mohorovičić discontinuity or Moho) is here taken to be at radius 6331 km (depth 40 km) under the continents but 6364 km (depth 7 km) under the ocean basins. For thermal properties at the same depths, see Table 7.2.

7.2 GENERAL MANTLE PROPERTIES

Mass:
  Lower mantle, 2.94×10^{24} kg
  Upper mantle, 1.07×10^{24} kg
Volume:
  Lower mantle, 6.04×10^{20} m^3
  Upper mantle, 2.93×10^{20} m^3
Moment of inertia, mantle + crust, 7.08×10^{37} kg m^2
Decompressed density:
  Lower mantle (high-pressure phases), 3977 kg m^{-3}
  Upper mantle, 3370 kg m^{-3}
Mean atomic weight, 21.045, by abundances in Table 7.3.

Viscosity has strong variations, laterally as well as radially. The following are representative values for selected regions:

Lithosphere, 10^{24} Pa s
Oceanic asthenosphere, $<4 \times 10^{19}$ Pa s, possibly much lower locally
Continental asthenosphere, $\sim 10^{20}$ Pa s
Upper mantle average, 10^{21} Pa s
Lower mantle, increasing with depth to 10^{23} Pa s (see note below on *Melting points in the mantle*)
D″ base and plume axes, locally 10^{18} Pa s

Electrical conductivity:

Upper mantle, 10^{-2} S m^{-1}
Lower mantle, 1 S m^{-1} up to 4 S m^{-1}
D″, local patches > 10 S m^{-1}

Thermal conductivity:

Upper mantle, 4 W m^{-1} K^{-1}
Lower mantle, 8 W m^{-1} K^{-1}

Melting points in the mantle. Being a mixture of minerals with different properties, mantle material does not have a single, simple melting point, but for many purposes the solidus temperature, at which the first liquid appears on heating, serves as the notional melting temperature, T_m. This increases more rapidly with compression than the ambient temperature, T, which closely follows an adiabat. The result is that T/T_m decreases with depth and this ratio controls the viscosity, which increases with depth through most of the mantle range. T_m varies from about 1400K at the top of the mantle to about 2300K at the top of the lower mantle and about 4000K at the bottom of the mantle (Figure 7.1).

There are several reasonably consistent estimates of mantle composition, mostly considering the 'bulk silicate earth', that is, the mantle and crust together, for example, Newsom (1995) and McDonough and Sun (1995). Table 7.3 gives a separate assessment for the mantle, with the crustal composition in Table 8.3. Within the uncertainties of this tabulation, there is no compositional distinction between the upper and lower mantles. The mean atomic weight is 21.045.

TABLE 7.2 THERMAL PROPERTIES AT DEPTHS LISTED IN TABLE 7.1

	Radius (km)	Temperature T(K)	Expansion Co-efficient α (10^{-6} K^{-1})	Specific Heat C_p (J K^{-1} kg^{-1})
D″	3480	3739	11.3	1203
	3600	2838	11.6	1191
	3630	2740	11.7	1190
Lower mantle	3800	2668	12.1	1191
	4000	2596	12.7	1192
	4200	2525	13.4	1193
	4400	2452	14.2	1195
	4600	2379	14.9	1196
	4800	2302	15.9	1198
	5000	2227	17.0	1201
	5200	2144	18.3	1203
	5400	2069	19.8	1206
	5600	1974	21.8	1209
	5701	1931	23.0	1214
Upper mantle	5701	2010	20.6	1200
	5821	1982	23.8	1208
	5971	1948	27.4	1217
	5971	1907	24.9	1202
	6061	1817	26.9	1210
	6151	1780	28.8	1210
	6151	1719	30.2	1205
	6251	1303	31.8	1197
Cont. Moho	6331	880	33.5	1206
Oceanic Moho	6364	475	27.0	1040

TABLE 7.3 ELEMENTAL COMPOSITION OF THE MANTLE

At. No.	Element	Mass Fraction (ppm)	Atomic Fraction (ppm)
1	H	60	1270
2	He	0.03	0.16
3	Li	1.5	4.6
4	Be	0.06	0.14
5	B	0.49	0.95
6	C	119	215
7	N	3	4.7
8	O	441 150	557 799
9	F	21.6	25
10	Ne	Trace	
11	Na	2699	2545
12	Mg	224 232	173 875
13	Al	21 952	17 639
14	Si	214 694	165 734
15	P	91.7	64
16	S	320	216
17	Cl	11.8	7.2
18	Ar	0.02	0.01
19	K	81	45
20	Ca	24 260	13 124
21	Sc	15.8	7.6
22	Ti	1196	542
23	V	92	39
24	Cr	2929	1221
25	Mn	1000	395
26	Fe	62 816	24 387
27	Co	105	39
28	Ni	2002	740
29	Cu	29	9.9
30	Zn	63	21

(Continued)

TABLE 7.3 *(Continued)* ELEMENTAL COMPOSITION OF THE MANTLE

At. No.	Element	Mass Fraction (ppm)	Atomic Fraction (ppm)
31	Ga	3.8	1.2
32	Ge	1.2	0.36
33	As	0.12	0.034
34	Se	0.03	0.008
35	Br	0.06	0.016
36	Kr	Trace	
37	Rb	0.4	0.098
38	Sr	20	4.9
39	Y	4.0	0.98
40	Zr	9.5	2.3
41	Nb	0.64	0.145
42	Mo	0.057	0.0125
43	Tc	Nil	
44	Ru	0.0043	0.00090
45	Rh	0.0014	0.00029
46	Pd	0.004	0.0008
47	Ag	0.01	0.0020
48	Cd	0.034	0.0064
49	In	0.016	0.0029
50	Sn	0.50	0.089
51	Sb	0.006	0.0010
52	Te	0.018	0.0030
53	I	0.004	0.00066
54	Xe	Trace	
55	Cs	0.007	0.0011
56	Ba	5.6	0.86
57	La	0.49	0.074
58	Ce	1.5	0.23
59	Pr	0.21	0.031

(Continued)

TABLE 7.3 *(Continued)* ELEMENTAL COMPOSITION OF THE MANTLE

At. No.	Element	Mass Fraction (ppm)	Atomic Fraction (ppm)
60	Nd	1.2	0.175
61	Pm	Nil	
62	Sm	0.40	0.056
63	Eu	0.16	0.022
64	Gd	0.58	0.078
65	Tb	0.11	0.015
66	Dy	0.67	0.087
67	Ho	0.15	0.019
68	Er	0.42	0.053
69	Tm	0.062	0.0077
70	Yb	0.43	0.052
71	Lu	0.07	0.008
72	Hf	0.27	0.032
73	Ta	0.03	0.0035
74	W	0.013	0.0015
75	Re	0.00025	0.00003
76	Os	0.0034	0.00038
77	Ir	0.0031	0.00034
78	Pt	0.0075	0.00081
79	Au	0.0009	0.000096
80	Hg	0.01	0.001
81	Tl	0.006	0.00062
82	Pb	0.15	0.015
83	Bi	0.05	0.005
84	Po	Th and U decay daughters	
85	At		
86	Rn		
87	Fr		

(Continued)

TABLE 7.3 *(Continued)* ELEMENTAL COMPOSITION OF THE MANTLE

At. No.	Element	Mass Fraction (ppm)	Atomic Fraction (ppm)
88	Ra	Th and U decay daughters	
89	Ac		
90	Th	0.108	0.0098
91	Pa	U decay daughter	
92	U	0.029	0.0026

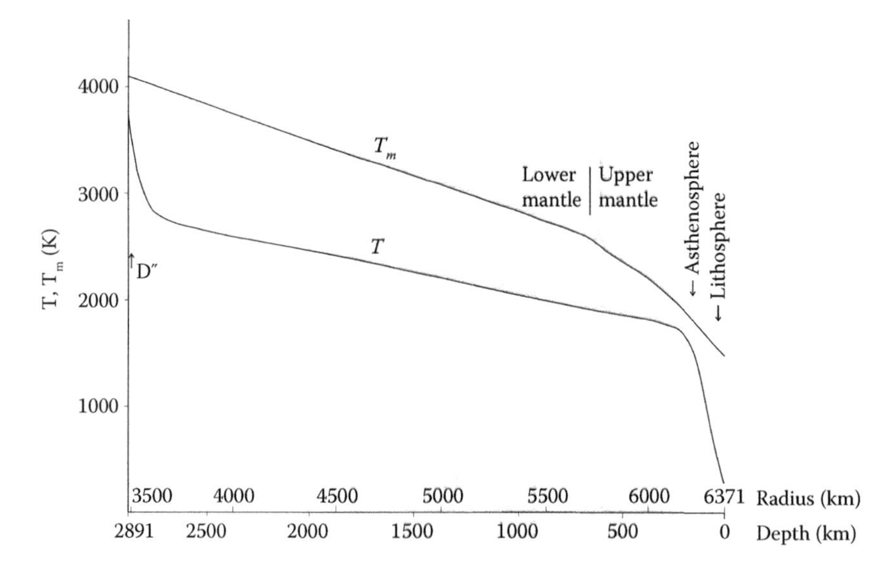

Figure 7.1 Variation in temperature and melting point (solidus temperature) with depth in the mantle.

7.3 MANTLE MINERALS AND PHASE TRANSITIONS

Mantle minerals are combinations of oxides, primarily SiO_2, MgO, FeO, CaO and Al_2O_3, with minor additions of others, in proportions and crystal structures that are pressure and temperature dependent. The mineral combinations are identified with depth ranges, between which there are phase transitions, with successively increasing densities represented by adiabatically extrapolated zero pressure values:

Uppermost mantle, from 7 km (oceanic areas) or 37 km (under continents) to 410 km, 3395 kg m^{-3}.
Transition zone, 410 km to 520 km, 3565 kg m^{-3}
520 km to 660 km, 3715 kg m^{-3}
Lower mantle, 660 km to ~ 2700 km, 4105 kg m^{-3}
D″, ~2700 km to 2890 km, 4160 kg m^{-3}

In interpreting these (extrapolated, zero pressure) density variations in terms of mineral structures, it is convenient to refer to a simplified model with SiO_2 and MgO only and then note the departures from this arising from inclusion of other elements. By this model, dubbed pyrolite (PYRoxene-OLivine), the uppermost mantle mineral structure, is approximated by a combination of four minerals:

~56% Mg_2SiO_4, forsterite (olivine, α-phase), density 3227 kg m^{-3}, mean at. wt. 20.099
~17% $MgSiO_3$, enstatite (orthopyroxene), density 3204 kg m^{-3}, mean at. wt. 20.078
~15% $MgSiO_3$, clinopyroxene, density 3188 kg m^{-3}, mean at. wt. 20.078
~12% $MgSiO_3$, garnet, density 3513 kg m^{-3}, mean at. wt. 20.078, but this occurs more often as pyrope, a garnet with the formula $Mg_3Al_2Si_3O_{12}$, density 3565 kg m^{-3} and mean at. wt. 20.156.

The coexistence of these minerals arises from differences in accommodating additions of other elements. All of them accept replacement of some Fe for Mg, with increased density. Clinopyroxene incorporates more of the other elements, particularly Ca, than forsterite or enstatite, and garnet includes Al. Having a higher density than the other three, although a similar atomic weight, garnet is favoured by pressure and absorbs the pyroxenes at increasing depth.

Transitions to denser phases are smeared over ranges in pressure by compositional variations and adjustments but are sharpest for the olivine structure. Combined with its compositional dominance, this means that the olivine transitions explain the seismologically observed mantle boundaries. The sequence of phases of Mg_2SiO_4, with extrapolated zero pressure densities, ρ_0, the transition pressures, P, at which they transform, the transition entropies, ΔS, and corresponding mantle depths, z, is listed in Table 7.4. Values in the table are derived by combining laboratory and seismological observations although discrepancies between them have been reported.

Structural changes in $MgSiO_3$ orthopyroxene are also important to the transition zone but are less clear-cut than forsterite transitions. Pyroxene

TABLE 7.4 Mg_2SiO_4 (FORSTERITE) PHASES

Structure	ρ_0 (kg m^{-3})	P (GPa)	ΔS (J kg^{-1} K^{-1})	z (km)
Forsterite (olivine)	3227	13.7	− 35	410
β-spinel (Wadsleyite)	3473	17.9	− 32	520
γ-spinel (Ringwoodite)	3548	23.3	+ 49	660
Mg–Si perovskite + periclase ($MgSiO_3$ + MgO)	3943	120	− 20	2600
Post-perovskite + periclase	4004			

transforms to an ilmenite structure deep in the transition zone but the pyroxenes are mainly replaced by garnet, incorporating Ca as well as Al, probably largely as the mineral majorite $(Mg,Fe,Ca)_3(Mg,Si,Al)_2Si_3O_{12}$. Over the pressure range 23–27 GPa, this transforms to a perovskite structure, with element ratios differing from the perovskite derived from olivine, and with rejection of Ca, which forms a third perovskite structure, basically $CaSiO_3$.

Most of the lower mantle is understood to be composed of the three perovskites plus (Mg,Fe)O ferropericlase (magnesiowüstite), but in the lowest region, D″, perovskite is converted to a denser structure (post-perovskite), with a dependence of the transition pressure on temperature that contributes to the heterogeneity of this region. Post-perovskite has the same orthorhombic crystal structure as $CaIrO_3$, which can be studied at low pressure.

Rejection of Fe by the Mg perovskite at about 100 GPa, with the appearance of a new 'H-phase' has been reported (Zhang et al. 2014), indicating that the lowest 600 km of the mantle has an Fe-depleted perovskite. It is not clear that the H-phase is involved because co-existing ferro-periclase, (Mg,Fe)O, accepts more Fe than the perovskite at all pressures.

7.4 ENERGY BALANCE OF THE MANTLE

Convection of the mantle is driven by upward heat transport itemised in Table 7.5. Heat transport and convective power:

96.3% convective, 3.7% by conduction

Convective power generation:

Lower mantle	2.28×10^{12} W
Upper mantle by upper mantle heat	0.86×10^{12} W
by lower mantle heat	4.56×10^{12} W
Total	7.70×10^{12} W

Notional thermodynamic efficiency 19.8%
(allowing for conductive heat transport)

Cooling rate:

Potential temperature	53 K/10^9 years
Deep lower mantle	85 K/10^9 years.

TABLE 7.5 HEAT TRANSFERS IN THE MANTLE, ALL VALUES IN TERAWATTS (10^{12} W)

Region	Heat Source	Heat Transfer	Net
Lower mantle	Heat from core	4.0	
	Radiogenic[a]	16.8	
	Tidal	0.2	
	Cooling	8.9	
	Heat to upper mantle	29.9	29.9
Upper mantle	Heat from lower mantle	29.9	
	Radiogenic[a]	6.1	
	Tidal	0.2	
	Cooling	2.6	
	Heat to crust		38.8

[a] 5.7×10^{-12} W kg^{-1}.

Chapter 8

The Crust

8.1 A VENEER DISPLAYING THE HISTORY OF THE EARTH

The word *crust* originally arose from Kelvin's idea of a solidified layer on a molten earth but is now understood to be a compositionally distinct veneer on the outer solid silicate part of the Earth (the mantle), from which it has been derived by volcanism. Two distinct crustal types are recognised, continental and oceanic, with average properties summarised in Table 8.1, although there are regional variations, summarised in Table 8.2, that depart strongly from the averages and, in some cases, appear as compromises between the two types. Both have depth variations approximated as layers, numbered 1 to 3 or 4 from the surface, and both are distinct, in composition and properties, from the underlying mantle.

There are several components to lithospheric buoyancies with values in Table 8.1. Cooled mature lithosphere (70–90 million years old) has shrunk thermally by about 2.1 km, increasing its density relative to the new lithosphere (~3300 kg m^{-3}) and giving it the driving force for subduction, with an effective mass increment per unit area of lithosphere (2100 × 3300) kg m^{-2} = 6.9×10^6 kg m^{-2}. The overlying oceanic crust (7 km with density contrast of 400 kg m^{-3}) has a mass decrement of 2.8×10^6 kg m^{-2}, relative to mantle material, leaving the oceanic lithosphere, including the crust, with a net mass increment $\Delta m = 4.1 \times 10^6$ kg m^{-2}. This causes a downward gravitational force $\Delta m.g = 4.0 \times 10^7$ N on the subducting slab per metre length of the subduction zone and per metre of its down-slope dimension. The average continental crust (37 km thick with a density contrast of 450 kg m^{-3}) has a mass decrement of 16.7×10^6 kg m^{-2} which, superimposed on the 6.9×10^6 kg m^{-2} increment of the cooled lithosphere, leaves a decrement of 9.8×10^6 kg m^{-2}, making it strongly buoyant.

TABLE 8.1 AVERAGE PHYSICAL PROPERTIES OF THE CRUST, INCORPORATING DETAILS FROM WHITE ET AL. (1992), MOONEY ET AL. (1998), AND BIRD (2003)

Property		Continental Crust	Oceanic Crust
Area (m^2)		2×10^{14}	3.1×10^{14}
Mass (kg)		21×10^{21}	7×10^{21}
Average thickness (km)		37	6.7 (excluding sediment)
Layer 1	Composition	Sediment	Sediment
	Density (kg m^{-3})	2000	1500–2000
	Thickness (km)	0 to 20	0–1
	P-wave vel. (km s^{-1})	<5.7	1.75 to 2
Layer 2 (upper)	Composition	Granitic	Basalt, with sea water percolation
	Density (kg m^{-3})	2670	2900–
	Thickness (km)	~10	1.7
	P-wave vel. (km s^{-1})	5.7–6.4	5.1
Layer 3 (middle)	Composition	Basaltic	Basalt/gabbro
	Density (kg m^{-3})	2900	2900+
	Thickness (km)	~10 to 15	5
	P-wave vel. (km s^{-1})	6.4–7.1	6.7
Layer 4 (lower)	Composition	Amphibolite, granulite	–
	Density (kg m^{-3})	~3000	
	Thickness (km)	7 to 10	
	P-wave vel. (km s^{-1})	7.1–7.6	
Underlying mantle	Composition	Ultra-basic	
	Density (kg m^{-3})	3300	
	P-wave vel. (km s^{-1})	8	
Age variation (10^6 years)		0–4000[a], mineral grains up to 4400	0–200, mean 65, average duration 92, spreading centre production 3.36 km^2/year

(Continued)

TABLE 8.1 ***(Continued)*** **AVERAGE PHYSICAL PROPERTIES OF THE CRUST, INCORPORATING DETAILS FROM WHITE ET AL. (1992), MOONEY ET AL. (1998), AND BIRD (2003)**

Property	Continental Crust	Oceanic Crust
Surface elevation (km), relative to sea level (see Figures 8.1 and 8.2)	−1 to +6[a]	−1 to −8
Mean density (kg m^{-3})	2850	2900
Lithospheric buoyancy (N m^{-2})[b]	+9.6 × 10^7	−4.0 × 10^7

[a] Age and elevation are negatively correlated because elevated land erodes fastest.
[b] Positive buoyancy indicates material that stays at the surface.

TABLE 8.2 CRUSTAL VARIATIONS USING DATA SELECTED FROM TANIMOTO (1995)

Crustal Type	Examples	Thickness (km)	Interpretation
Continental thick areas	Himalaya	Up to 70	Continental collision
	Andes	Up to 60	Subduction zone volcanism
	Alps, Caucasus, Rocky Mts.	40–50	Compression
	Shield areas	35–55	High-average density
Continental thin areas	Rifts (Red Sea, East Africa, Lake Baikal, Rio Grande, Rhine Graben)	<30	Crustal splitting
	Basin and Range, W. USA, and Mexico	25–30	Broad range of crustal extension
Oceanic thick areas	Ontong-Java Plateau	Up to 20	Intruded and extruded plume basalt
	Hawaiian swell		
	Japan	23–30	Oceanic crust converting to continental type
Oceanic thin areas	Slow-spreading ridges (<2 cm/year)	Down to 2	Limited lava at spreading centres. Layer 3 thin or missing
	Fracture zones/rifts	~4	Crustal splitting

8.2 AVERAGE CRUSTAL STRUCTURE AND ISOSTASY

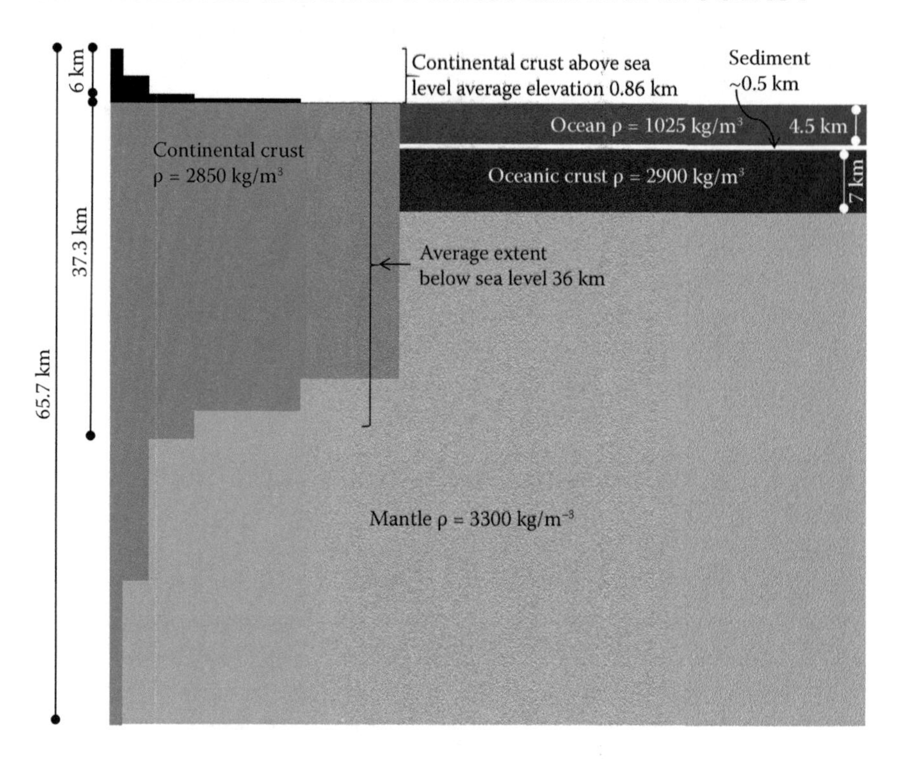

Figure 8.1 A simplified model of crustal structure, with densities and continental layer thicknesses adjusted for isostatic balance with the ocean average. The model is constrained to give an average surface elevation of continents of 0.8 km, relative to sea level (including submerged margins), with an average thickness of 36.8 km and an assumed uniform density. The distribution of surface elevation is illustrated in Figure 8.2.

8.3 HEAT BALANCE OF THE CRUST

From Table 7.5, heat flux, mantle to crust,	38.8×10^{12} W
Crustal radioactivity	8.0×10^{12} W
Heat flux, crust to oceans and atmosphere	$46.8 \pm 2.0 \times 10^{12}$ W

Unlike the mantle, on average, the crust is in thermal balance, neither warming nor cooling.

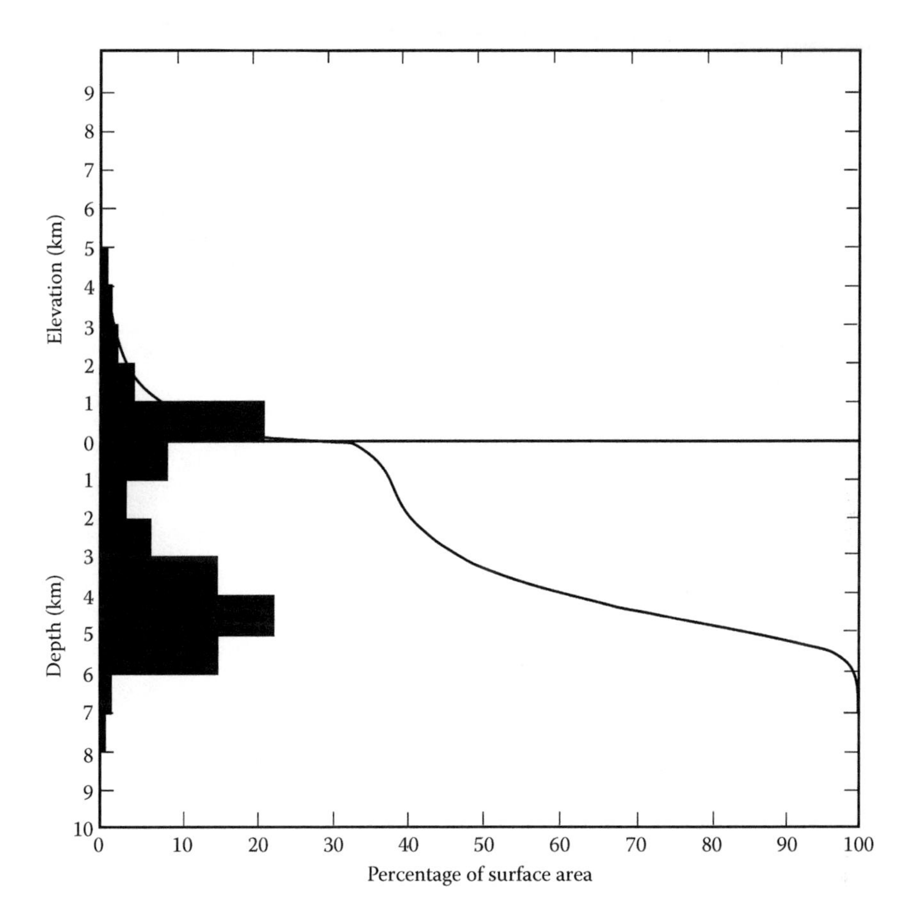

Figure 8.2 The hypsographic curve, representing the distribution of surface elevation of the solid Earth with a histogram of the areas in 1 km elevation intervals.

8.4 ELEMENTAL COMPOSITION OF THE CRUST

In Table 8.3, the upper continental crust refers to the uppermost 25%. The whole crust includes the oceanic crust but without the sediment. There are numerous estimates with significant disagreements. This compilation makes use of estimates by Newsom (1995), Stacey and Davis (2008), Rudnick and Gao (2014) and White and Klein (2014). Values are ppm by mass. The final column of Table 8.3 compares the elemental abundances of the upper continental crust with the mantle values in Table 7.3. As has often been noted, the radioactive elements, K, Rb, Th and U, are all concentrated in the crust. The significance of some other strong concentrations is considered in Chapter 23.

TABLE 8.3 ELEMENTAL COMPOSITION OF THE CRUST (PPM BY MASS), WITH RATIOS OF UPPER CONTINENTAL CRUST AND MANTLE VALUES

z	Element	Whole Crust	Upper Continental Crust	Upper Continental/ Mantle Ratios
1	H	1000	2000	33
2	He	1	3	100
3	Li	40	60	40
4	Be	1.6	2.1	35
5	B	8.8	17	35
6	C	2500	9000	76
7	N	40	180	60
8	O	457,000	470,000	1.07
9	F	500	800	37
10	Ne	trace	trace	–
11	Na	20,500	22,300	8.3
12	Mg	34,000	14,000	0.06
13	Al	79,000	80,000	3.6
14	Si	272,700	305,000	1.4
15	P	830	650	7.1
16	S	400	600	1.9
17	Cl	250	300	25
18	Ar	2.5	5	250
19	K	15,500	32,000	395
20	Ca	50,000	23,000	0.95
21	Sc	25	14	0.89
22	Ti	4000	3000	2.5
23	V	140	95	1.0
24	Cr	170	90	0.03
25	Mn	750	750	0.75
26	Fe	52,000	36,000	0.57
27	Co	27	17	0.16

(Continued)

TABLE 8.3 *(Continued)* ELEMENTAL COMPOSITION OF THE CRUST (PPM BY MASS), WITH RATIOS OF UPPER CONTINENTAL CRUST AND MANTLE VALUES

z	Element	Whole Crust	Upper Continental Crust	Upper Continental/ Mantle Ratios
28	Ni	75	47	0.02
29	Cu	30	27	0.93
30	Zn	67	67	1.1
31	Ga	18	18	4.7
32	Ge	1.5	1.4	1.2
33	As	2.2	2.0	17
34	Se	0.12	0.09	3
35	Br	0.8	1.6	27
36	Kr	trace	trace	–
37	Rb	40	84	210
38	Sr	270	30	1.5
39	Y	19	21	5.3
40	Zr	110	190	20
41	Nb	7	12	19
42	Mo	0.7	1.5	2.6
44	Ru	7×10^{-4}	3×10^{-4}	0.07
45	Rh	3×10^{-4}	2×10^{-4}	0.14
46	Pd	1.5×10^{-3}	5×10^{-4}	0.13
47	Ag	0.07	0.05	5
48	Cd	0.08	0.09	2.6
49	In	0.05	0.06	3.8
50	Sn	2	5	10
51	Sb	0.2	0.4	67
52	Te	0.001	0.001	0.06
53	I	0.7	1.4	350
54	Xe	trace	trace	–
55	Cs	1.6	4.9	700

(Continued)

TABLE 8.3 *(Continued)* ELEMENTAL COMPOSITION OF THE CRUST (PPM BY MASS), WITH RATIOS OF UPPER CONTINENTAL CRUST AND MANTLE VALUES

z	Element	Whole Crust	Upper Continental Crust	Upper Continental/ Mantle Ratios
56	Ba	350	600	107
57	La	15	30	61
58	Ce	50	60	40
59	Pr	4	7	33
60	Nd	15	25	21
62	Sm	3	5	13
63	Eu	1.0	1.0	6.3
64	Gd	2.5	4	6.9
65	Tb	0.6	0.7	6.4
66	Dy	3.3	4	6.0
67	Ho	0.7	0.8	5.3
68	Er	2.0	2.3	5.5
69	Tm	0.28	0.3	4.8
70	Yb	1.9	2	4.7
71	Lu	0.3	0.3	4.3
72	Hf	3	5	19
73	Ta	0.6	0.9	30
74	W	1.2	2	154
75	Re	2×10^{-4}	2×10^{-4}	0.8
76	Os	4×10^{-5}	3×10^{-5}	0.009
77	Ir	4×10^{-5}	2×10^{-5}	0.006
78	Pt	1.2×10^{-3}	6×10^{-4}	0.08
79	Au	3×10^{-3}	1.5×10^{-3}	1.7
80	Hg	0.03	0.05	5
81	Tl	0.6	0.9	150
82	Pb	9	17	113
83	Bi	0.1	0.15	3

(Continued)

TABLE 8.3 *(Continued)* ELEMENTAL COMPOSITION OF THE CRUST (PPM BY MASS), WITH RATIOS OF UPPER CONTINENTAL CRUST AND MANTLE VALUES

z	Element	Whole Crust	Upper Continental Crust	Upper Continental/ Mantle Ratios
84–89 and 91 U and Th decay series				
90	Th	4.5	16	148
92	U	1.2	4	138
Mean At. Wt.		20.96	20.11	

Chapter 9

The Oceans

9.1 GLOBAL OCEAN PROPERTIES

9.1.1 Sea Level Rise

The global average sea level, measured relative to the Earth's centre of mass, is rising by 3 mm/year (Figure 9.3). Due to a combination of ocean thermal expansion and melting of grounded ice, the total ocean volume is increasing by 2.9×10^{11} m^3/year. Greenland and Antarctic melting are the major contributors, and this will become increasingly so. Thermal expansion by ocean warming is significant at the present time, but its total potential effect is only tens of centimetres of sea level rise, whereas a complete melt of 'permanent' ice would cause a rise exceeding 60 m, decreasing to 50 m over thousands of years by isostatic adjustment. The present net greenhouse-induced heat input to the Earth, ~200 terawatts, is mainly absorbed by the oceans and, with their large heat capacity, this will continue for more than 100 years, but when the oceans reach thermal equilibrium with the atmosphere and no longer act as the major heat sink, the rate of sea level rise will exceed 4 cm/year.

9.2 SEA WATER PROPERTIES (FIGURES 9.1 THROUGH 9.3 AND TABLES 9.1 THROUGH 9.6)

The disparities in the abundance ratios in Table 9.3 emphasise that sea salt is not just a result of the evaporation of river water. Although there is no quantitative evidence that groundwater leaking into the oceans is dramatically different from river water, the salt introduced to the oceans by hydrothermal circulation at mid-ocean ridges is very different and appears to be the dominant source of ocean salt (Stacey and Hodgkinson 2013).

TABLE 9.1 MASS AND AREAS	
Total volume of sea water	1.4×10^{18} m^3
Mass	1.44×10^{21} kg
Areas	
Sea total	3.62×10^{14} m^2
Ocean basins	3.1×10^{14} m^2

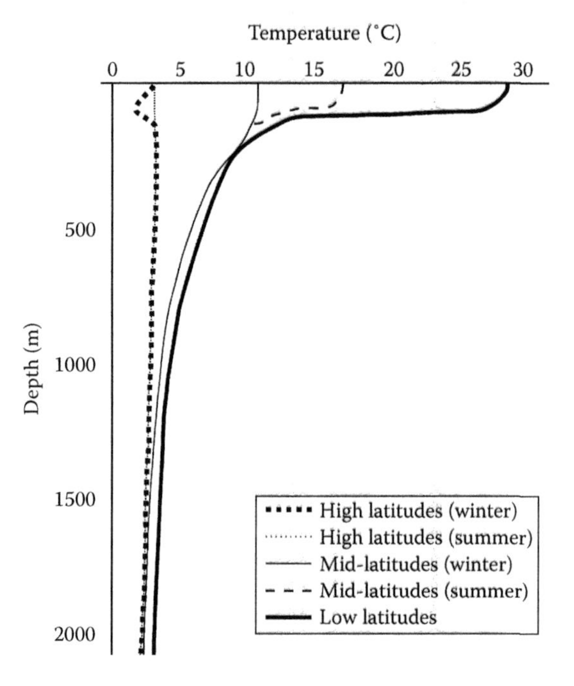

Figure 9.1 Temperature variations of the ocean with depth, latitude and season.

TABLE 9.2 GENERAL PROPERTIES		
Property	**0°C**	**20°C**
Viscosity (Pa s)	1.88×10^{-3}	1.08×10^{-3}
Thermal conductivity (W m^{-1} K^{-1})	0.570	0.600
Specific heat (J kg^{-1} K^{-1})	3985	3993
Bulk modulus (GPa)	2.15	2.34
Refractive index	1.3325	

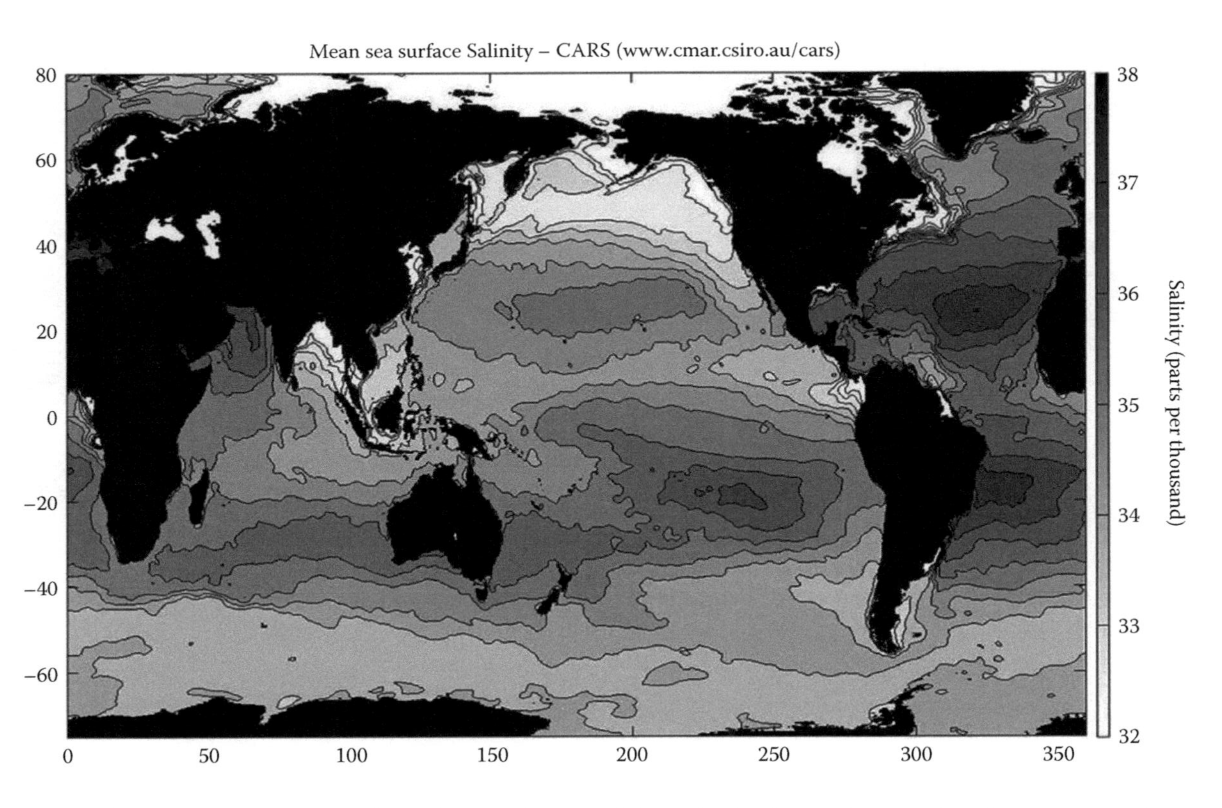

Figure 9.2 Global variations of sea surface salinity. Salinity variations diminish with depth and in the deep ocean the range is restricted 34.7‰ to 34.8‰. (Courtesy of Jeff Dunn, CSIRO Marine and Atmospheric Research.)

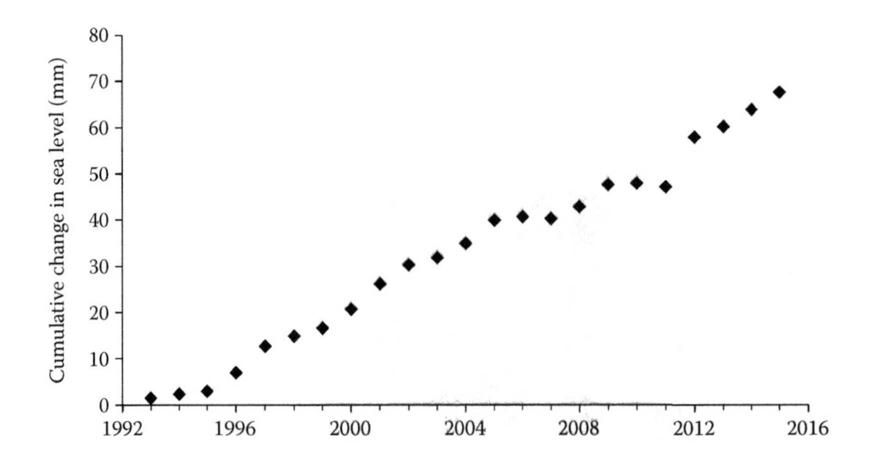

Figure 9.3 Annual average values of global mean sea level obtained by satellite altimetry (radar reflections from the sea). The ultimate reference point is the Earth's centre of mass. The gradient from a more detailed dataset [Beckley et al. (2010), GSFC (2013) and Combined TOPEX/Poseidon and Jason 1&2 project teams] is 3.21 mm/year.

TABLE 9.3 SEA WATER SOLUTES

		Abundance (ppm)	
Element	**Form**	**Oceans**	**Rivers**
Cl	Cl^-	19,353	7.8
Na	Na^+	10,781	6.3
Mg	Mg^{2+}	1280	4.1
S	SO_4^{2-}	905	3.7
Ca	Ca^{2+}	415	15
K	K^+	399	2.3
C	CO_2, CO_3^{2-}	27	11.4
O	O_2 gas	4.8	~10
	Compounds[a]	1970	~250
Si	SiO_3^{2-}	3.1	6.1
Fe	Fe^{2+}	0.04	0.37
U		0.0032	0.0005
Mg/Na		0.119	0.651
S/Cl		0.047	0.474

(Continued)

TABLE 9.3 *(Continued)* SEA WATER SOLUTES

Element	Form	Abundance (ppm)	
		Oceans	Rivers
Br	Br^-	67	0.02
N	N_2 gas	10	~9
	Nitrate	0.84	~1
Sr	Sr^{2+}	7.8	0.05
B	B^{3+}	4.4	0.01
F	F^-	1.3	0.1

Note: Abundances of elements expressed as parts per million by mass of sea water with an average solute concentration of 3.514% [data selected from an extensive table by Fegley (1995)] and a comparison with the global average river input.

[a] Sulphate, carbonate, silicate, nitrate.

TABLE 9.4 DENSITIES AND EXPANSION COEFFICIENTS OF FRESHWATER AND SEA WATER AT ATMOSPHERIC PRESSURE AND TEMPERATURES 0°C–40°C, WITH SEA WATER ACOUSTIC VELOCITY

Temp. (°C)	Freshwater ρ (kg m^{-3})	Freshwater α (10^{-6}K^{-1})	Sea Water (3.514% salinity)[a] ρ (kg m^{-3})	Sea Water (3.514% salinity)[a] α (10^{-6}K^{-1})	Acoustic Vel. (m s^{-1})
-1.9	–	–	1028.5	0	1441.4
0	999.87	-67	1028.4	27	1449.0
4	1000.00	0	1028.1	83	1465.6
5	999.92	16	1028.0	97	1470.6
10	999.73	88	1027.2	167	1489.8
15	999.13	151	1026.2	214	1506.7
20	998.23	206	1025.0	256	1521.4
25	997.07	256	1023.6	297	1534.3
30	995.67	302	1022.0	336	1545.5
35	994.06	344	1020.2	373	1554.9
40	992.25	381	1018.2	407	1562.7

[a] Variation in sea water density, ρ, with salinity, S (%), at T = 0°C to 40°C: $d\rho/dS$ = (8.0 – 0.014T(°C)) kg m^{-3} per 1% salinity increment at $S \approx 3.5\%$.

TABLE 9.5 PRESSURE VARIATIONS IN DENSITY, ρ, AND BULK MODULUS, *K*, OF SEA WATER (3.514% SALINITY) AT 4°C

P (MPa)	ρ (kg m^{-3})	*K* (GPa)
0	1028.13	2.149
20	1037.47	2.272
40	1046.40	2.396
60	1054.95	2.521
80	1063.15	2.644
100	1071.04	2.766

TABLE 9.6 ELECTRICAL CONDUCTIVITY, σ (S m^{-1}), OF SEA WATER AS A FUNCTION OF SALINITY AND TEMPERATURE

	Salinity (Mass %)			
Temp. (°C)	**3.0**	**3.5**	**3.514**	**4.0**
0	2.523	2.906	2.917	3.285
5	2.909	3.346	3.358	3.778
10	3.313	3.808	3.822	4.297
15	3.735	4.290	4.305	4.837
20	4.171	4.788	4.805	5.397
25	4.621	5.302	5.321	5.974
30			5.852	

Note: Measured values at salinities of 3.0%, 3.5% and 4.0% are interpolated to the global average of 3.514%, for the composition in Table 9.3. These are values are at atmospheric pressure. The pressure effect is slight.

9.3 TIDES AND TIDAL FRICTION

9.3.1 The Gravitational Forces

Deformation of the Earth by the gravitational forces of the Moon and Sun is described in terms of the gravitational potential due to the deforming body (mass m at distance R from the centre of the Earth) at any point on the surface (radius a) at angle ψ to the Earth–Moon or Sun axis.

$$W_2 = -\frac{Gma^2}{R^3}\left(\frac{3}{2}\cos^2\psi - \frac{1}{2}\right) \tag{9.1}$$

where G is the Newtonian gravitational constant. The gradient of this potential is the gravitational force at any point, with the geometric form represented in Figure 9.4, which causes ellipsoidal deformation.

At their mean distances, the factor (Gma^2/R^3) is 3.505 m^2s^{-2} for the Moon and 1.609 m^2s^{-2} for the Sun. The forces derived from this equation are those that would be exerted on a hypothetical rigid Earth, and the elastic response of the real Earth is summarised by three dimensionless numbers, k and h by A. E. H. Love and a third, l, by T. Shida, all generally referred to as Love numbers and defined as follows, with numerical values:

$k = 0.245$, referred to as the potential Love number, is the ratio of the additional potential caused by the redistribution of mass by tidal deformation of the Earth to the deforming potential represented by Equation 9.1. Calculation of the value from elasticity of the solid (oceanless) Earth gives $k = 0.298$.

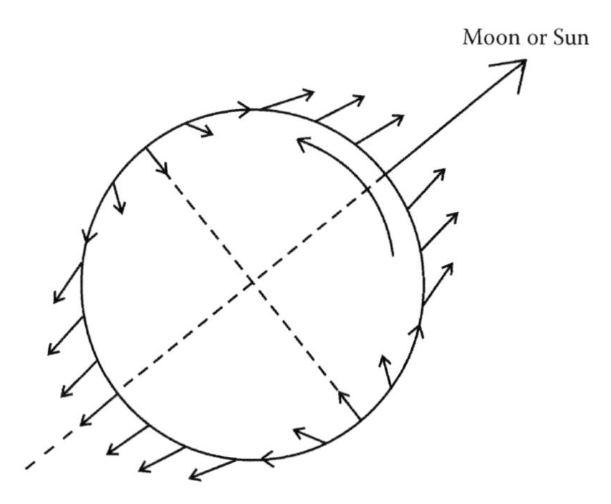

Figure 9.4 Tidal force at the surface of the Earth. This is the gradient of the tidal potential, W_2.

The difference is caused by the oceans for which the globally averaged tide is inverted, with highs appearing where an equilibrium tide would give lows;

$h = 0.603$ is the ratio of the height of the solid Earth tide to that of the deforming potential;

$l = 0.084$ (Shida's number) is the ratio of the horizontal component of the tidal displacement of the crust to the equilibrium displacement that would occur if the Earth were fluid.

Direct observation of individual Love numbers became possible with the availability of satellite data, but numerical values were obtained from surface data that give two combinations. Using the solid Earth value of k, they are as follows:

$\delta = 1 + h - (3/2)k = 1.156$, termed the gravimetric factor, is the ratio of observed gravity to that on a hypothetical undeformed Earth. In this expression, '1' represents the gradient of the deforming potential, W_2/g, h represents the tidal elevation at a point of observation of gravity, g, and the k term arises from the mass redistribution by tidal deformation.

$\eta = 1 + k - h = 0.695$, the tilt factor. This is observed astronomically as tidal tilt of the surface, relative to tilt of the deforming potential (the '1' in this expression).

Although Love numbers, especially k, are useful in global geophysics, they are imprecise because heterogeneity of the Earth and the complicated response of the oceans cause slightly different values to apply to different kinds of observation. In the case of k, this is indicated by applying a subscript, writing k_2 for ellipsoidal deformation, the subscript '2' indicating that it refers to a second-degree harmonic.

As illustrated in Figure 9.4, the tidal forces cause a deformation in the form of a prolate ellipsoid with its long axis oriented towards the Moon or Sun. At most latitudes, rotation of the Earth within this deforming envelope gives a dominantly semi-diurnal tide, with two highs and two lows per day. However, the rotational axis is inclined to the normal of the Moon and Sun orbits, resulting in diurnal components of tides that are prominent at high latitudes. The tides are complicated by ocean geometry, which locally modifies the relative amplitudes of different tidal components. Additionally, both diurnal and semi-diurnal tides are modulated by the monthly and annual cycles and by ellipticities of the orbits, giving sideband frequencies to the tidal periods, and there are small long-period tides, particularly that due to precession of the lunar orbit. The principal lunar and solar semi-diurnal periods are 12.42 hours and 12 hours, respectively, and the superposition causes a beat in tidal amplitude, with maxima (spring tides) and minima (neap tides) repeated every 14.75 days.

Without allowance for the complicated effects of marine tides, but including the factor $(1+k)$, the mean lunar tidal amplitude can be represented by the displacement, δa, of an equipotential surface, at radius a and the tidal contribution to gravity, δg, referred to the angle ψ between the radius to a point of observation and the Earth–Moon or Earth–Sun line, as in Equation 9.1.

$$\delta a/a = 3.5 \times 10^{-8} (3\cos^2\psi - 1) \tag{9.2}$$

$$\delta g/g = 7.0 \times 10^{-8} (3\cos^2\psi - 1) \tag{9.3}$$

The amplitude of the mean solar tide is 0.46 of these values and both vary with distance to the Moon or Sun according to the orbital ellipticities. The marine tide is far from equilibrium, being geographically highly variable with many local effects arising from delays and resonances caused by ocean basin geometry and sea-floor topography. The water load causes gravity variations, both directly and by deforming the solid ground, and for tidal corrections to gravity survey data, local information about gravity variations is needed.

9.3.2 Energy Dissipation (Tidal Friction)

Figure 9.4 indicates a prolate deformation of the Earth, with the elongation aligned with the Earth–Moon (or Sun) axis, but that is not the precise situation. The global tidal bulge is delayed by 12 minutes, corresponding to a 2.9° misalignment with that axis, as in Figure 9.5. The delay is a consequence of energy dissipation by the tides, predominantly the marine tide, but with a significant contribution by inelasticity of the solid Earth. The total tidal energy dissipation is 3.7×10^{12} W, of which ~15% is attributed to the solid tide. Consequences include slowing of the Earth's rotation, an increase in the length of day by 2.4 μs/year and expansion of the Moon's orbit by 3.7 cm/year.

Over geological time, tidal friction has caused a major change to the rate of the Earth's rotation and the Earth–Moon distance. The primordial rotational

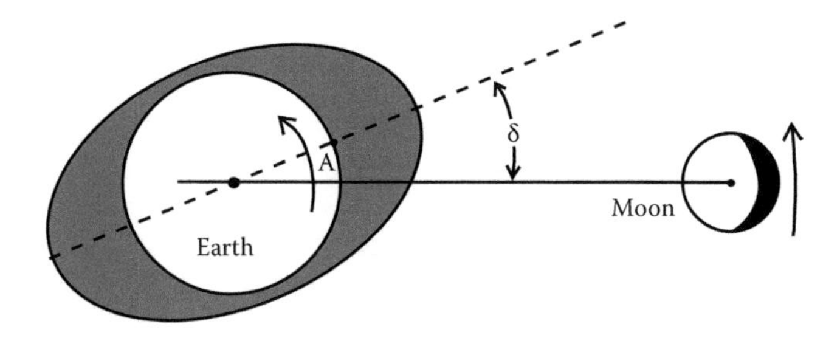

Figure 9.5 Tidal friction causes a delay in the tides and a misalignment of the tidal bulge, relative to the Earth–Moon axis. The angle δ is 2.9°, and the point A was aligned with the Moon 12 minutes ago.

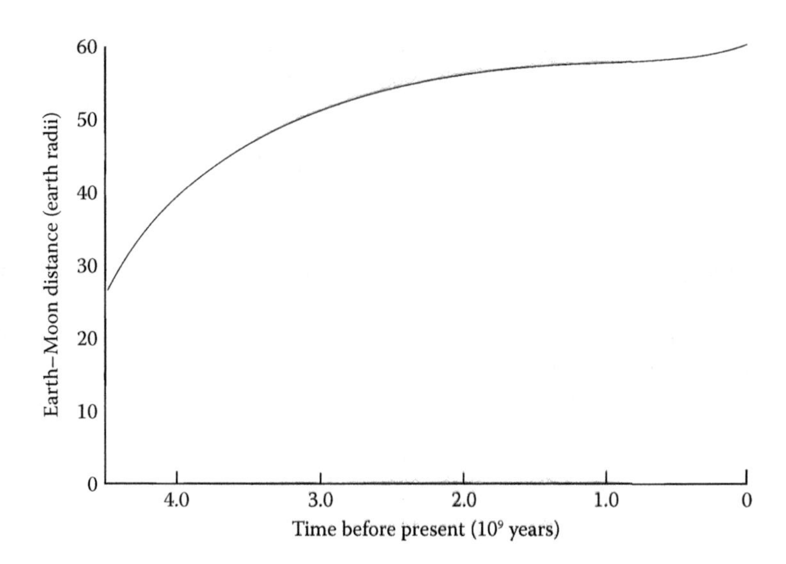

Figure 9.6 Earth–Moon separation over geological time inferred from tidal periods observed in ancient estuarine sediments.

period would have been in the range of 6 to 8 hours, with the lunar distance varying in the manner of Figure 9.6. Shorter-term variations in the rotation rate arising from irregularities in atmospheric and core motions are referred to in Section 5.3.

Chapter 10

Inland Salt Lakes

There are areas on all continents where topography and geological structures allow the accumulation of water with very restricted or no outflow to the oceans. Solutes from inflowing water are retained as the water itself evaporates and saline lakes are formed with various salinities (Table 10.1). Completely closed catchment areas, with no outflow, are termed endorheic, but this does not generally mean hydraulic isolation. In most cases, in addition to rainfall and river inputs, water both enters and leaves inland lakes as flow of groundwater, so that their salt compositions do not match the salts in rivers and streams flowing into them. This is well demonstrated by the largest of all salt lakes, the Caspian Sea, in which ionic ratios Na^+/Mg^{2+}, Ca^{2+}/Mg^{2+} and Cl^-/SO_4^{2-} differ from the ratios in the river input by factors of 2.33, 0.14 and 0.44, respectively. An interesting example of a lake that is apparently endorheic, but in which the water is fresh, is Lake Chad. In a dry area of high evaporation, bounded by four countries (Chad, Cameroun, Niger and Nigeria) all with irrigation ambitions, water is lost by groundwater flow into neighbouring wetlands, with a small river inflow sufficing to maintain a fresh lake. As the data in Table 9.3 show, there is also a compositional difference between the oceans and their river input, but in that case, the major reason is that salt is introduced by hydrothermal circulation of mid-ocean ridges.

Salt lakes are vulnerable to natural climatic variations and to human activity. Most occur in arid or semi-arid regions where evaporation is high and input low; complete drying up was the fate of many former lakes now apparent only as salt deposits. Some lakes are described as ephemeral, existing intermittently according to rainfall, a striking example being Lake Eyre in south-central Australia, which exceeds 9000 km^2 in area for a year or so once every few decades, but almost completely dries up in unfavourable times. Lake areas in Table 10.1 are maxima, with < signs to indicate that these may be former or transient areas. In some cases, river inputs have been diverted to irrigation projects, reducing lakes to token fragments, the most disastrous example being

TABLE 10.1 MAJOR SALT LAKES, INCLUDING THOSE IN THE SUB-SALINE CATEGORY (0.05% < SALINITY < 0.3%), NOT NECESSARILY ENDORHEIC

Lake	Boundary Countries	Area (km^2)	Salinity (%)	Ion Ratios	
				Mg^{2+}/Na^+	SO_4^{2-}/Cl^-
Caspian Sea	5 countries[b]	374,000	up to 1.4	0.233	0.605
Kara Bogaz[a]	Turkmenistan	<18,000	34	0.123	0.22
Balkhash	Kazakhstan	<18,000	0.05 to 0.7	19.2	1.6
Eyre	Australia	<9500	5 to 30	–	–
Titicaca	Peru, Bolivia	8300	0.08	0.19	1.08
Turkana (Rudolf)	Kenya, Ethiopia	<6400	0.2	Alkaline ($NaHCO_3$)	–
Issyk Kul	Kyrgistan	6200	0.6	2.6	1.4
Urmia	Iran	<5200	>30	–	–
Qinghai	China	<5700	1.3	–	–
Great Salt	USA	<8000	5 to 27	–	–
Van	Turkey	3700	2.4	0.36	0.42
Aral Sea	Kazakhstan, Uzbekistan	<3500	~1	–	–
Kivu	Rwanda, Congo	2300	Gas emission	0.22	0.43
Dead Sea	Jordan, Israel	<1000	34	1.08	0.002

[a] A satellite lake on the eastern shore of the Caspian Sea, receiving water from it by a narrow channel that was blocked for several years by a dam, now dismantled.
[b] Russia, Kazakhstan, Azerbaijan, Turkmenistan, Iran.

the Aral Sea in central Asia (Kazakhstan–Uzbekistan), where fishing has been replaced by dry salt which renders useless the agricultural land across which it blows.

Total salinity reflects the input–output competition. This applies to comparisons between lakes and to the seasonal and other variations in individual lakes, which have increasing salinity as their volumes decrease. It is measured as the total ionic content, generally expressed in milligrams per litre (parts per million), but in parts per thousand (‰) or per cent (%) for high concentrations. Different salinity ranges are categorised by type, distinguished by the biota adapted to them (Table 10.2). The sub-saline/saline boundary approximately marks the lower limit of salinity at which human taste identifies water as salty.

TABLE 10.2 SALINITY CATEGORIES

Salinity Type	Salinity Range (%)
Fresh	<0.05
Sub-saline	0.05 to 0.3
Saline	
Hyposaline	0.3 to 2.0
Mesosaline[a]	2.0 to 5.0
Hypersaline	>5.0

[a] This range brackets the average salinity of sea water, 3.5%.

Natural lakes cover the complete range from fresh to saturated with some constituents. Lakes with high concentrations are used to extract chemical industry feedstock from deposits or evaporation ponds on their boundaries and the various lake compositions determine which elements are usefully extracted. An interesting case is the Dead Sea, which is uniquely rich in bromine and is used for bromine extraction in plants on both the Jordanian and Israeli shores.

Chapter 11

Freshwater and Ice

Although there is an arbitrariness in the definition of freshwater used here (salinity $<0.05\%$, Table 10.2), for most purposes there is little difficulty in making the distinction because most of the water we consider is either well below or clearly above this limit. The only significant doubt is the fraction of groundwater that can be identified as fresh. The flow of groundwater into the sea is not well constrained by observations but is generally believed to be of order 10% of the river flow. However, much groundwater is saline and its contribution to ocean salt probably exceeds that of the river input. Table 11.1 summarises the distribution of the Earth's water. An estimate of the total water content of the Earth requires some information about the water in the mantle, which probably exceeds all the rest put together. Table 11.1 reports an upper limit, obtained from reported water contents of mineral inclusions in diamonds of very deep origin, which give more than 0.2% (Shirey et al. 2013), assuming this to be representative of the deep mantle. Water in Hawaiian lavas identified by Kokubu et al. (1961) as juvenile amounted to 0.09%. If this is a better representation of the mantle average, then the water content is 3.6×10^6 in units of 1000 km^3. The hydrogen content of the mantle listed in Table 7.3, derived from geochemical evidence of element partitioning, gives a value of 2.4×10^6 in the same units if it is assumed all to be in the form of molecular H_2O or as related H^+ and OH^- ions. Mantle rheology is controlled by water, or its separated ions, but does not give an estimate of concentration.

11.1 FRESHWATER LAKES

In a census of the Earth's freshwater, lakes are crucial, not just because they are a major store of water, but because they are windows on the water table and give some insight on the much larger store of groundwater and the extent to which that can be deemed fresh. Most studies of the global distribution of

TABLE 11.1 THE GLOBAL WATER INVENTORY

Condition	Location	Volume (units of 1000 km^3 = 10^{15} kg)
Saline	Oceans	1,340,000
	Lakes	86
	Ocean crust	20,000
	Continental groundwater	12,000
Fresh	Ice caps and glaciers	24,000
	Lakes and rivers	91
	Impoundments	3
	Wetlands	11,000
	Groundwater and permafrost	11,000
	Soil moisture	15
	Atmosphere	13
	Mantle[a]	up to 10^7

[a] Although not securely determined, water in the mantle is more abundant than has often been recognised, as discussed in the introductory paragraph to this chapter.

lakes have considered lake areas, with estimates of the global total lake area. Satellite surveys have simplified (and improved) this by properly including the very large number of small lakes, but do not distinguish freshwater from the saline lakes considered in Chapter 10. The conclusion is that the total area of surface water on land, excluding constructed reservoirs but including saline as well as freshwater, is about 4.2×10^6 km^2, 2.8% of the land area, with roughly equal contributions by each decade of the lake area range (10 km^2 to 100 km^2, 100 km^2 to 1000 km^2, etc.). Although the very small lakes make a major contribution to the total lake area, the depths are less than the depths of the largest lakes, which account for most of the total lake volume (Table 11.2) and this is, therefore, reasonably well constrained.

11.2 MAN-MADE RESERVOIRS

As with natural lakes, reservoirs constructed for various purposes cover a very wide size range from major hydroelectric facilities to small farm ponds, but none are as large as the biggest natural lakes and the few in the volume range of

TABLE 11.2 NATURAL FRESHWATER LAKES WITH VOLUMES EXCEEDING 100 km³

Lake	Bordering Countries	Area (km²)	Volume (km³)
Baikal	Russia	31,700	23,600
Tanganyika[a]	Tanzania, Congo, Burundi, Zaire	32,900	19,000
Superior	Canada, USA	82,100	12,000
Malawi (Nyasa)	Malawi, Tanzania, Mozambique	29,600	8000
Vostok	Antarctica	Subglacial	~5000
Michigan[b]	USA	57,800	4900
Huron[b]	Canada, USA	59,500	3200
Victoria	Tanzania, Uganda, Kenya	68,800	2750
Great Bear	Canada	31,300	2400
Ontario	Canada, USA	19,000	1700
Great Slave	Canada	28,600	1600
Ladoga	Russia	17,700	920
Erie	Canada, USA	25,660	490
Khövsgöl (Hovsgol)	Mongolia	2750	480
Winnipeg	Canada	24,500	295
Onega	Russia	9700	285
Nipigon	Canada	4848	250
Toba	Indonesia (Sumatra)	1130	240
Argentino	Argentina	1420	220
Athabasca	Canada	7850	205
Mistassini	Canada	2150	160
Vänern	Sweden	5600	155
Tahoe	USA	500	150
Albert (Nyanza)	Uganda, Congo	5300	130
Iliamna	USA (Alaska)	2600	120
Nettilling	Canada (Baffin Is.)	5300	115
Nicaragua	Nicaragua	8200	110

[a] Of the lakes listed here, Tanganyika comes closest to being classified as sub-saline and included in Table 10.1. It is sometimes locally brackish. Surface outflow is limited, evaporation is high and the lake is very long, allowing heterogeneity. Although the average solute concentration is about 0.05%, at the limit of the 'fresh' definition, it is dominated by bicarbonate ions, HCO_3^- with much less Na^+ or Cl^-, which would be of greater concern, and the water is regarded, and used, as fresh.

[b] These are often regarded as a single lake.

TABLE 11.3 MAN-MADE RESERVOIRS WITH STORAGE VOLUMES EXCEEDING 75 km^3

Lake	Country	Volume (km^3)
Kariba	Zambia, Zimbabwe	180
Bratsk	Russia	169
Nasser/Nubie	Egypt, Sudan	157 (132 in Egypt)
Volta	Ghana	150
Manicouagan	Canada	142
Guri	Venezuela	135

Table 11.2 are listed in Table 11.3. In listing them by water volume, the capacity at complete filling is reported but usage and variations in rainfall or input river flow mean that this overstates the actual water storage at any time, often by a factor of 2. All of the large dams have environmental effects and some face serious maintenance and safety issues. Taking into account the average fullness of man-made dams and ponds, the water storage reduces sea level by about 1 cm, an effect more than compensated by pumping groundwater.

There are 29 other reservoirs exceeding 30 km^3, but even these exclude some major hydroelectric installations, in particular Itaipu (Brazil, Paraguay), 29 km^3, on the Paraná River, which, in most years, generates more electrical power than the larger Three Gorges Dam installation (China) on the seasonally variable Yangtze River. Major hydroelectric stations are listed in Section 25.2.2.

11.3 RIVERS, STREAMS AND LAND DRAINAGE

Of the ~117,000 km^3/year of water that falls on land as rain or snow, 64% is returned to the atmosphere by evaporation and transpiration from vegetation. Most of the balance is conveyed to the oceans by rivers and streams, with a smaller amount (~5000 km^3) sinking in and becoming groundwater, which mostly reaches the oceans after a delay that may be very prolonged, even millions of years. Of the 37,000 km^3/year that flows more directly to the oceans, 50% is carried by the major rivers listed in Table 11.4. A disproportionately large fraction flows into the Atlantic, reflecting the fact that drainage basins connected to the Atlantic occupy 48% of the land area.

The discussion in Section 5.4.3 of gravity monitoring by the GRACE satellites points out that they are sensitive enough to detect, as localised removal

TABLE 11.4 RIVERS DISCHARGING MORE THAN 200 km³/year OF WATER INTO THE OCEANS, PLUS TWO OTHERS NOTABLE FOR PRESENT OR FORMER HEAVY SEDIMENT LOADS, WITH GEOGRAPHIC COORDINATES OF RIVER MOUTHS

Continent	River	Catchment Area (10^6 km²)	Average Discharge (km³/year)	River Mouth Coordinates
S America	Amazon	7.05	6600	0N 49W
Africa	Zaire (Congo)	4.01	1300	6S 12.5E
Asia	Ganga (Ganges)	1.64	1200	22N 91E
S America	Orinoco	0.88	1140	9S 62W
Asia	Chang Jiang (Yangtze)	1.81	950	31.5N 122E
S America	Rio de la Plata (Parana estuary)	3.10	695	35.5S 57W
Asia	Yenisey (Yenisei)	2.58	620	73N 80E
N America	Mississippi	3.20	536	29N 89W
Asia	Lena	2.49	533	73N 126E
N America	St Lawrence	1.03	530	49N 63W
Asia	Mekong	0.81	467	9.5N 106.5E
Asia	Irrawaddy	0.41	410	15.5N 95E
Asia	Ob	3.00	394	73N 73E
Asia	Amur	1.86	360	53N 141.5E
N America	MacKenzie	1.79	357	69.5N 135.5W
Asia	Zhujiang (Pearl)	0.45	300	22.5N 114E
Europe	Danube	0.82	255	45.5N 29.5E
Europe	Volga	1.38	254 (to Caspian Sea)	46N 48E
N America	Columbia	0.67	237	46.5N 124W
Africa	Zambeze (Zambezi)	1.33	223	19S 36.5E
S America	Magdalena	0.26	220	11S 74.5W
Asia	Indus	1.17	208	24N 67E
N America	Yukon	0.85	203	63N 165W
Asia	Huang He (Yellow)	0.74	81	38N 119E
Africa	Nile	3.26	75	31.5N 30–32E

of mass, groundwater depletion in major aquifers. This appears to be a growing problem in arid areas, although the global groundwater store is not diminishing.

11.3.1 Drainage Patterns

Land drainage is studied from several perspectives, which include flooding, hydropower and reservoir siting, erosion, sediment transport and ground stability, relationships to tectonic history, faults, local land structure and rock fabric, soil moisture and fertility. There have been numerous methods of quantifying stream patterns to develop systematic procedures applicable to different situations and, although there are almost as many methods as examples, four that have had multiple uses are illustrated in Figure 11.1 and detailed in Table 11.5. They all apply numbers to streams, starting with 1 for the smallest and incrementing as illustrated in Figure 11.1, at junctions. This requires decisions about categories of tributaries, for example, where a small stream joins a major one that has a hierarchy of tributaries higher upstream.

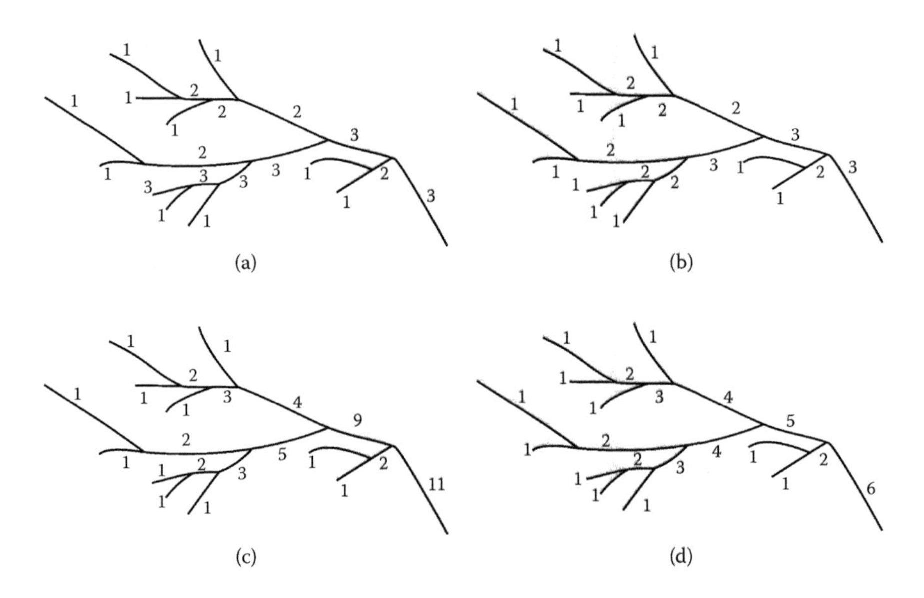

Figure 11.1 Stream numbering in four ordering systems: (a) Horton, (b) Strahler, (c) Shreve and (d) Hodgkinson et al., listed in Table 11.5.

TABLE 11.5 A COMPARISON OF USES/ADVANTAGES AND LIMITATIONS/ DISADVANTAGES OF THE FOUR STREAM ORDERING METHODS ILLUSTRATED IN FIGURE 11.1

Method Name	Numbering Scheme	Uses/Advantages	Limitations/ Disadvantages
(a) Horton 1945	Provides a simplified system with few orders through the network. Does not assign a new order at every new node.	Produces a natural stream ordering system with main streams and tributaries; able to identify main channel. Often combined with Strahler system for mathematical calculations of catchment parameters.	May cause orders to be ascribed to channels that lead to branched sections being designated high orders, regardless of size. May work best after applying Strahler ordering, then referred to as Horton–Strahler system.
(b) Strahler 1952, 1957	Does not assign a new order at every new node. A stream reach increases by one order at every point at which two streams of the same order join.	Has a mathematical background; can calculate 'bifurcation ratio' to identify flooding potential.	Assumes that flood is the accumulation of upstream content and does not consider additional local overland flow or back up of lower channels; may under-represent final discharge.
(c) Shreve 1966, 1967	Assigns a new stream order at every node. The new stream order is the sum of the orders of the adjoining reaches.	Is designed to mathematically consider magnitude of reaches and network discharge through the summing of orders at each connection or node.	'Magnitude' only considers upstream channel flow accumulation but does not consider additional overland flow directly into the channel; some stream orders may be 'missing'.

(Continued)

TABLE 11.5 *(Continued)* A COMPARISON OF USES/ADVANTAGES AND LIMITATIONS/DISADVANTAGES OF THE FOUR STREAM ORDERING METHODS ILLUSTRATED IN FIGURE 11.1

Method Name	Numbering Scheme	Uses/Advantages	Limitations/Disadvantages
(d) Hodgkinson et al. 2006	Assigns a new stream order at every node. New order number increases by 1, downstream, after every node.	Considers simple terms of 'magnitude' by increasing order at each node (as smaller streams join each trunk); includes every consecutive order; allows similar orders to be compared across the network for segment lengths, direction for each order and assessing network controls. Used as a topological or descriptive numbering system.	Designed for restricted use in topology. Not based on mathematical background.

11.4 SUSPENDED SOLIDS

11.4.1 Global Range

Standard definitions of sediment concentrations are listed in Table 11.6.

11.4.2 Turbidity

This is a measure of the concentration of suspended material in a liquid, apparent as cloudiness and quantified by optical observations. It is mainly applied to the assessment of water quality in natural situations (rivers, lakes and oceans) and in water supplies, but can be used in the description of any imperfectly transparent material. Conceptually, it is most simply determined either from the attenuation of light transmitted through a sample or the obscurity of an object viewed through it. For rapid, semi-quantitative field observations in sufficiently deep water, a Secchi disc of strongly contrasting black and white quadrants is lowered into the water until the disc is no longer visible.

TABLE 11.6 TOTAL SUSPENDED SOLIDS[a]

Suspended Sediment Concentration (Cs)	Typical Setting	Example
Very low (5–20 mg l^{-1})	Downstream of major lakes, flat humid regions, wetlands	Central Amazonia, Rhone Lacustre
Low (20–100 mg l^{-1})	Mountainous headwaters, low-relief basins draining erodible rocks	Loire, Sacramento
Medium (100–500 mg l^{-1})		Huai Mae Ya
High (500–2000 mg l^{-1})	Active mountain ranges	Alps, Caucasus, Andes, Mississippi
Very high (2000–10,000 mg l^{-1})	Active mountain ranges, highly erodible rocks, steep volcanic islands, semiarid regions	Rio Grande
Extremely high (10,000–50,000 mg l^{-1})	Absence of vegetative cover, medium to steep slopes, erosive seasonal rainfall, highly erodible material	Colorado, Nile, Huang He

[a] The total quantity of solid material is generally reported in mg per litre of water. These representative numbers can be compared with the global average mass of river sediment carried to the deep oceans, 540 mg/l (2×10^{13} kg/year in 37,000 km^3/year of water).

The observed Secchi depth is an inverse measure of turbidity that allows immediate qualitative comparisons between different natural environmental situations. For more quantitative measurements on small samples, the favoured instrumental method uses a nephelometer, which measures the scattering rather than attenuation of light. A light beam is incident on a sample, typically 2.5 cm in dimension, and the ratio of the intensity of light scattered at 90° to the sum of the intensities transmitted and reflected from the incident surface is recorded as the turbidity in nephelometric turbidity units (NTU). Inevitably, there are calibration questions and dilutions of formazin have been used as standards, but this is a noxious material and most observations are very variable, making precise observations unnecessary or even irrelevant. Dilutions of milk provide more readily accessible standards. For undiluted milk, turbidity varies from about 5000 NTU (full cream) to about 3000 NTU (reduced fat) and the scale is linear for lower turbidities obtained by dilution.

Standards of treated water supplies specify turbidity limits, typically ~1 NTU, but differing between administrations. Turbidities of river water are much higher than this, but depend on season and recent rainfall. Table 11.7 gives some indicative variations and examples.

TABLE 11.7 TURBIDITIES OF VARIOUS WATER SOURCES

Water Source	NTU
Processed drinking water	0.1 to 5
Thames	14 (average)
Upper Mississippi	15 to 37
Amazon	6 to 81
Congo	35
Brisbane	35 to 120
Ganges	54.9 to 404
White Nile	115
Blue Nile	7275

11.5 ICE

Including Antarctica and Greenland, approximately 11% of the Earth's land area is covered with permanent ice. Antarctica (14×10^6 km^2) accounts for nearly 90% of this and Greenland (1.1×10^6 km^2) 8%, with smaller glaciations on all major land masses except Australia. The total ice volume is 30×10^6 km^3, 68% of the global total of fresh water. Northern high latitudes have an additional seasonal snow and ice cover. If all the ice were to melt rapidly, sea level would rise by 76 m, reducing to about 62 m by isostatic adjustment over a few thousand years. A slow melt is ongoing and this is the major contributor to the present rate of sea level rise (3.2 mm/year, see Figure 9.3). Although Greenland has only about 11% as much ice as Antarctica, the rate of its melting is twice as fast. The melting of floating sea ice (Figure 11.2) is also faster in the Arctic,

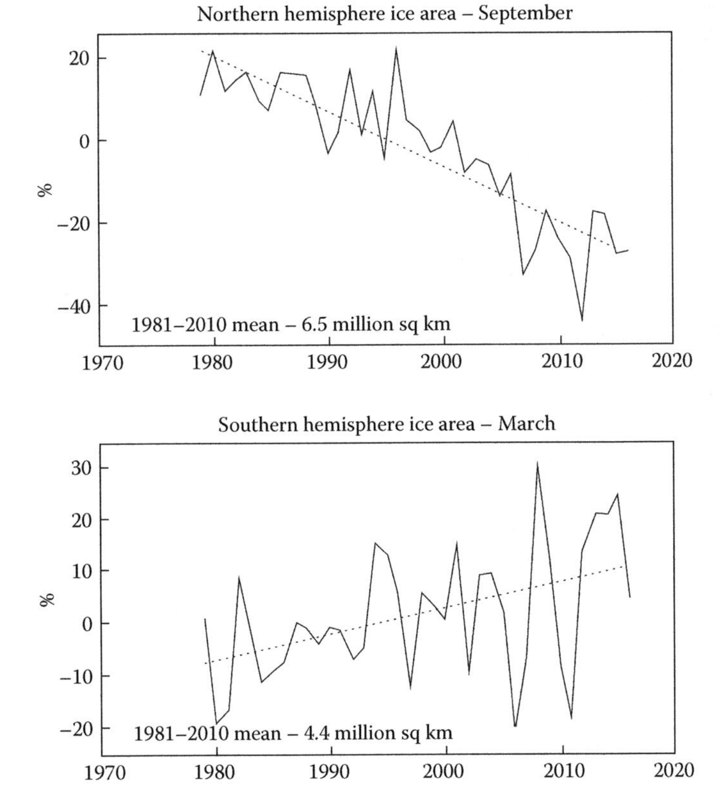

Figure 11.2 Percentage changes in sea ice over the period of satellite observations. (Courtesy of National Snow & Ice Data Centre, www.nsidc.org.)

but this does not contribute to the sea level rise, which is caused by the melting of land-based ice, with a smaller contribution by thermal expansion of the warming oceans. The complete Arctic ice cap is floating, but the Antarctic cap is almost completely grounded and the trend of the southern sea ice area in Figure 11.2 is slightly misleading both because it is thinning as its area increases, with a decrease in volume, and because it does not take account of the melting of grounded ice.

11.6 GLOBAL WATER EXCHANGES

Rain and snowfall: World total	504,000 km^3/year
On land	117,000 km^3/year
At sea	387,000 km^3/year
Evaporation and transpiration: From land	75,000 km^3/year
From sea	429,000 km^3/year
Land to sea flow: Rivers and streams	37,000 km^3/year
Groundwater 5000 km^3/year,	~500 km^3/year fresh
Groundwater pumping (human use)	~100 km^3/year
Average duration of atmospheric water	9.4 days
Average duration of ocean water (ocean volume/rainfall)	3250 years
Melted land-based ice to oceans	900 km^3/year
Subduction of water	~1 km^3/year
Average volcanic emission of water	<1 km^3/year

Chapter 12

The Atmosphere

12.1 GENERAL PHYSICAL PROPERTIES OF AIR AND THE ATMOSPHERE

Total atmospheric mass: 5.136×10^{18} kg

Moment of inertia: 1.413×10^{32} kg m^2

Temperature gradients:

Dry adiabat: $g/C_P = 9.75$ K/km

Tropospheric mean ≈ wet adiabatic gradient: 6.5 K/km

Mean molecular weight:

Dry air: 28.965

Water vapour: 18.1766

Density dry air:

at $P_0 = 101.325$ kPa, $T_0 = 273.15$ K: $\rho_0 = 1.291$ kg m^{-3}

$T_0 = 288$ K (15°C, mean surface T): $\rho_0 = 1.225$ kg m^{-3}

at other pressures and temperatures, P and T, $\rho = \rho_0 (P/P_0)(T_0/T)$, as for an ideal gas

Density of wet air, with a partial pressure P_w of water vapour:

$\rho_{wet} = (T_0/T)(P/P_0 - 0.372\, P_w/P_0)\, \rho_0$

where P_w is the product of relative humidity and saturation vapour pressure (Table 12.6).

Specific heat at constant pressure:

Dry air, $C_P = 1004$ J K^{-1} kg^{-1}

Water vapour, $C_P = 1830$ J K^{-1} kg^{-1}

Wet air, partial pressure of water vapour P_w: $C_P = (1004 + 826\, P_w/P)(1 - 0.372\, P_w/P)$ J K^{-1} kg^{-1}

Latent heat of evaporation, see Section 12.3.
Thermal conductivity: dry air at 15°C, $\kappa = 0.0252$ W m^{-1}K^{-1}, independent of pressure
Thermal diffusivity at 15°C and 101.325 kPa, $\eta = \kappa/\rho C_P = 20.4 \times 10^{-6}$ m^{2}s^{-1}
Dielectric constant, dry air at standard P, T: $k = 1 + 5.46 \times 10^{-4}$
Electrical resistivity in fine weather at 100 V/m: $\rho_e = 5 \times 10^{12}$ Ω m (for atmospheric electricity, see Section 12.5)
Magnetic susceptibility at standard P, T: $\chi = (\mu/\mu_0 - 1) = 3.74 \times 10^{-7}$
Thermal expansion coefficient: $\alpha = 1/T(\mathrm{K}) = 3.47 \times 10^{-3}$ K^{-1} at 15°C
Bulk modulus, K (1/compressibility)

Isothermal: $K_T = P$

Adiabatic: $K_S = (C_P/C_V)\,P = 1.402\,P$

C_P and C_V are specific heats at constant pressure and constant volume, respectively. For diatomic gases, such as N_2 and O_2, $C_P/C_V = 1.4$. The next most abundant atmospheric component is Ar, which is monatomic, with $C_P/C_V = 1.67$, but it only raises the atmospheric ratio to $C_P/C_V = 1.402$.
Speed of sound:

In air, $v = \sqrt{(K_S/\rho)} = 1.184 \times \sqrt{(P/\rho)}$

At constant T, $\rho \propto P$ and v is independent of pressure, but $\rho \propto 1/T$, as in the ideal gas equation, and $v \propto \sqrt{T}$

At 15°C and low altitudes, $v = 340.5$ to 341.2 m s^{-1}, depending on humidity (at 30°C, add 0.2 m s^{-1} per 10% humidity)

At modern aircraft altitudes, in the lower stratosphere, $v \approx 295$ m s^{-1}
Viscosity: 1.81×10^{-5} Pa s at 15°C, 1.52×10^{-5} Pa s at −40°C, independent of pressure.
Refractive index: see Section 12.4.

12.2 ATMOSPHERIC STRUCTURE AND COMPOSITION (TABLES 12.1 THROUGH 12.5)

Table 12.1 is a model giving pressure, P, temperature, T, and density, ρ, as functions of elevation, H. It best represents mid-season conditions at about 45° latitude. Geographical, seasonal and meteorological departures from it are most obvious at low elevations. The total extent of the atmosphere is generally taken to include the magnetosphere (Section 5.6). Table 12.2 lists constituents of dry air, with an added value for water. Selected values are converted to fractions of total Earth mass in Table 4.4, for comparison with other planets. Table 12.3 lists recent CO_2 values from Mauna Loa data, with earlier ice core data. CH_4 values do not have a regular trend. Southern hemisphere values are systematically lower for both, with a bigger difference for CH_4. Table 12.4 lists the isotopic ratios of atmospheric gases with selected values converted to ratios by mass in Table 4.6.

TABLE 12.1 INTERNATIONAL STANDARD ATMOSPHERE

H (km)	P (kPa)	T (K)	ρ (kg m^{-3})	Identified Layers and Properties
0	101.391	288	1.225	Scale heights (factor e change):
1	89.125	282	1.112	Pressure ~7.6 km
2	79.433	275	1.007	Density ~9.5 km
3	70.795	269	0.909	
4	61.660	262	0.818	0–11 km troposphere
5	53.703	256	0.736	
6	46.774	249	0.661	
8	35.481	236	0.526	
10	26.303	223	0.413	
15	12.023	217	0.195	11–48 km stratosphere
20	5.495	217	0.089	15–35 km ozone concentration
30	1.202	227	0.019	
40	0.288	250	0.0040	
50	0.079	271	0.0010	48–85 km mesosphere
60	0.022	247	3.1×10^{-4}	50–1000 km ionosphere
70	0.0052	220	8.3×10^{-5}	
80	0.00105	199	1.9×10^{-5}	~80 km noctilucent clouds
90	1.8×10^{-4}	187	3.4×10^{-6}	
100	3.2×10^{-5}	195	5.6×10^{-7}	95 km thermosphere
150	4.6×10^{-7}	634	2.1×10^{-9}	
500	3.0×10^{-10}	999	5.2×10^{-13}	500 km + exosphere, magnetosphere
1000	7.6×10^{-12}	1000	3.5×10^{-15}	

TABLE 12.2 COMPOSITION OF THE LOWER ATMOSPHERE

Constituent	Volume Fraction	Mass Fraction
N_2	78.08%	75.51%
O_2	20.95%	23.14%
H_2O	~0.5%	~0.3%
Ar	0.934%	1.29%

(Continued)

TABLE 12.2 *(Continued)* **COMPOSITION OF THE LOWER ATMOSPHERE**

Constituent	Volume Fraction	Mass Fraction
CO_2 (2015)	0.040%	0.061%
Ne	18.2 ppm	12.7 ppm
He	5.2 ppm	0.72 ppm
Kr	1.1 ppm	3.2 ppm
H_2	0.55 ppm	0.038 ppm
CO	125 ppb	121 ppb
Xe	87 ppb	394 ppb
O_3	~30 ppb	~50 ppb
Nitrogen oxides	~330 ppb	~490 ppb
NH_3	Up to 3 ppb	Up to 1.8 ppb
SO_2	~0.05 ppb	~0.11 ppb
CH_4 (2015)[a]	1.86 ppm	1.03 ppm
Heavier hydrocarbons	Up to 90 ppb	Up to 100 ppb
Chlorofluorocarbons	Up to 1.6 ppm	Up to 7 ppm

[a] Increasing by ~1%/year.

TABLE 12.3 ANNUAL AVERAGE CONCENTRATIONS OF CO_2 AND CH_4 IN THE NORTHERN HEMISPHERE ATMOSPHERE

	CO_2		CH_4
Date	ppm by Vol.	Rate of Change ppm/year	ppm by Vol.
1000–1700	281[a] (average)	~0	0.677[a] (average)
1800	282[a]	~0	0.730[a]
1900	287[a]	0.21	0.862[a]
1930	295[a]	0.49	1.002[a]
1960	316[b]	0.92	1.210[a]
1970	326[b]	1.11	1.352[a]
1980	338[b]	1.32	1.480[a]
1990	353[b]	1.56	1.731[a]
2000	370[a]	1.83	1.785[a]
2010	389	2.12	1.814
2015	401	2.28	1.831 (2013)

[a] Ice core data from Etheridge et al. (1998) and Dlugokencky et al. (2014).
[b] Air sample data from Keeling et al. (2001).

TABLE 12.4 ISOTOPIC RATIOS OF ATMOSPHERIC GASES BY VOLUME (FOR MASS RATIOS SEE TABLE 4.6)

Constituents	Ratio
D/H	1.558×10^{-4}
$^{3}He/^{4}He$	1.40×10^{-6}
$^{13}C/^{12}C$	0.0123
$^{15}N/^{14}N$	3.67×10^{-3}
$^{17}O/^{16}O$	3.727×10^{-4}
$^{18}O/^{16}O$	2.005×10^{-3}
$^{38}Ar/^{36}Ar$	0.1879
$^{40}Ar/^{36}Ar$	296.0

TABLE 12.5 ALTITUDE VARIATION IN THE COMPOSITION OF THE OUTER ATMOSPHERE AT ALTITUDES WHERE DIFFUSIVE SEPARATION OF CONSTITUENTS OCCURS

Altitude (km)	Composition (Volume %)					
	N_2	O_2	O	He	Ar	H
100	77	19	3.4	<0.05	0.8	<0.05
150	61	5.6	34	<0.05	0.1	<0.05
200	42	3.0	55	0.01	<0.05	<0.05
300	17	0.8	81	0.8	<0.05	<0.05
400	6.0	0.2	91	2.7	<0.05	<0.05
500	1.9	<0.05	90	8.2	<0.05	0.2
700	0.1	<0.05	55	43	<0.05	1.6
1000	<0.05	<0.05	5.7	88	<0.05	6.7

12.3 WATER VAPOUR PRESSURE AND LATENT HEAT

Table 12.6 gives values of vapour pressure, P_w, in equilibrium with liquid water at $T \geq 0°C$ and in equilibrium with ice at $T \leq 0°C$ and the latent heat of evaporation of water, L, or sublimation of vapour from ice. Reference atmospheric pressure at sea level is 101.3 kPa.

TABLE 12.6 VAPOUR PRESSURE AND LATENT HEAT

T (°C)	P_w (kPa)	L (kJ/kg)
-40	0.0129	2839
-30	0.0381	2839
-25	0.0635	2839
-20	0.1034	2838
-15	0.1654	2838
-10	0.2599	2837
-5	0.4016	2835
0	0.6117 (ice) 0.6112 (water)	2834 2501
2	0.7060	2496
4	0.8135	2491
6	0.9353	2487
8	1.0729	2482
10	1.2281	2477
12	1.4028	2473
14	1.5989	2468
16	1.8187	2464
18	2.0647	2459
20	2.3393	2455
22	2.6453	2450
24	2.9858	2446
26	3.3639	2442
28	3.7831	2437
30	4.2470	2433
32	4.7600	2429
34	5.3252	2425
36	5.9481	2421
38	6.6331	2417
40	7.3853	2412

Analytical approximations to the data in Table 12.6 are:
over the range $0°C \leq T \leq 40°C$
$P_w = 0.6112 \exp(0.07192\,T - 2.406\times10^{-4}\,T^2)$ for P_w in kPa and T in °C;
for vapour pressure over ice at $-40°C \leq T \leq 0°C$,
$P_w = 0.6117 \exp(0.08120\,T - 3.830\times10^{-4}\,T^2)$.
Vapour pressures over ice and supercooled water differ significantly.

12.4 REFRACTIVE INDEX OF AIR

Refractive index, $n = 1 + \delta$, where δ is a small quantity representing the departure of n from the vacuum value ($n = 1$). In the wavelength range of the visible spectrum (430 nm to 690 nm) for dry air at 0°C and standard atmospheric pressure ($P_0 = 101.32$ kPa), δ varies slightly with wavelength, λ, as in Table 12.7. At other pressures, P, and temperatures, T (Kelvin), δ is modified according to air density in the manner of an ideal gas, so the values in the table are multiplied by the factor $(P/P_0) \times (273.15/T)$. For air with a partial pressure P_w (kPa) of water vapour, δ is reduced by $0.23 \times 10^{-6} P_w/P_0$. P_w is the product of relative humidity and the saturation vapour pressure, as listed as a function of temperature in Table 12.6.

12.5 ATMOSPHERIC ELECTRICITY

The following are typical values of quantities that are variable.

Earth–ionosphere potential difference: 300,000 to 450,000 volts (Earth negative).

Effective capacitance: 0.11 farad

Charge: 4×10^4 Coulomb (maintained by lightning)

Fine weather field near the surface ~100 V/m

Fine weather current: 2000 amperes. Ohmic dissipation: ~1 GW

Unmaintained discharge time constant: 20 seconds

Charge transfer per lightning strike: typically 20 Coulombs, but 5 to 100 Coulombs have been reported.

Global rate of lightning strikes: 100 per second.

TABLE 12.7 WAVELENGTH DEPENDENCE OF THE REFRACTIVE INDEX OF AIR

λ (nm)	δ (10^{-6})
400	298.3
450	295.9
500	294.3
550	293.1
600	292.2
650	291.5
700	290.9

12.6 ENERGETICS OF THE ATMOSPHERE

12.6.1 Kinetics

Global rms wind speed: $(\int v^2 dm/M)^{1/2} \approx 14.3$ m s^{-1}

Kinetic energy: 5.3×10^{20} J

Rate of dissipation by friction with the Earth: ~4.5×10^{14} W

Relaxation time: ~14 days

Average global zonal (E–W) wind fluctuation is ≥2.5 m s^{-1}. Eastward wind is 1.7 m s^{-1} faster in January than in July, corresponding to a 0.9 millisecond annual variation in the length of the sidereal day (rotation relative to distant stars). Note: Integrated over sufficient time (several relaxation times) and all latitudes there can be no net torque between the atmosphere and Earth. The annual cycle reflects a latitude redistribution of zonal wind.

12.6.2 Thermal Balance

The solar constant (intensity of radiation at the mean distance of the Earth) is 1361.6 W m^{-2}, fluctuating by ±0.8 W m^{-2} over the ~11-year solar cycle, strongest at the maximum of solar activity. Longer-term variations are postulated as a cause of climate variations, but no trend has been observed over the 35 years of satellite data. The mean rate of solar energy at the top of the atmosphere is 174,000 TW, but the orbital ellipticity causes a 6.9% variation over the course of a year (strongest on about January 3, weakest on about July 4). However, although the intensity varies with distance, r, from the Sun as $1/r^2$, so does the orbital speed (Kepler's second law), and the radiation received is the same in all equal angular segments of the orbit. The annual total is the same for both hemispheres and climate differences must be explained as consequences of the seasonal differences in insolation and the continent–ocean distribution. It is convenient to consider four stages in the energy redistribution, (i) solar radiation in the atmosphere, (ii) energy exchange at the surface of the Earth, (iii) outgoing thermal radiation and (iv) smaller items involved in the greenhouse effect and thermal imbalance that are lost in the uncertainties of (i), (ii) and (iii). All components of this energy balance are given here in terawatts (TW) to emphasise comparison with the rate of human energy use (16 TW).

(i) Solar radiation

Received at top of atmosphere 174,000 TW

Reflected to space by the atmosphere, especially clouds 46,000 TW

Absorbed by the atmosphere 33,000 TW

Reaching Earth surface 95,000 TW
Reflected from surface 7,000 TW

(ii) Surface heat redistribution [major items excluding those in category (iv)]

Net solar heating of the surface 88,000 TW
Latent heat of evaporation and transpiration of water 39,500 TW
Thermally driven convection of heat from the surface 11,500 TW
Atmospheric thermal conduction 1000 TW
Thermal radiation from the surface 36,000 TW

(iii) Outgoing thermal radiation

Direct radiation from the surface to space 10,000 TW
Radiation from the surface absorbed by the atmosphere 26,000 TW
Conducted, convected and latent heat from the surface absorbed by the atmosphere 52,000 TW
Thermal radiation from the atmosphere 111,000 TW (total thermal radiation 121,000 TW)

(iv) Small items

Conducted into the continents 10 TW. [The heat flux into surface area A of a medium of density ρ, specific heat C_p and conductivity κ by a steadily increasing surface temperature $(\mathrm{d}T_0/\mathrm{d}t)$ imposed for time τ is $2(\rho C_p \kappa\tau/\pi)^{1/2} A(\mathrm{d}T_0/\mathrm{d}t)$].
Retained as atmospheric warming and increased water vapour 3 TW
Ocean heating 180 TW
Net melting of sea ice 15 TW
Melting of grounded ice 9 TW
Heat released by human activity 18 TW (includes 16 TW of energy use)
Geothermal heat 47 TW
Net photosynthetic energy retained by vegetation that is buried without decomposing 1.5 TW

Section V

Geological Activity

The Restless Earth

Chapter 13

Tectonics

13.1 THE SURFACE PLATES AND GEOMETRY

The evolution of surface features of the Earth, driven by mantle convection, is characterised by an arrangement of quasi-rigid lithospheric plates in relative motion at speeds of a few centimetres per year. Activities such as earthquakes and volcanic eruptions are concentrated at the boundaries between the plates, where the relative motion is concentrated. Although most boundaries, especially those where plates converge, involve deformation over tens or hundreds of kilometres, the concept of a set of plates that remain coherent and undeformed as they move, with growth or disappearance only at their boundaries, is a useful approximation. Progressive deformation within plates is commonly represented by independent micro-plates but, as the multiplicity of identifications increases, the usefulness of the plate concept decreases, and it is advantageous to restrict the perceived fragmentation to recognition of plates with clearly independent motions that relate to the underlying convective driving forces. This is the philosophy of the plate selection in Table 13.1 and Figures 13.1 and 13.2. Even so, the figures indicate a distinction between the seven major plates and smaller ones. The selection differs in minor ways from some others, most noticeably in the inclusion of Sunda as a plate that cannot be regarded as part of any of its neighbours and the exclusion of Somalia, which is not moving fast enough, relative to the rest of Africa, to be considered independent of it. Primary data are from a detailed analysis by Bird (2003), who listed a much larger number of plates, but for calculation of the areas in Table 13.1, most of the minor ones are counted as parts of those listed here. The size distribution illustrated in Figure 13.2 is not materially affected by this.

Motions of plates are presented in terms of Euler vectors, the mathematically general method of denoting displacements on a spherical surface. An Euler vector represents a displacement as a rotation about an axis through the centre of the Earth and is specified by the coordinates (latitude and longitude) of the intersection of the axis with the surface (its pole) and the

TABLE 13.1 PLATE AREAS AND VECTORS REPRESENTING THEIR MOTIONS RELATIVE TO THE PACIFIC PLATE

		Euler Vectors		
Plate	Area (10^6 km^2)	Pole Latitude	Pole Longitude	Angular Speed (deg./10^6 years)
Pacific	107.91	–	–	0 (reference)
Africa	77.63	59.160	-73.174	0.9270
Eurasia	60.00	61.066	-85.819	0.8591
Antarctica	58.15	64.315	-83.984	0.8695
North America	55.43	48.709	-78.167	0.7486
Australia	47.60	60.080	1.742	1.0744
South America	43.63	54.999	-85.752	0.6365
Nazca	16.38	55.578	-90.096	1.3599
India	12.95	60.494	-30.403	1.1034
Sunda	10.98	55.442	-72.955	1.1030
Philippine Sea	6.19	-1.200	-45.800	1.0000
Arabia	4.90	59.658	-33.193	1.1616
Caribbean	3.24	4.313	-79.431	0.9040
Cocos	2.93	36.823	-108.629	1.9975
Scotia	1.88	48.625	-81.454	0.6516
Juan de Fuca	0.26	35.000	26.000	0.5068

angle, or angular speed, of rotation about it. For the tectonic plates, there is no fixed reference and observations give rotations of plates relative to one another. The largest, and arguably the best observed, is the Pacific plate and this has been adopted as a reference, so that, as in Table 13.1, Euler vectors for other plates are quoted as measures of their motions relative to the Pacific plate. For a fundamental understanding of plate motion, this is slightly unfortunate because the Pacific plate is fast moving and the vectors give no intuitive indications of absolute plate motions. This shortcoming is partially mitigated by using as the reference the no-net-rotation frame, which sets the area-weighted average motion of all plates to zero, but this has two disadvantages. A revised estimate of any plate motion causes readjustment of the reference frame and therefore of the vectors for all plates. However, the no-net-rotation frame has led to an interesting result. In this frame, the rms speed of global plate motion is 3.8 cm/year.

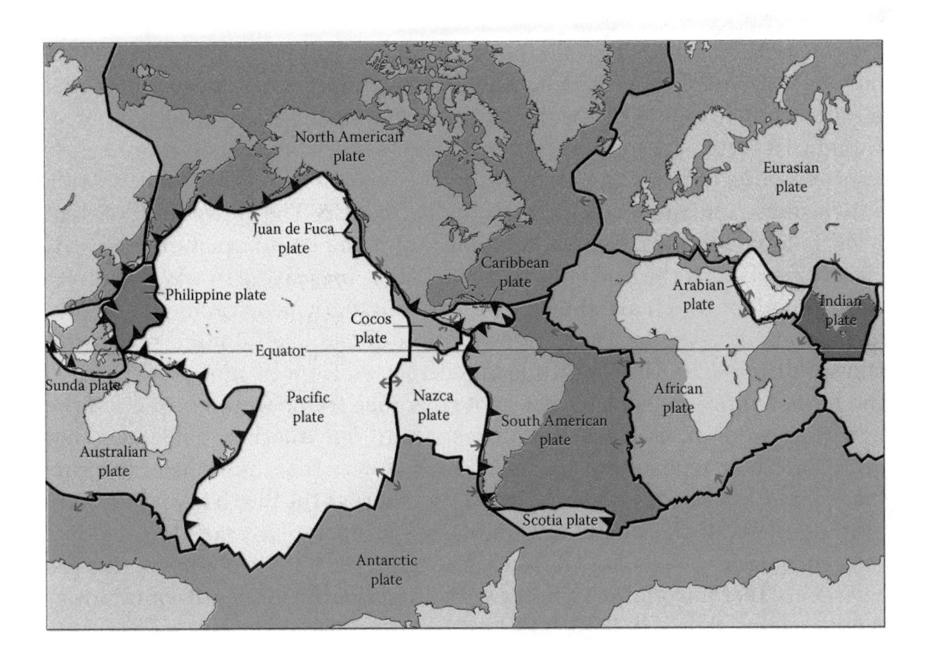

Figure 13.1 The tectonic plates. 'Sharks' teeth' indicate the directions of motion of down-going plates that are subducting under neighbouring plates. Directions of motions at spreading centres and transcurrent faults are marked by red arrows.

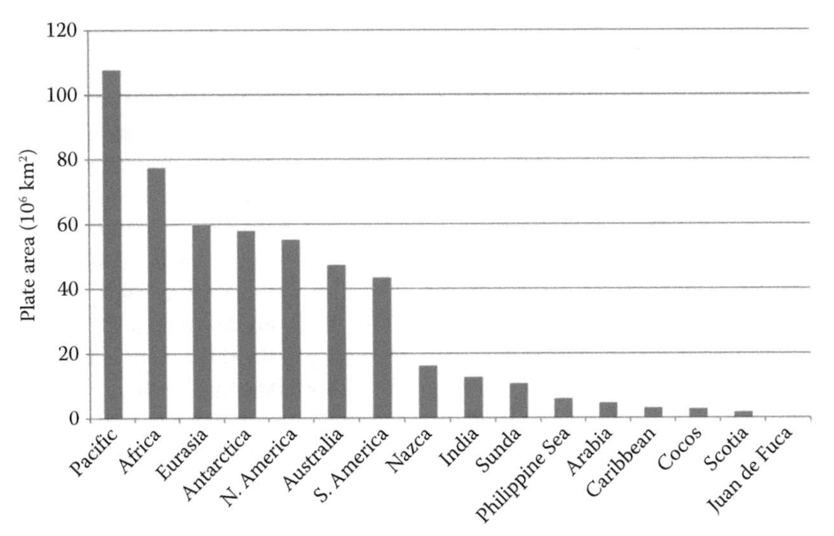

Figure 13.2 Plate areas arranged in order of size.

The other disadvantage of the no-net-rotation frame is that some plates, particularly the purely oceanic plates, move more easily over the deeper mantle than do continent-bearing plates, so that the frame does not correspond to a state of zero total torque on the lithosphere. This is more nearly approximated by referring plate motions to the 'hot spots', such as Hawaii, which mark volcanic sources deep in the mantle. The Hawaiian hot spot itself is a useful start because the motion of the Pacific plate across it is clearly marked by the line of islands, with ages increasing in a WNW direction, and all of which are shown by palaeomagnetic measurements to have formed at the fixed latitude, 19°N. Motion of the Pacific plate relative to Hawaii differs in both magnitude and direction from its motion relative to the no-net-rotation frame. A hot spot reference frame would be ideal if the hot spots were all stationary with respect to one another, but that is not quite so. Their relative motions are much slower than motions of the surface plates, but not zero. Their distribution across the Earth is very uneven, making difficult the identification of an appropriate average to take as a reference frame. So, for the time being at least, the Pacific plate remains the reference. Tectonic plate geometry with an indication of relative motions at boundaries is shown in Figure 13.1.

13.2 TECTONIC ENERGY AND MANTLE VISCOSITY

Tectonic processes are driven by thermal convection of the mantle, with the energy balance summarised by the data in Section 7.4. The rate of production of fresh oceanic crust is 3.36 km^2/year from 67,000 km of spreading ridges (plus some back arc spreading), corresponding to an average spreading rate of 2.5 cm/year each way from a ridge. The total area of deep ocean is 3.1×10^8 km^2, so the rate of production corresponds to a mean ocean floor lifetime of 92 million years. Production is balanced by the rate of subduction at 51,000 km of convergence zones but, since only one plate subducts at each boundary, the average subduction speed is 6.6 cm/year. The mean thermal shrinkage of mature subducting lithosphere is 2.1 km and if it were all to sink through the entire upper mantle (to 660 km depth), the rate of gravitational energy release would be about 5×10^{12} W, and correspond closely to the thermodynamic calculation of upper mantle convective energy in Section 7.4. With this dissipation, viscous control of the observed subduction speed indicates an effective upper mantle viscosity of $\sim 1.2 \times 10^{21}$ Pa s. Plate motion across the surface is much less dissipative and implies an asthenospheric viscosity $\sim 2 \times 10^{19}$ Pa s. As in Section 7.4, there is much less convective energy in the lower mantle, because both heat flux and the temperature ratios that determine thermodynamic efficiency decrease with depth, and a simple analytical approximation gives the variation with depth, z, of the convective energy generation per unit volume:

$$1.5 \times 10^{-8}(1 - z/2890\ \text{km})\ \text{W m}^{-3} \tag{13.1}$$

The variation with depth of the lower mantle viscosity is not effectively constrained by observations, but could reach 10^{24} Pa s a few hundred kilometres from the bottom, decreasing at greater depth to 10^{18} Pa s at the core–mantle boundary, this being the viscosity of plume material originating there. The total convective energy dissipation by the plumes is ~1.6×10^{12} W, independent of the convective energy of the mantle referred to above.

13.3 FAULT STRESSES AND ORIENTATIONS AND SEISMIC FIRST MOTIONS

13.3.1 Faults as Responses to Stress

At any point in a solid medium, the stress pattern can be resolved into three mutually perpendicular principal (normal) stresses, which, in the convention of rock mechanics, are taken as positive for axial compression, with σ_1, σ_2 and σ_3 as the maximum, intermediate and minimum values, respectively. The average, $P = (\sigma_1 + \sigma_2 + \sigma_3)/3$, is the ambient (lithostatic) pressure and differences are referred to as deviatoric stresses, which are all small enough to be analysed by standard, linear elasticity theory, although P may be much larger at depth in the Earth. The orientations of the principal stresses responsible for each type of surface faulting, the Anderson faulting criteria, are indicated in Figure 13.3a. The dip angles of the thrust and normal fault planes are controlled by Coulomb's faulting criteria, which depend on the effective frictional coefficients, μ, across the planes, as conventionally represented by the Mohr failure envelope diagram (Jaeger 1971) (Figure 13.4). A low value, $\mu \approx 0.2$, indicative of fault lubrication, is consistent with average observed dip angles of 39° for thrust faults and 50° for normal faults, although for thrust faults they vary from 10° to 60°.

13.3.2 Seismic First Motions

The requirement that a medium be in static equilibrium both before and after an earthquake, means that there is no net torque on it either before or after the shock and therefore that this is true also for the pattern of stress release. To avoid the implication of accelerating rotation, the pattern must be described as that of a balanced double couple, with the torque represented by the release

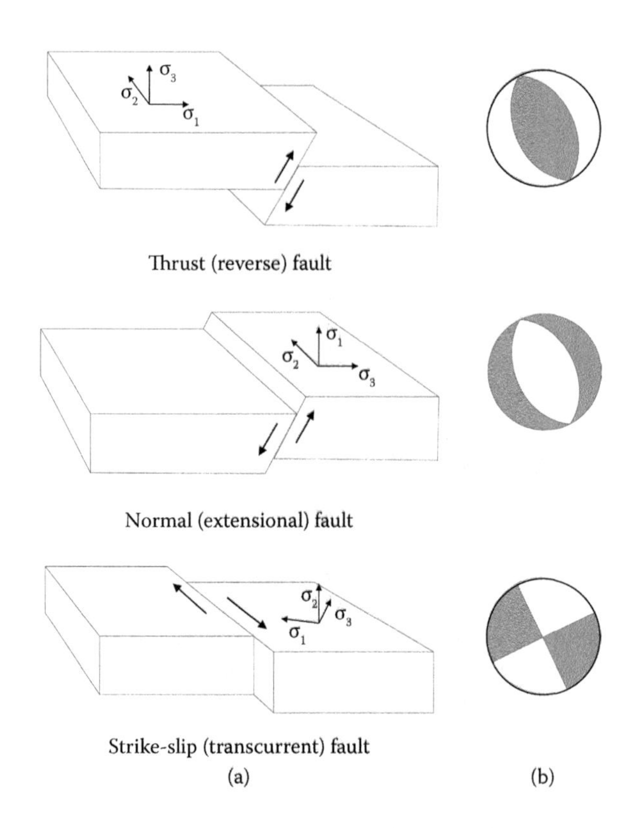

Figure 13.3 (a) The three basic types of fault displacement, with directions of the principal stresses that cause them (σ_1 maximum, σ_2 intermediate and σ_3 minimum), not necessarily all compressive. In the strike-slip case, σ_1 and σ_2 may be interchanged according to the angle relative to the strike of the fault. (b) 'Beach ball' diagrams of the signs of first motion compressional waves in the hemispheres below earthquakes, resulting in the illustrated fault displacements, shaded for initial compression and white for rarefaction.

of stress across the fault plane balanced by an equal and opposite torque across an auxiliary plane perpendicular to it. The result is a pattern of elastic wave radiation of quadrantal form, that is, initial compression in a pair of opposite quadrants and initial rarefaction in the other quadrants as in Figure 13.3b. With allowance for the refraction of seismic waves, initial pulses observed at stations distributed about an earthquake allow the orientation of the fault plane and the direction of fault movement to be identified, but with the ambiguity that the fault plane and the auxiliary plane are not distinguishable

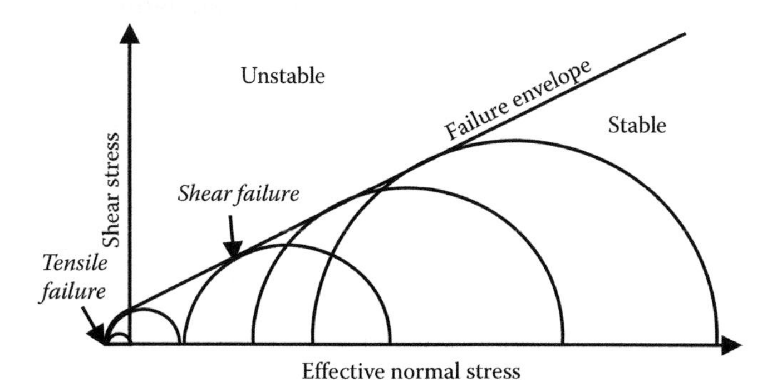

Figure 13.4 The Mohr failure envelope diagram, showing the best-fit line that represents the loci of shear and normal stresses at failure and delineating stable and unstable states of stress for a stressed rock.

without additional information. The distinction may be apparent for faults that are large enough to be outlined by aftershocks. Another common situation is asymmetrical seismic radiation from a fault movement that is not synchronous over the whole area but is unilateral in the sense of propagating as a patch of active displacement from one end to the other. This gives sharper and stronger compression in the direction of fault propagation, a Doppler effect of a moving source.

Breaks start at initiating points (foci or hypocentres) and spread at speeds that are less than, but up to 90% of, the speeds of shear waves in the faulted media. Localised damage, tsunamis and generation of the elastic waves that are recorded around the Earth occur because the breaks spread rapidly. The earthquakes considered in Sections 14.1 and 14.2 are of this type. More recently recognised are slow earthquakes, discussed in Section 14.4, that do not cause these effects.

Chapter 14

Earthquakes

14.1 MEASURES OF SIZE: MAGNITUDES, MOMENTS, ENERGIES, FAULT DIMENSIONS AND FREQUENCY OF OCCURRENCE

Earthquakes are localised hiccups in the convection of the mantle and occur where the material is too cool to deform smoothly and breaks discontinuously. Its temperature is below about half of the melting temperature on the absolute (Kelvin) temperature scale. It includes the crust but is restricted to a small fraction of the mantle, the lithosphere, roughly the uppermost 100 km, and the subduction zones, where lithospheric slabs descend into the mantle, retaining sufficient coolness to depths up to 700 km. Seismic energy is derived from the convective motion and is only a fraction of the convective energy of the whole mantle, itemised in Section 7.4, but represents a concentration of it. Earthquake energies are generally discussed in terms of magnitudes, M, of which there are some minor variants, mostly based on original work on surface waves from Californian earthquakes by C. F. Richter. The Richter surface wave magnitude, M_S, is the logarithm of the ratio of the amplitude of a seismic wave (in microns) to its period (in seconds), normalised to observations at 100 km from the source and adjusted for source depth. Since this ratio is a measure of the elastic strain energy density of a wave, there is a direct relationship between M_S and the energy, E, of an earthquake, which, with allowance for duration of the wave train, is

$$\log_{10} E\,(\text{joules}) = 1.5\,M_S + 4.8 \tag{14.1}$$

This is the energy radiated as seismic waves and does not include energy dissipated as heat and pulverisation of rock along a fault.

The conventionally determined Richter magnitude uses wave amplitudes recorded on seismometers with natural periods of about 20 seconds, which do not respond effectively to the long periods of waves from very large earthquakes. A more fundamental measure of the size of an earthquake is its moment, M_0, defined as the integral over the fault surface, S, of the product of fault slip, b, and rigidity modulus, μ, of the medium.

$$M_0 = \int \mu b dS \tag{14.2}$$

This quantity is determined from the very low-frequency spectral power of the seismic waves, now routinely measured, and is the basis of a revised magnitude scale, M_W. This is adjusted to coincide with the M_S scale for magnitudes below 6 (for which there is no problem with M_S) by the relationship

$$M_W = (2/3)\log_{10}M_0 - 6.07 \tag{14.3}$$

with M_0 in Newton metres. This means that Equation 14.1 can be used with M_W instead of M_S, extending the validity of the equation to the largest magnitudes.

The number of earthquakes per year with magnitudes M or greater is given by a simple relationship known as the Gutenberg–Richter 'law' after its originators. With numerical values by Kanamori and Brodsky (2004) for the global total of earthquakes, this is

$$\log_{10} N = (8.0 \pm 0.2) - (1.00 \pm 0.3)M \tag{14.4}$$

The second coefficient (1.00) is found to be the same, within uncertainties of observation, for local and regional earthquakes as well as the global catalogue. By the global numbers in Equation 14.4, one $M \geq 8$ event would be expected per year and one $M \geq 9$ shock per 10 years but the numbers cannot be relied on for the largest events. Combining Equations 14.1 and 14.4, we can calculate the total energy within any magnitude range, drawing attention to the fact that, unless an upper bound is imposed on the magnitudes to which Equation 14.4 applies, infinite total energy is implied. The form of a roll-off that needs to be applied to the 'law' cannot be inferred from observations because, for the largest earthquakes, numbers are too small to give reliable statistics and the largest earthquakes dominate the energy calculation. Estimates of the annual average total seismic energy release are hardly more than guesses, but there is a particular interest in this as the fraction of the global total convective/tectonic energy that is expressed in earthquakes. One event equivalent to the largest well-recorded shock, Chile 1960 ($M_w = 9.5$), per year would be more than 5% of that global total.

Earthquake intensity, I, is a qualitative, partly subjective scale of earthquake shaking, based on a mix of human reactions and structural damage, as in Table 14.2. Every earthquake has a range of intensities diminishing with distance from its epicentre, but varying with local geology. The most common use of intensity scales is by insurance assessors. There is a rough general correlation between the maximum intensity of an earthquake, I_{max}, and its magnitude, M_W, as listed in Table 14.1. Table 14.2 is a modified version of the Mercalli intensity scale of

TABLE 14.1 AVERAGE DIMENSIONS OF EARTHQUAKES AND FAULTS FOR $M_W \geq 1$, WITH AN APPROXIMATE RELATIONSHIP TO MAXIMUM INTENSITY, I_{MAX} (TABLE 14.2), AND NUMBER PER YEAR, n

Magnitude (M_W)	Moment (N m)	Energy (J)	Fault Area	Max. Displacement[a]	I_{max}	n
1	4.0×10^{10}	2.0×10^{6}	570 m^2	2.4 mm	I	10^7
2	1.3×10^{12}	6.3×10^{7}	5700 m^2	7.5 mm	II	10^6
3	4.0×10^{13}	2.0×10^{9}	0.057 km^2	2.4 cm	III	10^5
4	1.3×10^{15}	6.3×10^{10}	0.57 km^2	7.5 cm	IV–V	10^4
5	4.0×10^{16}	2.0×10^{12}	5.7 km^2	24 cm	VI	10^3
6	1.3×10^{18}	6.3×10^{13}	57 km^2	0.75 m	VII	100
7	4.0×10^{19}	2.0×10^{15}	570 km^2	2 m	VII–VIII	10
8	1.3×10^{21}	6.3×10^{16}	5700 km^2	6 m	IX–X	1
9	4.0×10^{22}	2.0×10^{18}	57,000 km^2	20 m	X+	~0.05
9.5	2.3×10^{23}	1.1×10^{19}	180,000 km^2	40 m		

[a] These averaged numbers allow for unequal fault dimensions (length/width) characteristic of large subduction zone earthquakes. For shocks on transcurrent faults (e.g. San Francisco, 1906), the dimension ratio may be much larger and the scaling of displacement with magnitude is weaker.

TABLE 14.2 EARTHQUAKE INTENSITY SCALE

Intensity	Effects
I	Not normally felt.
II	Felt by a few people in very quiet situations, particularly in upper floors of buildings.
III	Felt by many people indoors, but often not recognised as an earthquake, and not noticed outside.
IV	Crockery and windows rattle. Similar to passing of a heavy vehicle.
V	Unstable objects may be overturned. Sleepers awoken.
VI	Slight damage to building plaster, etc. Furniture may be displaced.
VII	Serious damage to poorly constructed buildings, cracking of brickwork.
VIII	Serious damage, even collapse, of all but specially constructed buildings, unreinforced walls and chimneys. Furniture overturned.
IX	Serious damage, even of well-constructed buildings.
X	Most buildings and infrastructure destroyed or rendered inoperable.
XI, XII	Arise from geological focussing, with local accelerations exceeding gravity.

earthquake shaking, itself a modification in the 1880s by Giuseppe Mercalli of an earlier Rossi–Forel scale. There are several rival scales, some cited as modified Mercalli scales, so that his name is most often cited. Some give 12 intensity levels instead of the 10 listed here, but levels XI and XII are rarely considered and, if reported, are generally consequences of local effects.

14.2 HISTORICALLY SIGNIFICANT EVENTS

As widely noted but unexplained, very large earthquakes cluster in time. This is illustrated in Figure 14.1, which includes all earthquakes with $M > 8.5$ since 1950, identified by letters A to M: A, Assam–Tibet (8.7) 1950; B, Kamchatka (9.0) 1952; C, Andreanov Islands (8.7) 1957; D, Valdivia–Chile (9.5) 1960; E, Kuril Islands (8.5) 1963; F, Alaska (9.2) 1964; G, Rat Islands, Alaska (8.7) 1965; H, Sumatra (9.2) 2004; I, Sumatra (8.6) 2005; J, Sumatra (8.5) 2007; K, BioBio (Maule), Chile (8.8) 2010; L, Tohuku, Japan (9.0) 2011; M, Sumatra (8.6) 2012.

14.2.1 Earthquakes of Particular Scientific Interest

1700. Coastal Washington State, USA, and British Columbia, Canada. $M \sim 9$ (estimated). A very large subduction zone earthquake identified from Japanese tsunami records, but not previously recognised. Geological investigations

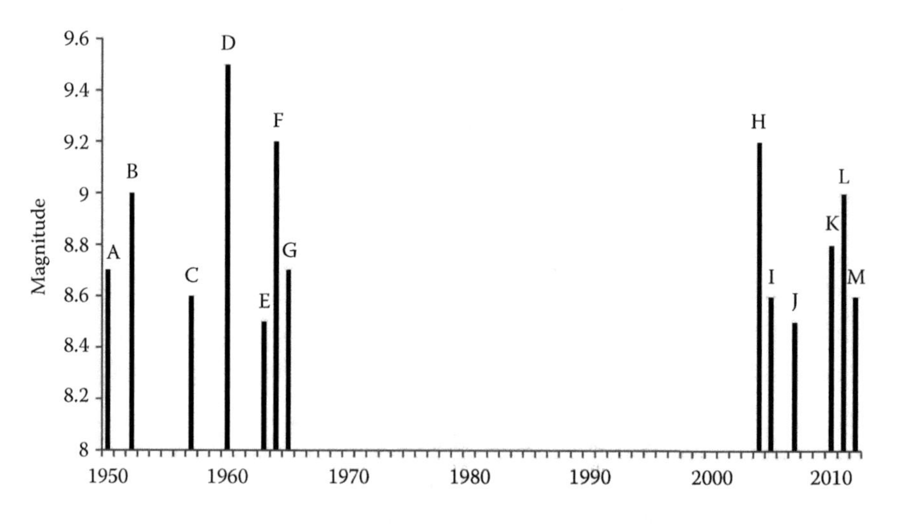

Figure 14.1 Major earthquakes ($M_W > 8.5$) since 1950, identified by letters A to M in the text.

confirmed that there had been a major disturbance at that time and a local tribal story identified the event as having occurred at night, which is consistent with the Japanese tsunami timing.

1755. Lisbon, Portugal. $M > 8.7$. A subduction zone earthquake, accompanied by a massive tsunami during church services in Europe. A magnitude estimate by Gutenberg and Richter (1954) was based on reports of the amplitudes of swing of church chandeliers across Europe.

1811–1812. New Madrid, Missouri, USA. A series of shocks with three having magnitudes variously estimated to be in the range 7.5 to 8.0. These were mid-continent events, remote from plate boundaries, inviting surprise that such large events could occur outside the recognised zones of major seismicity. They are now identified as lying on a band of continuing, less dramatic events that are attributable to deep crustal adjustment to past tectonic activity.

1906. San Francisco. $M = 7.9$. A geodetic survey of the area before the earthquake was repeated afterwards, giving a detailed record of strike-slip displacements and focussing attention on the elastic rebound theory of the earthquake mechanism. This became central to seismological studies, although it led to an overemphasis of the significance of strike-slip displacements in understanding the ultimate causes of earthquakes.

1931. Hawkes Bay, New Zealand. $M = 7.9$. Drew attention to level changes that can occur during subduction zone earthquakes. A total of 40 km^2 of sea floor became dry land.

1960. Chile. $M = 9.5$. The largest earthquake recorded by modern instrumentation. It was preceded by very large foreshocks as well as a deep subduction zone creep event ('slowquake') and generated a tsunami that caused widespread damage and loss of life across the Pacific, especially in Hawaii and Japan as well as Chile.

1964. Alaska. $M = 9.2$. One of the two major earthquakes (with Chile 1960) used to produce new seismological models of Earth structure from free oscillation records.

1994. Bolivia. $M = 8.2$ at 647 km depth. The deepest known earthquake of comparable magnitude. Such very deep events give vital information about the structure of the Earth by giving clearer seismic records than are possible from shallow earthquakes that are affected by structural complexities near the surface. In particular, low-degree components of the free-oscillation spectrum are excited with much less high-frequency content in the records.

2004. Sumatra. $M = 9.2$. Notable for a massive tsunami in the Indian Ocean, not widely recognised as tsunami-prone. Tsunami precautions and prediction systems were concentrated in the Pacific.

2011. Tohoku, Japan. $M = 9.0$. Consequential damage from a massive tsunami included destruction of nuclear power reactors at Fukushima, on the Pacific coast, with a large area rendered uninhabitable and fishing disallowed by radioactive pollution.

2013. Okhotsk Sea, off far-east Russia. $M = 8.3$ at 609 km depth. A very large, deep event, with strong low-frequency seismic radiation.

14.3 SEISMIC WAVES

Earthquakes start with an initial fault displacement at a triggering point referred to as the focus, or hypocentre, and spread from there. The surface points above the foci are the epicentres and maps of seismicity, showing the distribution of activity, are plots of epicentres. Most fault displacements are rapid, effectively jerks, and the response of the medium is to radiate elastic waves, which are recorded at instruments around the globe. The waves travel faster than the fault propagation (although generally not much faster), so that remote observations see first pulses that are caused by the initial fault displacements and these are used both to locate the sources (identifying epicentres) and fault geometry. This means using the orientations and polarities (compression, rarefaction or shear) of the pulses to infer the fault planes and directions of movement, represented by 'beach ball' diagrams of focal mechanisms as in Figure 13.3b. Although this process emphasises the fastest (first-arriving P, primary, compressional) waves, several wave types, detailed in Table 14.3, are used in exploring details of the Earth's internal structure.

14.4 SLOW EARTHQUAKES AND NON-VOLCANIC TREMOR

For many years, very slow fault displacements have been observed at restricted surface locations, one being a limited range of the San Andreas Fault in California, where they have been known as aseismic creep events. As a surface phenomenon they are rare, but at depths of tens of kilometres, they are now recognised as normal fault behaviour, especially in subduction zones. Being closely related to more familiar seismic activity on the same faults, they are referred to as slow earthquakes, slowquakes or silent earthquakes. The slowquakes attracting most interest are those occurring on major faults below the seismogenic zones of 'fast earthquakes'. Although they are slow, taking hours to months to occur, compared with seconds to minutes for 'fast' events, they are discrete increments in fault displacement, presumed to require triggering or initiation processes, as do 'fast earthquakes'. They resemble 'fast' events more closely than they do to the steady state creep, observed in high temperature

TABLE 14.3 TYPES OF SEISMIC WAVE

Wave Type	Particle Motion			Wave Speed (km/s) or Period (minutes)
Body waves	P	Compressional	Direction of wave motion	8.07 to 11.27 km/s (core) 8.11 to 13.72 km/s (mantle)
	S	Shear	Transverse to wave motion	3.61 to 3.65 km/s (inner core). No outer core S wave 4.49 to 7.33 km/s (mantle)
Surface waves	Rayleigh	Dispersive waves (speed varies with frequency or wavelength)	Vertical ellipses	Up to 7.4 km/s at 18-min period
	Love		Horizontal ellipses	Up to 5 km/s at 12-min period
Free oscillations	Spheroidal	Standing Rayleigh waves	Combined radial and circumferential motion	Periods up to 53.9 min
	Toroidal	Standing Love waves	Circumferential oscillations	Periods up to 44.2 min

laboratory experiments (as in Figure 21.3) and responsible for the convective motion of the bulk of the mantle. Slowquakes evidently occur in material with rheological behaviour intermediate between steady state creep and the fracturing of faults in shallow layers. Since they indicate stress release at depth, it is presumed that they enhance the stress in the shallower seismogenic zone, where it accumulates to cause major shocks. This concern applies particularly to the Cascadia subduction zone on the NW coastal fringe of North America (British Columbia to northern California), where deep slowquakes occur regularly (at 10- to 14-month intervals), and the rare shallow shocks can be very big, as in the case of the first entry in Section 14.2.1.

Although, in many cases, slowquakes are small but frequent, this is not always the case. At the other extreme, evidence of a deep slow-slip event with a fault displacement equivalent to an earthquake of magnitude ~8 was reported to precede the Chile 1960 $M = 9.5$ shock and repeated events with $M \sim 7$ occur deep under Wellington, New Zealand. The time taken by slow events is highly variable and there may be no clear demarcation between slow and 'normal' events. Also, the expression 'silent earthquakes' is questioned and the alternative 'quiet earthquakes' is suggested because they are accompanied by tremor, irregular bursts of vibration similar to those observed with volcanic activity and attributed to fluid movement. The same explanation probably applies to non-volcanic tremor, most obviously in subduction zones, where it occurs in subducted crust with hydrated minerals and probably also in sedimentary pore water. Tremor is easier to observe, and use as evidence of slow-slip events, than is the static surface deformation monitored by geodetic methods.

14.5 SEISMIC TOMOGRAPHY

Tomography is a word adopted from medical imaging with multi-directional X-ray beams. A patient is rotated, to be presented in different orientations to the beams, and measured variations in the intensities of transmitted X-rays are used to identify local heterogeneity in their absorption. The seismological analogue uses the travel times of seismic waves on intersecting ray paths to image variations in wave speed. The result, a three-dimensional picture of the Earth's internal structure, now attracts more attention than the average radial variation in properties represented by the model in Tables 6.1 and 7.1. Development of a model from travel time data is an inverse process, starting with travel times for an initial (assumed) model, such as the radial model in Table 7.1, and repeatedly adjusting it to improve the fit to observations. For a detailed model requiring a large number of observations, the process is very computer-intensive. There are some limitations and approximations. A travel time is the time of flight for the fastest path between two points and involves diffraction around the heterogeneities that are investigated, enlarging images of high-velocity features and obscuring low velocities. With sufficiently detailed records, this problem can be reduced by a full wave solution, but that greatly increases the computational complexity and is not widely used. Resolution is limited also by the seismic wavelengths, which can only resolve features larger than themselves, so that the long wavelengths needed for deep penetration can see only broad-scale features. Tomographic inversions assume isotropic elastic properties, with waves having the same speeds in all directions at any point. This is unlikely to be precisely correct anywhere in the mantle, but it is not clear how seriously the models are compromised by this assumption.

Tomographic velocity 'anomalies' can be interpreted in terms of both temperature and compositional variations where both P and S wave data are clear.

TABLE 14.4 TEMPERATURE DEPENDENCES OF SEISMIC VELOCITIES, V_P (COMPRESSIONAL), V_S (SHEAR) AND V_φ (BULK) IN THE LOWER MANTLE

Radius (km)	$(\partial \ln V_S/\partial T)_P$ (10^{-5} K^{-1})	$(\partial \ln V_P/\partial T)_P$ (10^{-5} K^{-1})	$(\partial \ln V_\varphi/\partial T)_P$ (10^{-5} K^{-1})
3480	-2.473	-1.001	-0.125
3600	-2.561	-1.050	-0.141
3800	-2.714	-1.136	-0.164
4000	-2.870	-1.229	-0.210
4200	-3.042	-1.334	-0.259
4400	-3.228	-1.449	-0.309
4600	-3.451	-1.595	-0.385
4800	-3.703	-1.762	-0.473
5000	-3.990	-1.960	-0.584
5200	-4.341	-2.209	-0.683
5400	-4.760	-2.518	-0.919
5600	-5.283	-2.922	-1.184
5700	-5.543	-3.154	-1.509

The temperature effect is subject to a straightforward interpretation (Table 14.4), in which it is seen that V_S is more temperature sensitive than V_P. However, most reports compare V_S with the so-called bulk sound velocity, V_φ

$$V_\varphi = (V_P^2 - (4/3)\, V_S^2)^{1/2} = (K_S/\rho)^{1/2} \tag{14.5}$$

because the greater contrast makes clearer how much of the observed velocity variations must be interpreted as compositional, arising from variations in the proportions of different minerals and the abundances of iron in them.

Chapter 15

Volcanism

Volcanic activity is very diverse, with three fundamentally different causes and corresponding effects. Qualitatively different phenomena are observed at ocean ridge spreading centres, at subduction zones and at isolated hot spots, such as Hawaii. At any time, there are typically about 50 volcanoes in various states and stages of eruption and a much larger number described as dormant (but not extinct). Eruptions are assigned volcanic explosivity indices (VEIs, Table 15.1), numbers that are measures of the volumes of pyroclastic material emitted, on a logarithmic scale, that is, 8 for volumes exceeding 1000 km^3, 7 for 100–1000 km^3, and so on down to 2. This scale is attributed to a 1982 proposal by C. Newhall and S. Self of USA, but is just a minor rescaling of the volcanic intensity scale of Tsuya (1955) of Japan. It is not to be confused with an index of volcanic explosiveness proposed in 1927 by K. Sapper, a number between 0 and 100 that is the percentage of the total eruptive material that is fragmented (pyroclastic).

15.1 SPREADING CENTRES

The 3.4 km^2/year of fresh ocean floor that is generated by the sea-floor spreading process at 67,000 km of ocean ridge is capped by a layer of basaltic rock averaging about 6 km thick. This is produced by partial melting of the rising ultrabasic asthenospheric material. It means a more or less steady emission of 20 km^3/year of basaltic lava at the ridges. Individual eruptions at local points have been identified but the process is passive and concealed by ocean several kilometres deep.

15.2 SUBDUCTION ZONES

The ocean-floor lithosphere, with its basaltic crustal layer and accumulating sediment, survives, on average, for 92 million years before plunging into the

TABLE 15.1 VOLCANIC EXPLOSIVITY INDEX

VEI	Ejecta Volume (Bulk)	Classification	Style	Plume	Frequency	Injection	Examples
0	<10,000 m^3	Hawaiian	Effusive	<100 m	Constant	Tropospheric negligible Stratospheric none	Kilauea
1	>10,000 m^3	Hawaiian/ Strombolian	Gentle	100 m–1 km	Daily	Tropospheric minor Stratospheric none	Stromboli
2	>1,000,000 m^3	Strombolian/ Vulcanian	Explosive	1–5 km	Weekly	Tropospheric moderate Stratospheric none	Cumbre Vieja
3	>10,000,000 m^3	Vulcanian/Pelean	Catastrophic	3–15 km	Few months	Tropospheric substantial Stratospheric possible	Nevado del Ruiz
4	>0.1 km^3	Pelean/Plinian	Cataclysmic	>10–25 km	≥1 year	Tropospheric substantial Stratospheric definite	Eyjafjallajö-kull Pelée
5	>1 km^3	Plinian	Paroxysmic	>20–35 km	≥10 years	Tropospheric substantial Stratospheric significant	Fuji Vesuvius
6	>10 km^3	Plinian/Ultra-Plinian/Ignimbrite	Colossal	>35 km	≥100 years	Tropospheric substantial Stratospheric substantial	Pinatubo Krakatoa
7	>100 km^3	Ultra-Plinian / Ignimbrite	Super-colossal	>40 km	≥1000 years	Tropospheric substantial Stratospheric substantial	Tambora Thera
8	>1000 km^3	Supervolcanic	Mega-colossal	>50 km	≥10,000 years	Tropospheric substantial Stratospheric substantial	Yellowstone Toba

Note: The dust veil index (DVI) was developed to assess the effect of volcanic eruptions on climate, based on the principle that the volume of ejected material directly affects radiation in the atmosphere and to the surface. The use of the DVI has lapsed with introduction of the volcanic explosivity index (VEI).

mantle along the 51,000 km of subduction zones. Sea water percolation during the millions of years of cooling and cracking causes hydration of the crustal minerals and infusion of sea-salt elements, so that the subducting crust, and possibly the uppermost mantle component, has a high volatile content, largely water. This lowers its solidus temperature, so that, as it is heated with increasing depth, melting occurs when it has descended to about 100 km, roughly the top of the asthenosphere. It produces an acid (Si-rich) magma, typically of andesitic composition, with a high volatile content. The resulting subduction zone volcanism is often explosive, with violent release of the volatiles, and can be very destructive. Following is a brief selection of major explosive eruptions that have changed the course of history by occurring at critical locations or times, had widespread, perhaps global, environmental effects or invited particular scientific interest.

15.2.1 Notable Subduction Zone Eruptions

Yellowstone, Wyoming, USA. Most recent major eruption ~640,000 years BP, VEI-8. This is an interesting hybrid volcanic system, being both a subduction zone volcano and a hot spot. Although, on this basis, it may appear to have features in common with the hot spot volcanoes in Section 15.3, its eruptive style is characteristic of the violently explosive 'supervolcanoes' such as Toba (following paragraph). Essential differences between these eruptive styles are the viscosity of the magma and the abundance of volatiles, especially water. In the Yellowstone situation, the volcanism may be driven primarily by the deep plume (hot spot) material, but the water would have been derived from hydrated minerals in a subducted slab.

Toba, north-central Sumatra, ~73,000 years BP, VEI-8. By most estimates, the largest eruption for more than the last million years. It left a crater 3000 km^2 in area, now a lake, although this is probably an amalgamation of at least three calderas, with the last eruption responsible for a third of the area. Apart from intense local effects, the eruption spread ash across the Indian Ocean and southern Asia, many metres thick in nearer parts, such as the Malayan Peninsula, with a total volume of at least 800 km^3. It would have released a very large volume of noxious gases, especially SO_2, causing cooling of global extent and inviting speculation of environmental stress sufficient to constrain human evolution, although contrary evidence has been presented (Lane et al. 2014).

Campi Flegrei, Southern Italy, ~39,000 years BP, VEI-8. A 175 km^2 caldera west of Naples, much of it submerged, is identified as the source of extensive ignimbrites and an ash layer over large parts of Europe and the Mediterranean, amounting to at least 200 km^3. The caldera is marked by 24 craters, tuff rings and cinder cones and is considered to be a supervolcano. The area was in the grip of the last ice age at the time of the eruption, which compounded the environmental stress, making much of it uninhabitable for early humans and most animals for an extended period. The Neanderthal population was concentrated in southern Europe and never recovered. There was also a population of *Homo sapiens* that would have suffered the same fate but was more widespread and able to migrate back to re-occupy the area. They were often, but probably unjustly, blamed for the disappearance of the Neanderthals. Volcanic activity in the area continues, although without eruptions as dramatic as that of 39,000 BP.

Thera/Santorini, southern Aegean Sea, 1628 BC, VEI-7. A major eruption in a populated area, destroying the island at the centre of the Minoan civilisation, leaving a crater in place of most of the island and spreading ash over a wide area. A branch of the Minoan culture on the island of Crete, to the south, was effectively destroyed by a tsunami generated by the eruption (Section 16.3). The Minoan civilisation never recovered and political power in the Mediterranean shifted to mainland Greece.

Vesuvius, Italy, AD 79, VEI-5/6. A major eruption in an area of the Roman Empire with a literate population. Written accounts of events preceding and during the main eruption, especially in letters by Pliny the Younger, whose uncle, Pliny the Elder, died in the eruption, made it a landmark in the history of volcano studies. Pliny's description identified a particular type of eruption and the term 'Plinian' eruption is now applied to others of the same type. It is a characteristic of many subduction zone volcanoes, involving eruptive columns reaching the stratosphere and producing explosive releases of gas-charged clouds of hot ash that spread out over the surface at high speeds, destroying everything in their paths. Such a phenomenon, termed a pyroclastic flow, is often also referred to as a nuée ardente (see Pelée, below). There is confusion over the precise date of the Vesuvius eruption of AD 79 and it is probable that there were two distinct events. The first pyroclastic flow, in August, destroyed Pompeii and neighbouring towns on the Bay of Naples, with their inhabitants, and a much deeper layer of ash (up to 25 m) was deposited in November, so that the towns were concealed from view and effectively forgotten until rediscovered by modern archaeology.

Tambora, Sumbawa Island, Indonesia, 1815, VEI-7. Arguably the biggest volcanic eruption with recorded, detailed observations. Before the eruption, it was a large, typically conical stratovolcano but was reduced to a caldera with a rim 1400 m lower than the original peak and a caldera floor of 1300m lower still. A period of global cooling with cold summers, crop failures and famine followed the eruption and is attributed to it. Climatic effects of volcanism are referred to in Section 15.4.

Krakatoa, Sunda Strait, between Java and Sumatra, Indonesia, 1883, VEI-6. A massive eruption generating tsunamis that destroyed many coastal villages (Section 16.3). Although smaller than the Tambora eruption, its atmospheric effects were widely recognised, by spectacular sunsets and a general cooling, alerting the scientific community to the significance of volcanic effects on climate. The Krakatoa observations showed that pyroclastic flows can travel tens of kilometres across water.

Pelée, Martinique, West Indies, 1902, VEI-4. Notable for the very complete destruction of the town of Saint Pierre, by a pyroclastic flow, which was highly directional and almost focussed on the town, as documented by reports of survivors from surrounding areas and ships offshore. From the observed effects, it is evident that the flow was particularly hot and very fast, causing comprehensive structural damage before igniting everything combustible. This event drew attention to the mechanics of pyroclastic flows, with hot gas embedding fuming ash, making it dense enough to rush downhill, although commonly also explosively driven initially by the side of the mountain blowing out, as in this case and also Mount St. Helens (USA) in 1980. The Pelée event led to the origin of the French term 'nuée ardente', glowing cloud. This eruption was preceded, by a few hours, by a similar eruption of the Soufrière volcano on St. Vincent, two islands south of Pelée, on the chain of volcanic islands, with the residents of St. Lucia, the island between them, nervously watching their own volcano.

Pinatubo, Philippines, 1991, VEI-6. The largest eruption monitored by satellites, giving particularly comprehensive coverage of atmospheric effects. The eruption cloud injected about 20 million tonnes of SO_2 into the stratosphere, where it was discernible for more than 3 years and caused significant cooling in the Northern Hemisphere. Its progressive decrease with time appeared to be reasonably approximated by an exponential decay with a half-life of about 9 months. Although the troposphere was cooled, the stratosphere was warmed for about 2 years by the destruction of ozone, which is a greenhouse gas that has a negative (cooling) effect at that altitude.

15.3 FLOOD BASALTS, HOT SPOTS AND MASS EXTINCTIONS

Flood basalts are products of extreme eruptions of basalt from extended fissures, and their identification with mass extinctions of species indicates that they can be more environmentally damaging than explosive eruptions. There are two reasons for this. They are more voluminous and they occur over longer time periods, ensuring a global spread of their effects. The 1783 eruption of Laki volcano, Iceland, was a well-documented example of a smaller fissure eruption, partly because of the report of its climatic effect by a prominent commentator, the American statesman Benjamin Franklin, who was the ambassador in Paris at the time. Although it was very small on the scale of flood basalts, it caused unseasonably cold weather across the Northern Hemisphere, with widespread crop failures and starvation. A layer of SO_2 in a Greenland ice core reinforced the presumption that the principal culprit was the SO_2 emission. Scaling up the Laki flows to the major flood basalts leaves no doubt that the effects of flood basalt eruptions were dramatic. Not only were some individual flows very large, tens of thousands of cubic kilometres, and would have been fuming for many years, they were also repeated at intervals that were probably thousands of years, allowing intervening periods of strong warming by the volcanic CO_2 between sudden onsets of global cooling. Ocean acidification probably contributed to the marine extinctions.

Table 15.2 lists the Phanerozoic flood basalts on land that have been identified with mass extinctions, tectonic events (continental break-up) and/or continuing hot spot volcanism. Earlier flood basalts include especially the extensive McKenzie formation in northern Canada (1200 million years old) but cannot be related in this way to other events. Marine basalt formations that appear similar, but with less dramatic atmospheric effects, include the massive Ontong Java Plateau (S–W Pacific). Hot spot volcanism lacking a known connection with flood basalts or extinctions occurs in central Africa (Nyamuragira and Nyirogongo volcanoes). A characteristic of flood basalts is that they are layered series of individually massive flows from extended fissures, generally accumulating in less than a million years to total volumes of millions of cubic kilometres, much more voluminous than other volcanic features. Extinction events are not all equally severe, but the two most dramatic (end Permian and end Cretaceous) give the clearest indications of the nature of these events. For a review of the subject, see Bond and Wignall (2014).

15.4 ENVIRONMENTAL EFFECTS OF VOLCANISM

Apart from the heat released to the oceans in the sea-floor spreading process, volcanic activity makes a trivial contribution to the terrestrial heat flux and that is only 0.05% of the solar energy reaching the surface. Except very locally, volcanic heat is inconsequential and global environmental effects of volcanism are caused by the ash and gas emitted to the atmosphere, particularly by what reaches the stratosphere, where it can remain for months or years, especially for emissions by tropical volcanoes that are caught up in rising, warm, tropical air. Dust and gas from major eruptions may be identified after one or more circulations of the Earth. The duration of atmospheric residence of ash particles,

TABLE 15.2 RELATIONSHIPS OF FLOOD BASALTS TO EXTINCTIONS AND HOT SPOTS

Date (10^6 years BP)	Flood Basalt Province	Location	Extinction Event	Tectonic Event	Continuing Hot Spot	Comments
511	Kalkarindji	NW Australia	Mid-Cambrian			Jourdan et al. (2014)
260	Emeishan	SW China	End Capitanian (Guadalupian stage of Permian)			Dwarfed by following Siberian event
250	Siberia	N Asia	End Permian			The most nearly complete extinction
201	Central Atlantic	Brazil, Morocco	End Triassic	Atlantic opening		Now widely separated basalt trapps
183	Karoo-Ferrar	Africa, Antarctica	Pliensbachian-Toarcian	Africa–Antarctica split	Marion Island	A part of Gondwana break-up
135	Paraná-Etendeka	Brazil, Namibia		S Atlantic opening	Tristan Island	
~100	Rajmahal	Eastern India			Kerguelen Island	Ninety-east ridge connection
65	Deccan	Central India	End Cretaceous		Reunion Island	Second most severe extinction event
56	North Atlantic	Greenland, N Europe	End Palaeocene	N Atlantic opening	Iceland	Hot spot superimposed on mid-Atlantic ridge
34	Afro-Arabia	Ethiopia, Yemen	End Eocene	Africa–Arabia split		
16	Columbia	NW USA			Yellowstone	Hot spot superimposed on a subduction zone

and sulphuric acid droplets that condense from oxidised SO_2, is affected by atmospheric turbulence, and by aggregation of particles and droplets, but the speed of free fall, controlled by size, is a measure of the minimum time. It is not well estimated from Stokes' law calculations of terminal velocities of spheres in a viscous medium, but is approximated by numbers in Table 15.3 of fall out times from the stratosphere of particles with a range of sizes. The lower density is applicable to the acid droplets, but is a reasonable approximation also for many ash particles.

The most environmentally significant atmospheric gas is SO_2, because the sulphuric acid droplets that it forms are strong reflectors of sunlight and cause surface cooling. The average duration of atmospheric SO_2 following a sudden stratospheric injection is best estimated from observations following the 1991 Pinatubo eruption (Section 15.2.1, last entry), which showed an approximately exponential decay with a 9-month half-life. This is shorter than the 15 months required for exchange of atmospheric constituents between hemispheres, so that brief eruptions at high latitudes have climatic effects only in one hemisphere, whereas tropical eruptions affect both. Very prolonged eruptions at any latitude have global consequences. The spectacular sunsets widely observed for months after major eruptions require large volumes of the very fine particles to have reached the stratosphere.

Although CO_2 is more abundant than SO_2 in volcanic emissions, its effect is much less noticeable for at least two reasons; it does not form particles or droplets that reflect sunlight and it is a common atmospheric constituent anyway. As noted in Table 24.13, the annual average volcanic emission of CO_2 is about 0.2% of the emission by fossil fuel use.

TABLE 15.3 VOLCANIC PARTICLE FALL OUT

Grain/Droplet Size (μm)	Time for Free Fall from 15 km	
	Density 1000 kg m^{-3}	Density 2500 kg m^{-3}
1000 (1 mm)	60 minutes	25 minutes
300	5 hours	2 hours
100	20 hours	8 hours
30	4 days	2 days
10	17 days	7 days
3	80 days	33 days
1	1 year	5 months
0.3	5 years	2 years

15.4.1 Volcanic Ash and Aircraft

The number of reports of damage to aircraft resulting from encounters with volcanic ash is of order 100, 30% of them described as 'serious' and 10% involving engine failure. Although there have been no known crashes, there have been some close calls. Alarm in the aviation industry and an international resolve to establish a worldwide warning system followed an incident in June 1982, when British Airways Flight 9 flew into ash from Galunggung volcano in Indonesia. All four engines failed but three of them were restarted after an extended glide to low altitude and the plane landed without casualties, in spite of negligible vision through its sand-blasted windscreen. Concern intensified in December 1989, when KLM Flight 867 flew into ash from Redoubt volcano in Alaska, with an essentially complete repeat of the 1982 event. In both cases, the cost of aircraft repairs was a large fraction of the total replacement cost. The end result is an international monitoring system, involving both ground and satellite observations, linked to nine warning centres around the world, each responsible for a particular volcanic region. Any area where the density of ash is estimated to have reached 4 mg per cubic metre of air is declared as prohibited airspace. The most extensive aviation shutdown occurred in April–May, 2010 when ash from Eyjaflallajökull volcano in Iceland drifted across a large part of Europe. In addition to flight diversions, several other precautionary flight cancellations confirm that the warning system is fully operational.

Chapter 16

Tsunamis

16.1 WAVE PROPAGATION

Marine waves generated by sudden water displacements are referred to by the Japanese word tsunami, which has a literal translation 'harbour wave'. This expression arises from local resonant amplification of tsunamis in some harbours. Tsunamis are examples of what are technically shallow water waves, that is, waves of wavelength much greater than the water depth. Unlike the familiar ocean waves, their propagation speed (Table 16.1) is independent of wavelength or frequency, but this is not always obvious. It depends on depth and, in water of variable depth, refraction/diffraction effects cause apparent dispersion, with lower frequencies (longer wavelengths) arriving first at great distances.

For any particular tsunami, wavelength is proportional to speed and contracts as it approaches shallow water, with a corresponding increase in amplitude or height, but the vulnerability of particular sections of coastline arises as much from the focussing effect of refraction by sea-floor topography.

Three causes of tsunamis are recognised: submarine earthquakes, volcanic eruptions and landslides/submarine slumping (which may be triggered by earthquakes). Their effects are subtly different. Volcanic eruptions and most landslides or slumping events are effectively point sources, with small horizontal extent compared with the scale over which the waves propagate. Although they generate waves of greatest local amplitude, the amplitude falls off more rapidly with distance than for waves generated by major subduction zone earthquakes on reverse faults that may be 1000 km or more in horizontal extent and cause destruction and deaths thousands of kilometres away across oceans. Some historically important tsunamis follow.

TABLE 16.1 WAVE SPEEDS IN WATER OF UNIFORM DEPTH MUCH LESS THAN THE WAVELENGTH		
Depth	**Speed (m/s)**	**Speed (km/h)**
5 km	220	707
3 km	170	617
1 km	99	356
300 m	54	195
100 m	31	113
50 m	22	80

16.2 EARTHQUAKE-TRIGGERED TSUNAMIS

For all except the AD 360 event in the following list, the earthquakes are represented in Figure 14.1 and Section 14.2.1, which give details.

AD 360, Eastern Mediterranean. Although there are reports of violent shaking, there are none from the epicentral region of what must have been a major shock on the submarine Hellenic fault, which is recognised to extend along a trench south of Crete. A detailed report from Alexandria describes effects that were typical of major tsunamis, in this case extending for many kilometres inland across the flat Nile delta. The major settlements of Greco-Roman culture along the North African coast never recovered.

1700 Washington State – British Columbia.

1755 Lisbon.

1952 Kamchatka. This event was triggered by event B in Figure 14.1. It destroyed several towns, including Severo-Kurilsk, which has never been rebuilt.

1960 Chile. Event D in Figure 14.1.

1964 Alaska. Event F in Figure14.1.

2004 Sumatra. Event H in Figure 14.1.

2011 Tohoku. Event L in Figure 14.1.

16.3 VOLCANICALLY GENERATED TSUNAMIS

The two events listed here appear with the subduction zone eruptions in Section 15.2.1.

BC 1628 Thera/Santorini.

AD 1883 Krakatoa.

16.4 LANDSLIDE/SLUMP-GENERATED TSUNAMIS

~BC 6200, Norwegian Sea. This was caused by a major slump, or possibly three related events, in which more than 3000 km^3 of sediment slid into the deep Atlantic from the edge of the continental shelf off Norway. Tsunami deposits many metres above sea level have been identified in both Norway and Scotland. The size of the slump appears greater than is explained by normal accumulation of sediment and was plausibly an accumulation of ice-rafted debris dropped at the edge of the northern ice cap during the last ice age. Earthquake triggering is possible but with no evidence.

1958 Lituya Bay, Alaska. An earthquake triggered the collapse of 30 million cubic metres of rock into an inlet of the bay from the side of an adjacent mountain. The resulting wave is estimated to have exceeded 500 m in height when it crossed the opposite bank of the inlet before discharging into the main bay. This is the largest known wave height.

1998 Aitape area, north New Guinea. The tsunami, which followed an earthquake of magnitude 7.1 north of the island, exceeded 15 m in height, destroying coastal villages and the local economy. It was too big to be attributed directly to the earthquake and was a consequence of a major slump triggered by the earthquake.

Slumping may contribute to the generation of many of the tsunamis that are attributable to earthquakes. These include the 2004 Sumatra event, for which the record indicated a secondary, delayed source, consistent with slumping.

Chapter 17

Weathering, Erosion, Crustal Recycling and Regeneration

17.1 SOME DEFINITIONS AND SPECIALIST JARGON

Weathering is the chemical and physical degradation of rock, causing it to lose coherence or chemical identity. It is commonly a preliminary to, but by definition does not include, *erosion* which is the physical removal or displacement of material. Chemical agents of weathering (water, acids, salts, oxygen) do not necessarily act independently of physical agents (thermal changes, including freezing and melting of water, penetration by plant roots). Both may be accompanied by erosion processes, driven by wave action, stream flow, wind, heavy rainfall or gravitational instability of slopes. However, particularly in the case of chemical weathering, the residua may not erode but remain in situ as weathered rock (such as laterite). The vulnerability of a rock to erosion is referred to as its *erodibility*. This increases with weathering, but it is not a specific rock property as it depends on the *erosivity* (effectiveness) of eroding influences to which it is subjected and land, slope, vegetation and soil cover. Although these concepts are not reliably quantified, in principle with a suitable choice of units, erosion will occur if the product of the erosivity of an eroding agent and the erodibility of a target rock exceeds a specified value. However, these concepts are not readily applied to large-scale erosion processes, such as landslides, for which a general term *mass wasting* is used (Section 17.5).

Some common words have specific meanings when applied to erosion processes, particularly if they are transient consequences of current or recent heavy rain. *Splash* is the displacement of particles by impacting raindrops. Its effect is to render loosely aggregated rock or soil more vulnerable to erosion by water movement. *Rill erosion* is the transport of loosened particles in shallow, ephemeral channels. If they develop further, with repeated heavy rainfall, the process becomes *gully erosion* in more permanent channels. These effects are most noticeable on recently cultivated or overgrazed land and in arid areas.

If the land structure does not make it vulnerable to channel formation, persistent rain may cause *sheet erosion*, transporting loosened particles, organic material and nutrients across an extended area, leaving a stony surface, perhaps with exposed roots.

17.2 A GLOBAL PERSPECTIVE

The rate of sediment deposition in the oceans and shallow seas increased by a factor ~2 as agriculture became widespread, but has decreased more recently as major dams were constructed on sediment-bearing rivers. Major reservoirs formed by damming rivers are listed in Table 11.3 and a more extensive list of hydroelectric stations using them is given in Table 25.3. As the dams silt up and progressively become less useful, sediment transport to the sea may increase again. Notable changes have occurred in the Nile Delta, where agriculture used to rely on an annual influx of silt, now blocked by the Aswan Dam, and in the wetlands at the mouth of the Colorado River, where the diminished water flow is recognised to have serious environmental effects. Many small dams have become useless and are being removed, but the long-term future of big ones is unclear. The Aswan Dam, in particular, is expected to be useful for no more than 100 years. The sequence of changes causes some confusion to estimates of the long-term rate of continental erosion derived from thicknesses of marine sediment and sediment loads of rivers. The total sediment load carried by the world's rivers is about 20×10^9 tonnes/year, but 80% is deposited on wetlands, deltas, shallow shelves and continental margins. The deep ocean sedimentation rate is about 4×10^9 tonnes/year of solid (rocky) continental material, not including biological products that incorporate soluble products of erosion, especially calcium. Viewed on a geological time scale, it is a rapid transfer of continental material to the ocean floors. The mass of land above sea level is 3×10^{17} tonnes and this much is transferred to the deep oceans in 75 million years. However, large areas of all continents have ages exceeding 2000 million years. These numbers focus attention on the tectonic processes that maintain the continent–ocean balance. The ancient continental cratons are worn down and flat. The steep land that erodes rapidly is geologically youthful, a product of volcanism or crustal compression. Even compression can be regarded as crustal recycling; regeneration must be attributed to processes broadly described as igneous (Section 17.6).

The rate at which continental crust is eroded (and replaced) requires an assessment of the extent to which it is weathered by chemical interactions that have implications for global chemistry. Hydration of minerals on land as well as the sea floor and the recycling of sea salt in the subduction–volcanism cycle are obvious examples but present no serious conceptual problems. The oxidation of minerals during weathering, especially of iron-bearing minerals, but also aluminium, requires more detailed examination because recycling is less apparent. Oxygen is absorbed from the atmosphere in converting Fe^{2+} ions to Fe^{3+} as igneous rocks are weathered and it is also consumed in oxidising volcanic gases, but there is no obvious return path. The rate at which oxygen is consumed in these processes is not very accurately determined but would certainly exhaust the atmospheric oxygen in less than 50 million years, plausibly within 15 million years. It is maintained by two processes, photosynthesis, with permanent burial of reduced carbon, and photolysis of atmospheric water vapour, releasing oxygen and allowing the escape of hydrogen to space. Photosynthesis is now dominant, but cannot always have been so. The oxygen balance is more delicate than is generally realised.

17.3 EROSIVE EFFECT OF RAINFALL

The erosivity of rain depends strongly on raindrop size. The droplets in cloud and mist are smaller than 0.1 mm and fall noticeably only in very still air. The size classified as drizzle is generally in the range 0.2 mm to 0.5 mm diameters. Drops larger than 0.5 mm fall fast enough to be recognised as rain. Small drops are maintained spherical by surface tension, but this is less effective with larger drops (up to ~5 mm), which become elliptical in response to air resistance to their fall. Their erosive effects depend on impacting mass as well as impact speed, which is 2 m/s for 0.5 mm drops but 9 m/s for 5 mm drops in still air and increased by wind. Since free fall speed increases with drop size and both increase with rainfall intensity (as defined in Table 17.1), heavy rain is most effective.

17.4 RIVER AND STREAM TRANSPORT OF SEDIMENT

The ultimate destiny of eroded material depends on the effectiveness of water transport. The slowing of water flow downstream allows redeposition of sediment, but in some situations further erosion may occur. Simple transportation of sediment with neither sedimentation nor further erosion is rare. Figure 17.1 summarises the factors controlling this competition. This figure is necessarily approximate because it does not allow for factors such as particle density, water depth and the influence of stream/river floor roughness on turbulence. Sediment load is often estimated from turbidity (Section 11.4.2).

TABLE 17.1 TYPICAL DESCRIPTIONS OF RAINFALL INTENSITY, WHICH VARY GEOGRAPHICALLY

Description	Rainfall
Slight drizzle	Trace to 0.3mm/h
Moderate drizzle	0.5–0.5mm/h
Heavy drizzle	>1 mm/h
Light rain	<2.5 mm/h
Moderate rain	2.5–7.6 mm/h
Heavy rain	7.6–50 mm/h
Torrential rain	>50 mm/h

Note: The word drizzle also has a definition in terms of droplet size, but as this correlates with intensity of rainfall no confusion arises.

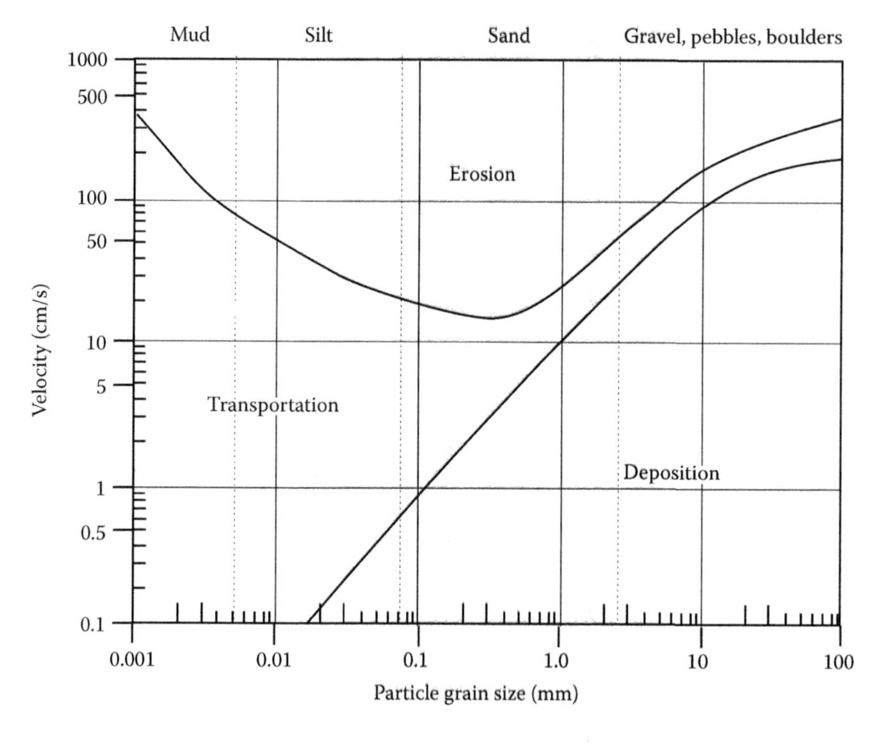

Figure 17.1 Effects of water speed and particle size on the competition between erosion, sediment transport and active sedimentation in moving water.

17.5 MASS WASTING

Mass wasting is a collective term for a range of erosion processes in which material moves in bulk, in the manner of a landslide, and not as independent particles. Distinct events are generally triggered by external influences, such as earthquakes, volcanic eruptions or heavy rain, always in combination with gravitational instability. Energy for the motion is derived gravitationally from the weight of moving material and not normally from triggering effects or external influences, such as water flow. Although this means that there is a net downward component of the motion, in some cases, consequential horizontal displacements can be much larger, essentially because large masses are involved and the momentum generated is not quickly dissipated. Table 17.2 summarises event types and gives details of examples. The general term 'landslide' can refer to a range of phenomena that are here identified by different descriptive words, but the categories are not all distinct.

TABLE 17.2 TYPES OF MASS-WASTING EVENT: DESCRIPTIONS, FEATURES AND EXAMPLES

Descriptions and Features	Examples
Slide *Landslide*: Sudden mass movement downslope of earth, debris and rock. Caused by gravitational instability. May be triggered by rainfall or earthquake.	Vajont Dam, Italy, 1963. 260,000,000 m^3 rock and debris fell into the reservoir at 110km/h displacing 50,000,000 m^3 water over the top of the dam in a 25-m high wave that destroyed five villages and caused around 2000 fatalities.
Rockslide: Landslide of rocky and stony material, often falling due to failure along a fault or bedding plain.	Frank, Alberta, Canada, 1903. 82 million tonnes of limestone slid down Turtle Mountain and buried part of the town, causing 76 deaths.
Slump: Movement of a loosely consolidated mass a short distance downslope. May be *translational* or *rotational* if the motion is constrained by the bedding plane or shape of underlying rock. Triggered by earthquakes or freeze–thaw weakening.	Limbe, Cameroon, 2001. Multiple translational slumps of highly weathered volcanic rock after prolonged heavy rain. Many houses destroyed, 24 deaths. Holbeck Hall, Yorkshire, UK, 1993. A progressive rotational slump of 10^6 tonnes of cliff after heavy rain. Cliff hotel destroyed.

(Continued)

TABLE 17.2 *(Continued)* TYPES OF MASS-WASTING EVENT: DESCRIPTIONS, FEATURES AND EXAMPLES

Descriptions and Features	Examples
Earthflow: Slow movement of material, water saturated over periods of high precipitation.	Portuguese Bend, California. The ongoing movement, accentuated by construction in the area that began in 1956, resulting in abandonment of homes.
Submarine slide: Thick sedimentary layers that have accumulated on steep slopes offshore of rapidly eroding land are vulnerable to major slides of soft material. Seismic and tsunami records commonly indicate secondary or delayed signals consistent with seismically triggered slips that may not be otherwise observed. The 2004 Sumatra–Andaman magnitude 9.2 earthquake was a particular example.	North Atlantic, ~8200 years ago. The largest documented slide occurred on the edge of the continental shelf off Norway, when ~3500 km^3 of sediment slid into the deep ocean (the *Storegga slide*), leaving tsunami deposits on elevated land in Norway and Scotland. The trigger is not known. Aitape, New Guinea, 1998. A submarine landslide followed a magnitude 7.1 earthquake, causing a 15 m tsunami, which destroyed coastal settlements with >2000 deaths.
Rockfall: Rock detaching and moving, sliding, toppling or freely falling from cliff faces under gravity. May be a consequence of weakening caused by weathering, with falling rocks dislodging additional material.	Lituya Bay, Alaska, 1958. About 30 million m^3 of rock fell from around 900 m into Gilbert Inlet. The resulting wave removed all trees and vegetation from elevations up to 524 m above sea level around the bay. Resulted in 5 deaths.
Avalanche: Rock and debris entrained in a slide of ice and snow, which accumulates increased volume as it moves down a steep slope.	Galtür, Austria, 1999. Ski resort town ruined. Thirty-one deaths.
Sturzstrom: A landslide involving a thick layer of rock and soil that acquires enough momentum to carry it a long way horizontally, distributing debris over a wide area.	St. Helens Volcano, Washington State, USA, 1980. During a major eruption, the north face of the mountain blew out, distributing 2.8 km^3 of rock over many kilometres and blocking a river. Heart Mountain, Wyoming. Parts of a 1000 km^3 landslide moved up to 50 km.

(Continued)

TABLE 17.2 *(Continued)* TYPES OF MASS-WASTING EVENT: DESCRIPTIONS, FEATURES AND EXAMPLES

Descriptions and Features	Examples
Debris flow: Water-laden mass of various sized rocks, soil, sand and mud, fast moving and able to entrain additional material. Sediment concentration may exceed 50% of the volume.	Vargas, Venezuela, 1999. Four days of heavy rainfall triggered landslides on steep slopes that led to massive debris flows. Seven hundred dwellings were destroyed or damaged with severe disruption of infrastructure. More than 8000 deaths.
Mudflow: Rapidly moving fine material able to travel long distances down low-slope angles entraining various particle sizes. Typically occurring in highly saturated rock and soil.	Oso, Washington, USA, 2014. After 45 days of heavy rain, an estimated 5×10^6 m^3 of mud was mobilised, destroying many houses and causing 43 fatalities.
Lahar: This is a mud and debris flow caused by rupture of a volcanic lake during an eruption or extreme rainfall on volcanic ash on steep slopes.	Nevado del Ruiz volcano, Colombia, 1985. The eruption melted glaciers and snow, generating four lahars that led to over 23,000 fatalities in nearby towns. Mount Pinatubo. In the years following the 1991 eruption, dozens of people have died from lahars.
Jökulhlaup: A kind of lahar. An Icelandic word for a mud and debris flow driven by a sub-glacial volcanic eruption.	Iceland, 1727. Three deaths and loss of livestock and buildings.
Pyroclastic flow: A cloud of hot ash and gas driven by a lateral eruption from a volcano. It may travel for tens of kilometres and settle while still hot, forming ignimbrite.	Details of two well-documented examples (Vesuvius and Pelée) are given in Section 15.2.1.
Creep: Very slow downslope movement by deformation of material.	Typical evidence includes tilted gravestones, fence posts and telegraph poles; cracking of man-made structures; and trees bending at the base as they grow vertically during continued creep. May cause infrastructure damage. Engineering consideration for construction and buried pipes.
Solifluction: Creep in water-saturated material.	Gradual soil thinning on slopes that may cause lobe-shaped structures of transported sediment on hillsides.

17.6 IGNEOUS REGENERATION OF THE CRUST

17.6.1 Continental Crust

The continents are maintained by igneous activity, volcanic and plutonic. Estimated rates of the processes listed in Table 17.3 must, in the long term, collectively come close to balancing erosion. However, in the present (agricultural) era, erosion has accelerated and is removing sediment from land that has been made vulnerable by human activity as well as from land that is erosion-prone because it is elevated. The numbers refer only to sediment that reaches the deep oceans, directly or indirectly, and does not include sediment deposited in low-lying wetlands, estuaries, deltas and continental shelves, where it is considered to remain continental. The state of balance in Precambrian times, when volcanism was more vigorous but the bare continents would have eroded more rapidly, is unclear.

17.6.2 Oceanic Crust

The sea-floor spreading process produces 20–25 km^3/year of basaltic sea floor crust (3.4 km^2/year of thickness 6–7 km) more or less steadily at mid-ocean ridges, but its existence is transient (mean lifetime 92 million years) and after hydration and chemical exchange with ocean solutes, it becomes feedstock for subduction zone volcanism. A very small fraction is added directly to the continents.

TABLE 17.3 THE RATES OF CRUSTAL WASTING AND REGENERATION BY VARIOUS PROCESSES

Process	Product	Continental Growth Rate (km^3/year)
Sea-floor spreading	Basalt	~0.1
Subduction zone volcanism	Acid rocks	3
Hot spots, flood basalts	Basic rock	1.5
Erosion, pre-agricultural	Sediment removed to deep ocean	–4
Erosion, present day		–8[a]
	Net Continental Growth	
Pre-agricultural		+0.6 ± 1
Present day		–3.4 ± 1

[a] The reduction in sediment transport by modern dams is here discounted.

Chapter 18

Geological Time and the Fossil Record

18.1 DEFINITIONS

Geological units are divided in various ways, depending on the characteristics considered and the classification methods. The term stratigraphy describes the ways in which strata are defined. Tables 18.1, 18.2 and 18.3 identify the commonly used stratigraphic terms. Geochronologic units are periods of time, whereas chronostratigraphic units are geological materials. Geochronology is a tool used in chronostratigraphy. Chronostratigraphic units are geological markers or periods of time marked by fossil occurrences that indicate relative ages. Geochronology determines the age of a rock or fossil using markers including fossils, isotopic dating, luminescence dating and palaeomagnetic dating and chemical indicators. The ages determined may be either relative or absolute depending on the methods.

TABLE 18.1 LITHOSTRATIGRAPHIC TERMS

Term	Meaning
Bed	Smallest recognisable stratigraphic unit/layer
Member	A distinct part or subdivision of a formation
Formation	The primary unit of lithostratigraphy
Group	Two or more formations sharing distinct characteristics
Supergroup	Two or more groups sharing similar characteristics

TABLE 18.2 BIOSTRATIGRAPHIC TERMS

Term	Meaning
Index fossil	A fossil of a creature that was geographically widespread, rapidly evolving, abundant in the rock record, well preserved and easy to identify.
Zone	Stratum characterised by a range of fossils, representing the time between the appearance of species at the base and the next zone (after Oppel 1856).
Assemblage	A group of (three or more) fossils of creatures that coexisted.
Lineage zone	A zone that hosts parts of the evolutionary lineage of a fossil group.
Interval zone	The top and bottom of which mark the first and last occurrence of a taxon.
Range zone	Interval that represents the range of occurrence of a taxon.
Assemblage zone	Strata characterised by an assemblage distinguishing the zone from surrounding strata.
Abundance zone	Strata in which the abundance of a taxon or a group of taxa is significantly greater than in the adjacent strata.

TABLE 18.3 CHRONOSTRATIGRAPHIC AND GEOCHRONOLOGIC TERMS

Chronostratigraphic Terms	Corresponding Geochronologic Terms
Eonothem	Eon
Erathem	Era
System	Period
Series	Epoch
Stage	Age
Chronozone	Chron

18.2 GEOLOGIC RECORD

Geological time was originally recorded as a sequence of fossils left by biological life as it evolved. A quantitative scale was added when radiogenic dating methods developed. Figure 18.1 presents the named periods with end dates.

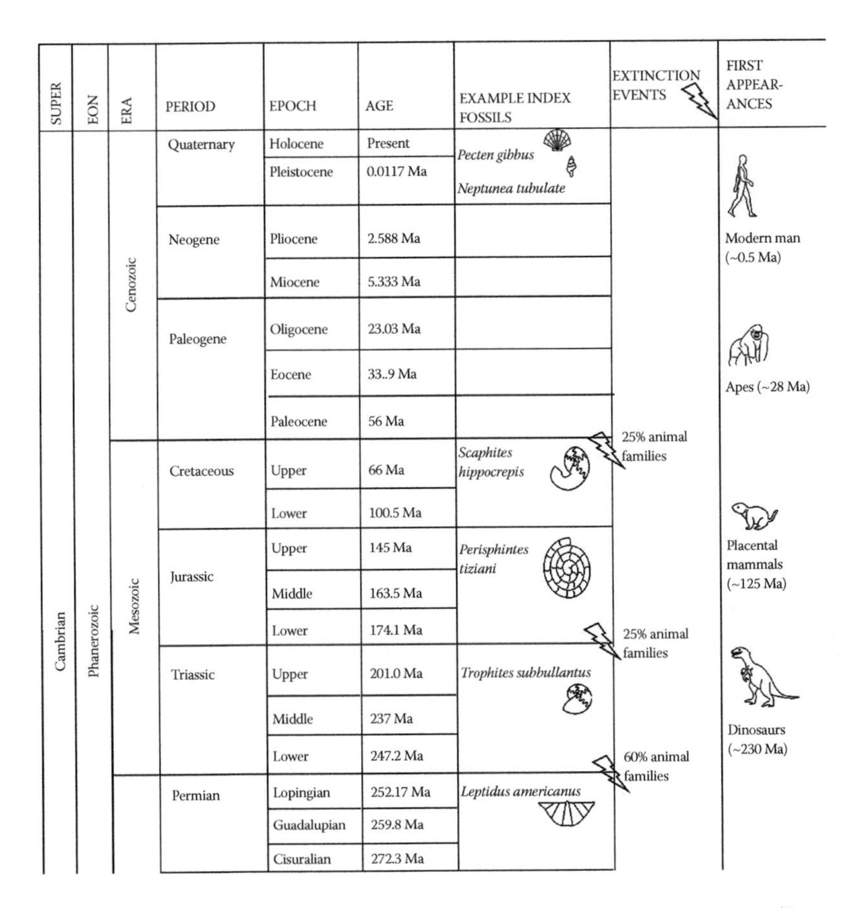

SUPER	EON	ERA	PERIOD	EPOCH	AGE	EXAMPLE INDEX FOSSILS	EXTINCTION EVENTS	FIRST APPEAR-ANCES
Cambrian	Phanerozoic	Cenozoic	Quaternary	Holocene	Present	*Pecten gibbus*		
				Pleistocene	0.0117 Ma	*Neptunea tubulate*		
			Neogene	Pliocene	2.588 Ma			Modern man (~0.5 Ma)
				Miocene	5.333 Ma			
			Paleogene	Oligocene	23.03 Ma			
				Eocene	33..9 Ma			Apes (~28 Ma)
				Paleocene	56 Ma		25% animal families	
		Mesozoic	Cretaceous	Upper	66 Ma	*Scaphites hippocrepis*		
				Lower	100.5 Ma			
			Jurassic	Upper	145 Ma	*Perisphintes tiziani*		Placental mammals (~125 Ma)
				Middle	163.5 Ma			
				Lower	174.1 Ma		25% animal families	
			Triassic	Upper	201.0 Ma	*Trophites subbullantus*		
				Middle	237 Ma			Dinosaurs (~230 Ma)
				Lower	247.2 Ma		60% animal families	
			Permian	Lopingian	252.17 Ma	*Leptidus americanus*		
				Guadalupian	259.8 Ma			
				Cisuralian	272.3 Ma			

Figure 18.1 The geologic time scale. *(Continued)*

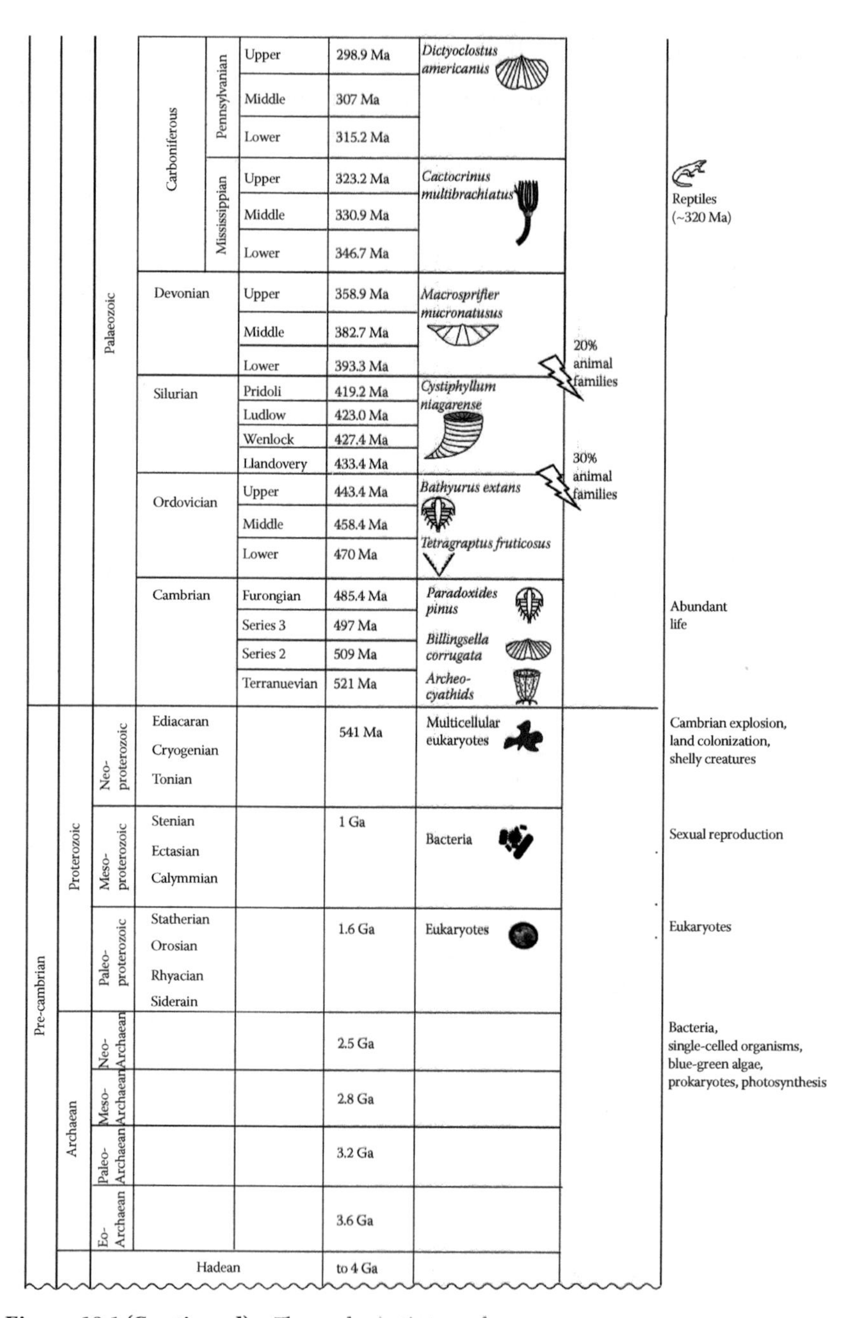

Figure 18.1 (Continued) The geologic time scale.

Section VI

Rocks and Minerals

Chapter 19

Mineral Types and Characteristics

19.1 DESCRIPTIVE TERMS

Various features of minerals are used to describe and identify them. In a hand specimen, diagnostic properties, defined in the following subsections (Tables 19.1 through 19.8 and 19.11 through 19.20), are colour, streak, Mohs hardness, fracture, cleavage, habit, lustre and specific gravity.

19.1.1 Colour of Mineral in Hand Specimen

As the word implies, a hand specimen of rock or mineral is typically of a size that will fit in the hand. To be useful, it must have a freshly broken surface that exposes unweathered grains. Colour is simply a general description of light reflected from the surface, which is often different from the colour seen in unpolarised transmitted light (in thin-section under a microscope).

19.1.2 Mohs Hardness

The scale proposed in 1812 by the mineralogist Friedrich Mohs is a relative scale of scratch resistance of minerals judged by the fact that harder materials visibly scratch softer ones in hand specimens. Type specimens are used to specify the hardness scale from 1 (softest) to 10 (hardest) (Table 19.1) and common, readily available materials are generally used (Table 19.2).

19.1.3 Streak

If a mineral specimen is scratched on an unglazed porcelain plate ('streak plate'), the colour of the resulting fine powder is its 'streak', which may be different from both the colours in reflected light and in thin section. Minerals with Mohs hardness exceeding 6.5 cannot normally produce a streak.

TABLE 19.1 THE MOHS HARDNESS SCALE

Mohs Hardness	Index Mineral
1	Talc
2	Gypsum
3	Calcite
4	Fluorite
5	Apatite
6	Orthoclase feldspar
7	Quartz
8	Topaz
9	Corundum
10	Diamond

TABLE 19.2 COMMON ITEMS USED FOR TESTING HARDNESS

Hardness	Material
2–3	Finger nail
3	Copper coin
5–6	Knife blade
6	Broken glass
7	Steel file

19.1.4 Fracture

The shape and texture of a fractured mineral or rock surface is its fracture type, generally one of the standard types in Table 19.3. Unlike cleavage (Section 19.1.5), which is the tendency of a mineral to break along a surface defined by its internal lattice or crystal structure, fracture describes less regular breakage.

19.1.5 Cleavage

A mineral's cleavage describes is its tendency, when subject to stress, to break along smooth planes self-selected by weak atomic bonds. It may occur in multiple orientations or faces, depending on the mineral. Mineralogists refer to the orientation of a cleavage plane as its direction (normal to the plane). The logic of this can be seen by referring to Miller indices (Section 19.2.1).

TABLE 19.3 STANDARD DESCRIPTIONS OF FRACTURED SURFACES	
Fracture Type	**Description**
Conchoidal	Curved breakage, resembling the curves of a sea shell or broken, thick glass
Uneven	A rough and uneven surface
Splintery	Elongated splinters, often sharp
Earthy	Crumbly
Even	Smooth surface
Hackly	Rough, jagged points
Brittle	Sharp cracks and edges, large defects
Fibrous	Thin, elongated threads
Micaceous	Displaying thin layers, as in mica

TABLE 19.4 CLEAVAGE DIRECTIONS	
Cleavage	**Examples**
In one direction (planar)	Muscovite (mica generally)
In two directions	Feldspar (at right angles)
In three directions	Halite (cubic cleavage, 3 cleavages at right angles); calcite (rhombohedral, 120° and 60°)
None	Quartz (irregular breakage)

TABLE 19.5 CLEAVAGE QUALITY	
Quality	**Description of Broken Surface**
Perfect	Smooth, no roughness
Very good/good	Smooth surface with minor residual rough areas (usually where faces meet)
Poor/very weak	Smooth edge (where faces in different planes meet) is not easy to detect, leaving a mainly rough surface, but planes can be identified
Indistinct	Very rough edges, but some planar surfaces may be noticeable
None	No particular direction or surface is evident

TABLE 19.6 CLEAVAGE HABIT

Habit	Description
Basal	A plane such as mica
Cubic	Exhibited by minerals that are isometric, forming cubes as in galena (PbS)
Octahedral	Minerals that are isometric, forming octahedra and cleaving triangular wedges, retain the octahedral shape of minerals such as fluorite
Prismatic	Cleavage forms thin, prismatic crystals perpendicular to the primary cleavage plane in minerals such as orthoclase
Pinicoidal	Tabular or prismatic elongated crystals, in minerals such as barite
Rhombohedral	Producing small rhombohedra, in minerals such as calcite

19.1.6 Mineral Habit

TABLE 19.7 SOME OF THE COMMON (MOST OBVIOUS) TERMS USED TO DESCRIBE THE GENERAL APPEARANCES OF MINERALS IN HAND SPECIMENS

Habit	Description
Acicular	Needle or blade-like
Amorphous	No identifiable shape
Bladed	Flat, slender, blade-like
Botryoidal	Cluster of rounded, conjoined shapes, like a bunch of grapes
Columnar	Conjoined column shapes
Cruciform	A cross-like structure
Crystalline	One or more well-defined conjoined crystals
Cubic	Cube-shaped
Dendritic	Complex branching shapes, often through a matrix
Drusy	Numerous crystal tips
Equant or blocky	Roughly spherical/ball-shaped or cubic/boxy in shape. Can be rhombohedral
Euhedral	Well-developed crystal faces

(Continued)

TABLE 19.7 *(Continued)* SOME OF THE COMMON (MOST OBVIOUS) TERMS USED TO DESCRIBE THE GENERAL APPEARANCES OF MINERALS IN HAND SPECIMENS

Habit	Description
Fibrous	Thin conjoined, often flaky fibres
Foliated	Flat layers like leaves of a book
Friable	Crumbly, loosely cohesive, falls apart easily
Globular	Nearly spherical form
Grainy	Appearance of small grains
Granular	Matrix containing aggregate of crystals
Lamellar	Thin flat layer
Lath	Tabular, often rectangular shaped
Massive	Cluster or mass of distinct forms
Microcrystalline	Multiple small, well-defined conjoined crystals
Nodular	Irregular group of spherical nodules
Platy	Similar to foliated and tabular, flat, layered
Prismatic	Elongate, prism-like
Spherical	Rounded, sphere-like
Tabular	Shapes that are thin in one direction, like platy and foliated
Twinned	Adjoined crystals that are mirror images

19.1.7 Lustre

The mineral lustre (Table 19.8) refers to the way in which light is reflected from a surface. There are no clear boundaries between types of lustre.

19.1.8 Specific Gravity

Specific gravity is synonymous with density reported in grams per cubic centimetre (g/cm^3), being the weight relative to an equal volume of water. Measurements are made at atmospheric pressure and ambient temperature (specifically 20°C if high precision is required). The mineral with the lowest specific gravity is amber (1.05 to 1.30 g/cm^3), and the naturally occurring mineral/element with the highest specific gravity is gold (19.30 g/cm^3).

TABLE 19.8 COMMON TERMS USED TO DESCRIBE MINERAL LUSTRE

Lustre	Description
Adamantine	Uncommon, typically highly reflective surfaces, as in diamond
Dull (or earthy)	Very low or no reflection, often with a coarse surface such as clay
Greasy	Resembles grease, for example, opal
Metallic and sub-metallic	Metallic lustre resembles the surface of polished metal as in pyrite; sub-metallic less reflective, as in spheralite
Pearly	Gives the appearance of pearls, as in muscovite
Resinous	Having the appearance of resin or plastic, as in amber
Silky	Aligned, smooth appearance of silk fibres; for example, asbestos
Vitreous	Glass-like reflective surface, very common; for example, quartz
Waxy	Resembles wax; for example, jade

19.2 CRYSTAL SYSTEMS

Crystals can be classified by systems that describe their atomic lattices or structures. The system is defined in terms of the relative lengths of the unit vectors, *a*, *b* and *c*, of the lattice unit cell and the angles between them. There are seven crystal systems, defined according to their symmetry axes, planes of symmetry and crystallographic axes (the vectors *a*, *b* and *c*). The orientation of each axis is designated + or –. The angles between the positive axis ends are designated α, β and γ. As in the convention of geometry,

- α lies between *b* and *c*.
- β lies between *a* and *c*.
- γ lies between *a* and *b*.

19.2.1 Miller Indices

In 1839, W. H. Miller introduced a notation, now universally adopted, to represent the orientations of planes and axes in a lattice or crystal structure by three integers, usually designated *h*, *k*, *l*. In the case of a plane, these are specified by the reciprocals of the distances from a convenient origin to the intersections of the plane with the three principal axes. If, for example, the distances are 1, 2 and 3 units, the reciprocals are 1, 1/2, 1/3 and the smallest set of integers with the same ratios is $h, k, l = 6, 3, 2$, which completely specifies the orientation of the plane. Other planes with similar ratios with respect to any of the axes are given the collective generic notation {632} in curly brackets, but for specific axes or planes the notation is (632) and if one of the intercepts is negative a superscript minus sign is added, as in $(6\bar{3}2)$. Parallel planes have the same generic indices. A plane or line that has no intercept with an axis is deemed to have an intercept at infinity and assigned an index zero. Directions of axes, or lines, are represented similarly but in square brackets for the generic notation, as in [632]. Most of the indices of interest are simple numbers. Thus, [100] represents the orientation of a cube edge and [111] denotes a cube body diagonal. In the simple case of cubic structures, an axis with particular indices is normal to planes with the same indices.

Referring to cleavage planes (Section 19.1.5), in identifying the 'direction' of a plane by its Miller indices, although it is two-dimensional, its orientation is completely specified without considering directions of pairs of axes in the plane.

19.2.2 Bravais Lattices and X-Ray Diffraction

Table 19.9 identifies the seven crystal systems defined by crystal shape. Although the shapes result from the arrangements of constituent atoms on regular lattices, there is no 1:1 correspondence between crystal shape and atomic lattice structure; several different lattices may produce a common crystal shape, but they are distinguishable by X-ray diffraction and may have significantly different physical properties. Symmetry principles specify 14 fundamental types of lattice, as first enunciated in 1848 by the French crystallographer A. Bravais. The relationship between Bravais lattices and corresponding crystal systems is illustrated in Figure 19.1 for the case of cubic crystals, for which there are three different Bravais lattices. The body-centred structure is two superimposed simple cubic lattices displaced from one another by half a body diagonal and the face-centred structure is four superimposed simple cubic lattices each displaced from the others by half face diagonals (the broken lines in Figure 19.1). Similar variations give two tetragonal lattices, four orthorhombic lattices, two monoclinic but only one each for the hexagonal, triclinic and trigonal systems.

The cubes are repeated indefinitely in three directions, sharing their corners, so that each cube has eight corners and each corner is shared by eight cubes. The individual points are not necessarily single atoms but repeating units that may represent molecules.

X-ray crystallography identifies crystal structures by diffraction of X-rays from planes of identical atoms in their Bravais lattices. This means that in a

TABLE 19.9 THE SEVEN CRYSTAL SYSTEMS

System	Symmetry Axes	Number of Planes of Symmetry	Crystal Image	Example
Hexagonal	Three equal with fourth perpendicular to them of a different length	7		Emerald, graphite
Isometric or cubic	Three, all perpendicular and of equal length	9	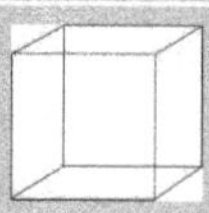	Diamond, pyrite
Monoclinic	Three unequal, two inclined to one other and the third perpendicular to their plane	1	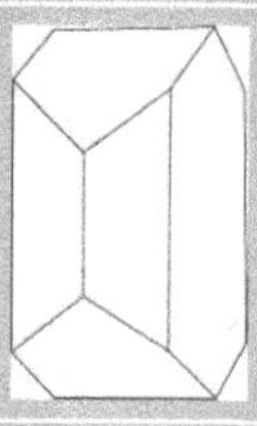	Jade, gypsum
Orthorhombic	Three perpendicular of different lengths	3	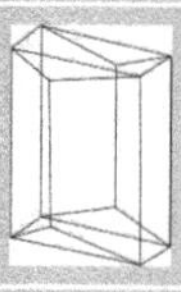	Peridot, topaz
Tetragonal	Three, all mutually perpendicular, two equal and the third one shorter	5	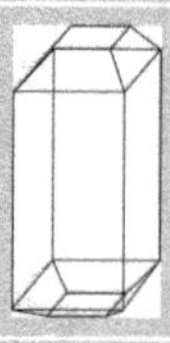	Zircon, rutile
Triclinic	Three unequal with oblique angles	0	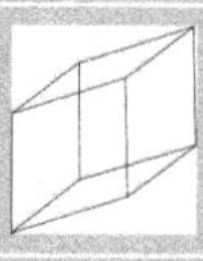	Turquoise, labradorite
Trigonal (often described as a subset of hexagonal)	Four axes, three of equal lengths in a plane with 120° intersections; the fourth perpendicular to the plane	3	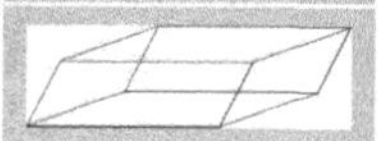	Quartz, calcite

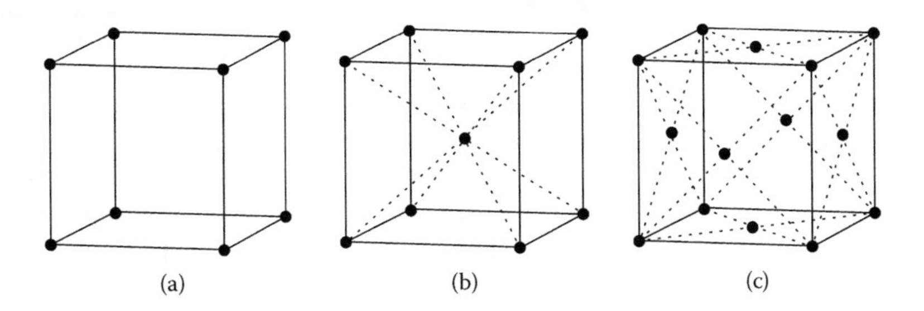

Figure 19.1 Lattice structures for the three types of cubic crystal: (a) simple cubic, (b) body-centred cubic and (c) face-centred cubic.

body-centred cubic element, such as iron at laboratory temperature, with all atoms in Figure 19.1b the same, there are planes of identical atoms in various orientations, which include both the corners and the body centres of the illustrated cube and its neighbours. The pattern of orientations of all these planes and the spacing of equivalent planes determine the corresponding X-ray diffraction pattern, which is characteristic of the body-centred cubic structure. However, the arrangement of atoms in a caesium chloride crystal can be represented by the same figure but with the corners caesium and the body centres chlorine or vice versa. This is not recognised as body-centred structure but as a superposition of two simple cubic structures, one of caesium and the other chlorine. The Bravais lattice is simple cubic with two atoms (Cs + Cl) at each lattice point and that is how X-rays see it. Another simple compound that illustrates the principle is sodium chloride (NaCl), which, if all atoms appeared identical, would look simple cubic, but has a face-centred Bravais lattice, with two atoms (Na + Cl) at each lattice point. The lines of atoms in a regular simple cubic array are alternately Na and Cl, giving superimposed Na and Cl face-centred cubic lattices.

19.3 SILICATES

Silicates are compounds of silica, SiO_2, with other elements, especially Mg, Fe, Al, Ca (as oxides). Except at the high pressures of the lower mantle, their structures are controlled by the strong preference of the Si ions for tetrahedral bonding, with each Si atom bonded to four oxygen atoms, all at equal angles to one another, and the $(SiO_4)^{2-}$ tetrahedra linked to one another by sharing oxygen ions. The low-pressure crystal form of SiO_2 (quartz) is a three-dimensional array of tetrahedra, each sharing oxygen ions with neighbours in all directions. Each Si ion is bonded to four O neighbours and each O ion is shared by two

Si ions, making the overall chemical formula $(SiO_2)_n$. The tetrahedral bonding results in an open crystal structure with two important consequences: (i) it allows atoms of a range of other elements to fit between the tetrahedra and contribute to the bonding that holds them together in mineral crystals, and (ii) the tetrahedral structure does not withstand strong compression and the high pressures of the lower mantle force the Si atoms into sixfold coordination with oxygen instead of fourfold. The result is closer atomic packing and higher density, producing coesite and stishovite as high pressure forms of quartz and $(Mg,Fe)SiO_3$ with a perovskite structure, which is the high pressure form of pyroxene and the most abundant lower mantle mineral. At a given temperature and pressure, a particular silica polymorph may be stable, and its presence in a rock identifies the conditions of its formation. Under different conditions, however, a polymorph may break bonds to become a new stable form (reconstructive transformation). Alternatively, small adjustments to a crystal lattice such as the angle between bonds, but without breaking any of them, may lead to a new structure. This is known as displacive transformation. Given suitable changes of conditions and sufficient time, polymorphs may further transform or revert to previous polymorphs.

The diversity of low-pressure silicates is a consequence of the several ways in which SiO_4 tetrahedra can be linked (polymerisation) and the ways in which other elements can provide linkages. Table 19.10 lists the principle linkage types and Tables 19.11 through 19.15 list the properties of examples of each type of silicate structure. Silicate polymerisation controls the viscosity of the melt from which the minerals form. Silicates with the highest degree of polymerisation (multiplicity of bonds) are 'framework' silicates (Table 19.11), and the degree reduces through the series of silicate structures in Tables 19.11 through 19.15. Orthosilicates are not polymerised because the negative charges of the anions are neutralised by the cations of additional elements.

Basaltic magma has low Si content and low degrees of polymerisation, resulting in low viscosity. It erupts easily and flood basalts flow for hundreds of kilometres. Rhyolitic (Si rich) magmas, with high degrees of polymerisation and high viscosity, may crystallise before reaching the surface. The highly viscous nature of the magma traps gas bubbles that cause explosive eruptions. Bowen's reaction series (Figure 20.7) illustrates the series of minerals that form as magma cools and the relationship between increasing polymerisation and increasing stability with consequent increasing resistance to weathering of the resulting minerals. Similar tables follow for carbonates (Table 19.16), phosphates and borates (Table 19.17), sulphates (Table 19.18), oxides, hydroxides and halides (Table 19.19) and some native elements (Table 19.20).

TABLE 19.10 STRUCTURES RESULTING FROM ALTERNATIVE WAYS OF COUPLING SiO_4 TETRAHEDRA

Silicate Structure	Atomic Arrangement	Examples
Framework silicates (tectosilicates)	3D array of tetrahedra bonded together in all directions	Feldspar, silica group (Table 19.11)
Sheet silicates (phyllosilicates)	Tetrahedra share three oxygen atoms with neighbours, forming 2D sheets and leading to strong cleavage planes	Serpentine, mica (Table 19.12)
Chain silicates (inosilicates)	Tetrahedra share two or three oxygen atoms with neighbours and form chains	Pyroxene, amphibole (Table 19.13)
Ring silicates (cyclosilicates)	Two oxygen atoms shared and structure arranged in a ring	Beryl (Table 19.14)
Orthosilicates	Consisting of isolated SiO_4 tetrahedra, bonded with other elements	Garnet (Table 19.15)
Aluminium silicates	Each of the two silica tetrahedra shares an oxygen atom with an aluminium silicate tetrahedron	Kyanite (Table 19.15)

TABLE 19.11 SELECTED FRAMEWORK SILICATES (TECTOSILICATES)

Group/Mineral			Colour in Hand Specimen	Streak	Hardness (Mohs)	Fracture	Cleavage with Miller Indices of Planes	Habit	Lustre	Crystal System	Specific Gravity (g/cm^3)
Feldspar group	Alkali feldspar	Microcline $KAlSi_3O_8$	Blue–green, green, grey, yellow	White	6–6.5	Uneven	{001} perfect, {010} good	Blocky, prismatic, crystalline	Vitreous	Triclinic	2.56
		Orthoclase $KAlSi_3O_8$	Colourless, white, yellow, pink, orange, blue, green–grey	White	6	Conchoidal to uneven	{001} perfect, {010} good	Blocky or tabular	Vitreous to dull	Monoclinic	2.55–2.6
	Plagioclase	Albite $NaAlSi_3O_8$	White, grey, bluish-green	White	6.0–6.5	Conchoidal to uneven	{001} perfect, {010} very good, {110} imperfect	Blocky, tabular, platy	Vitreous, pearly	Triclinic	2.6–2.65
		Anorthite $CaAl_2Si_2O_8$	Grey, white, pink	White	6	Conchoidal to uneven	{001} perfect, {010} good, {110} poor	Blocky, tabular	Vitreous to dull	Triclinic	2.72–2.76

(Continued)

TABLE 19.11 *(Continued)* SELECTED FRAMEWORK SILICATES (TECTOSILICATES)

Group/Mineral			Colour in Hand Specimen	Streak	Hardness (Mohs)	Fracture	Cleavage with Miller Indices of Planes	Habit	Lustre	Crystal System	Specific Gravity (g/cm³)
Feldspathoids		Sodalite $Na_8(Al_6Si_6O_{24})Cl_2$	Royal/azure blue, violet, green, pink, yellow, white veins	White	5.5–6.0	Brittle, conchoidal to uneven	{110} poor	Massive	Vitreous or greasy	Isometric hextetrahedral	2.29
Silica group		Quartz SiO_2	Colourless, milky, white, purple, pink, blue, brown, black, orange, yellow or banded	White	7	Conchoidal	{0110} indistinct/ very weak	Variable (drusy, grainy, bladed, nodular, prismatic)	Glassy to vitreous, waxy to dull	Trigonal	2.6–2.7
		Chalcedony SiO_2	Varies, often banded	White	6.0–7.0	Uneven, conchoidal	None	Variable, can form pseudomorphs of organic material	Vitreous, waxy, dull, greasy	Trigonal	2.59–2.61

TABLE 19.12 SELECTED COMMON SHEET SILICATES (PHYLLOSILICATES)

Group	Mineral	Colour in Hand Specimen	Streak	Hardness (Mohs)	Fracture	Cleavage with Miller Indices of Planes	Habit	Lustre	Crystal system	Specific gravity (g/cm^3)
Serpentine	Chrysotile $Mg_3(Si_2O_5)(OH)_4$	Green, grey–green	White	2.5–3.0	Fibrous	Fibrous	Acicular	Greasy, waxy, silky	Monoclinic-prismatic	2.53–2.65
Serpentine	Antigorite $(Mg,Fe)_3Si_2O_5(OH)_4$	Dark green, green, grey, blue–grey, brown, black	White	3.5–4	Brittle, conchoidal	{001} perfect	Fibrous, flat, tabular, lamellar	Vitreous, greasy, waxy, silky	Monoclinic	2.55
Mica	Muscovite $KAl_2(AlSi_3O_{10})(F,OH)_2$	White, grey, pale brown, pale green	White	2.0–2.5	Brittle, uneven, micaceous	{001} perfect	Tabular, lamellar, foliated, platy	Vitreous, silky, pearly	Monoclinic	2.83–3.0
Mica	Biotite $K(Mg,Fe)_3(AlSi_3O_{10})(F,OH)_2$	Dark green, dark brown, black, yellow, white	Grey	2.5–3.0	Uneven, micaceous	{001} perfect	Lamellar, foliated, platy	Vitreous, pearly	Monoclinic	3.1
Mica	Glauconite $(K,Na)(Fe^{3+},Al,Mg)_2(Si,Al)_4O_{10}(OH)_2$	Blue–green, green, yellow–green	Light green	2	Micaceous	{001} perfect	Lamellar, foliated, platy	Dull	Monoclinic	2.4–2.95
Mica	Chlorite $(Mg,Fe)_3(Si,Al)_4O_{10}(OH)_2{\cdot}(Mg,Fe)_3(OH)_6$	Green, grey–green, white, black	Light green to grey	2–2.5	Uneven	{001} perfect	Foliated	Vitreous, pearly, dull	Monoclinic	2.6–3.3

(Continued)

TABLE 19.12 *(Continued)* SELECTED COMMON SHEET SILICATES (PHYLLOSILICATES)

Group/Mineral		Colour in Hand Specimen	Streak	Hardness (Mohs)	Fracture	Cleavage with Miller Indices of Planes	Habit	Lustre	Crystal system	Specific gravity (g/cm^3)
Clay	Kaolinite $Al_2Si_2O_5(OH)_4$	White, grey, yellow, pale brown	White	2–2.5	Uneven, micaceous	{001} perfect	Massive, platy, microcrystalline	Waxy, pearly, dull, earthy	Triclinic	2.63
	Montmorillonite/Smectite $(Ca,Na,H)(Al,Mg,Fe,Zn)_2(Si,Al)_4O_{10}(OH)_2 \cdot H_2O$	White, pink, yellow, green, blue, red	White	1–2	Uneven	{001} perfect	Massive, lamellar, globular	Dull, earthy	Monoclinic	1.7–2
	Illite $(K,H_3O)(Al,Mg,Fe)_2(Si,Al)_4O_{10}[(OH)_2,(H_2O)]$	Silver–white, grey–white, grey–green	White	1–2	Uneven, micaceous	{001} perfect		Pearly, dull	Monoclinic	2.6–2.9
Others	Talc $Mg_3Si_4O_{10}(OH)_2$	Pale green, grey, white, yellow, pale brown	White	1	Uneven, flat surface	perfect in one direction {001}	Massive, foliated	Vitreous, pearly	Monoclinic	2.75
	Epidote $Ca_2(Al_2,Fe)(SiO_4)(Si_2O_7)O(OH)$	Grey, yellow–green, brown–black, black, yellow	Grey-white	7	Uneven	1 plane {001} Perfect	Fibrous, massive, prismatic	Vitreous	Monoclinic	3.45

TABLE 19.13 CHAIN SILICATES (INOSILICATES)

Group	Mineral	Colour in Hand Specimen	Streak	Hardness (Mohs)	Fracture	Cleavage with Miller Indices of Planes	Habit	Lustre	Crystal System	Specific Gravity (g/cm³)
Pyroxene	Augite $(Ca,Na)(Mg,Fe,Al,Ti)(Si,Al)_2O_6$	Green, dark brown, black	Colourless to light green	5.0–6.0	Splintery, uneven	2 planes, at 87° and 93°	Prismatic, acicular, dendritic	Vitreous, dull	Monoclinic	3.2–3.56
Amphibole	Hornblende $Ca_2(Mg,Fe,Al)_5(Al,Si)_8O_{22}(OH)_2$	Brown, green, green–black	White	5.5–6.0	Sub-conchoidal	2 planes	Columnar, fibrous, granular, massive	Vitreous, pearly	Monoclinic	3.24
Amphibole	Glaucophane $Na_2(Mg_3Al_2)Si_8O_{22}(OH)_2$	Grey, blue–black, azure blue	Grey–blue	6–6.5	Brittle, conchoidal	2 planes, {110} good, {001} good	Columnar, fibrous, granular, massive	Vitreous, pearly	Monoclinic	2.99–3.13

TABLE 19.14 RING SILICATES (CYCLOSILICATES)

Group/ Mineral	Colour in Hand Specimen	Streak	Hardness (Mohs)	Fracture	Cleavage with Miller Indices of Planes	Habit	Lustre	Crystal system	Specific gravity (g/cm^3)
Beryl $Be_3Al_2(SiO_3)_6$	Colourless, blue, green, yellow, pink	White	7.5–8.0	Brittle, conchoidal	1 plane {001} imperfect	Columnar, crystalline, massive	Vitreous, resinous	Hexagonal	2.77

TABLE 19.15 ORTHOSILICATES, ALUMINIUM SILICATES

Group/Mineral		Colour in Hand Specimen	Streak	Hardness (Mohs)	Fracture	Cleavage with Miller Indices of Planes	Habit	Lustre	Crystal system	Specific gravity (g/cm³)
Garnet	Pyrope $Mg_2Al_2(SiO_4)_3$	Red, dark red, white to purple	White	7–7.5	Conchoidal	None	Granular, massive, crystalline	Vitreous	Cubic	3.56–3.78
	Almandine $Fe^{2+}{}_3Al_2Si_3O_{12}$	Brown, red, red–orange, purple-red, black	White	7–7.5	Conchoidal	None	Granular, massive	Resinous, vitreous	Isometric, hexoctahedral	4.2
	Andradite $Ca_3Fe_2(SiO_4)_3$	Yellow, yellow–green, dark green, brown–red, brown, grey, black	White	6.5–7	Conchoidal to uneven	None	Granular, massive, crystalline	Resinous to dull	Cubic	3.85
Aluminium silicate	Kyanite Al_2SiO_5	Blue, white, grey–yellow, pink, green, black	White	4.5–5	Splintery	2 planes: perfect {100}, imperfect {010}, 70° angle	Columnar, bladed, fibrous	Vitreous to pearly	Triclinic	3.5–3.6

(Continued)

TABLE 19.15 *(Continued)* ORTHOSILICATES, ALUMINIUM SILICATES

Group/Mineral		Colour in Hand Specimen	Streak	Hardness (Mohs)	Fracture	Cleavage with Miller Indices of Planes	Habit	Lustre	Crystal system	Specific gravity (g/cm^3)
	Zircon $ZiSiO_4$	Brown, red, green, grey, colourless	White	7.5	Uneven	{110} indistinct	Tabular			4.66–4.85
	Staurolite $Fe^{2+}{}_2Al_9O_6(SiO_4)_4(O,OH)_2$	Brown, brown–black, brown–yellow, brown–red	Grey	7–7.5	Sub-conchoidal	{010} distinct	Usually twinned crystals	Dull, vitreous	Monoclinic	3.64–3.82
Other orthosilicates	Olivine $(Mg,Fe)_2SiO_4$	Green, olive green, black–green, brown–black	White	6.5	Brittle, conchoidal	{001} good, {010} distinct	Massive, granular	Vitreous	Orthorhombic	3.3–3.8
	Topaz $Al_2SiO_4(F,OH)_2$	Pale blue, yellow, yellow–red, red, colourless	White	8	Brittle, uneven	{001} perfect	Massive, crystalline, prismatic	Vitreous	Orthorhombic	3.55

19.4 CARBONATES, PHOSPHATES, BORATES AND SULPHATES

TABLE 19.16 CARBONATES

Mineral	Colour in Hand Specimen	Streak	Hardness (Mohs)	Fracture	Cleavage with Miller Indices of Planes	Habit	Lustre	Crystal System	Specific Gravity (g/cm^3)
Aragonite $CaCO_3$	White, grey, yellow–white, colourless	White	3.5–4	Sub-conchoidal	{010} distinct	Fibrous, columnar	Vitreous	Orthorhombic	2.93
Calcite $CaCO_3$	White, pink, brown, yellow, colourless	White	3	Brittle, conchoidal	{1011} perfect	Massive, crystalline	Vitreous	Trigonal	2.71
Dolomite $CaMg(CO_3)_2$	Grey, red–white, brown–white, white	White	3.5–4	Brittle, conchoidal	{1011} perfect	Blocky, crystalline, massive	Vitreous	Trigonal	2.84
Magnesite $MgCO_3$	Grey–white, yellow–white, brown–white, white, colourless	White	3.5–4.5	Brittle, conchoidal	{1011} perfect	Massive, fibrous	Vitreous	Trigonal	2.98–3.2

(Continued)

TABLE 19.16 *(Continued)* CARBONATES

Mineral	Colour in Hand Specimen	Streak	Hardness (Mohs)	Fracture	Cleavage with Miller Indices of Planes	Habit	Lustre	Crystal System	Specific Gravity (g/cm^3)
Siderite $FeCO_3$	Grey, yellow–grey, green–grey, brown, yellow–brown	White	3.5	Brittle, conchoidal	{0111} perfect	Botryoidal, massive, tabular	Vitreous	Trigonal–hexagonal	3.87–3.94
Malachite $Cu_2CO_3(OH)_2$	Black–green, dark green, green	Light green	3.5	Uneven	{201} perfect	Botryoidal, massive	Vitreous	Monoclinic	3.6–4.03

TABLE 19.17 PHOSPHATES AND BORATES

Mineral/Metal		Colour in Hand Specimen	Streak	Hardness (Mohs)	Fracture	Cleavage	Habit	Lustre	Crystal System	Specific Gravity (g/cm^3)
Phosphates	Apatite $Ca_5(PO_4)_3(F,Cl,OH)$	Brown, blue, violet, green, colourless	White	5	Conchoidal	{0001} indistinct	Massive, euhedral	Vitreous, resinous	Hexagonal	3.2
	Monazite $(Ce,La)PO_4$	Red–brown, brown, yellow, green, pink, grey	White	5.0–5.5	Brittle	{100} distinct, {010} poor	Platy, prismatic	Vitreous, greasy, resinous	Monoclinic	5.04
Borates	Borax $Na_2B_4O_7 \cdot 10H_2O$	Blue, grey, green, grey–white, colourless	White	2–2.5	Conchoidal, brittle	{100} perfect, {110} perfect	Massive, prismatic, tabular	Greasy	Monoclinic	1.71
	Ulexite $(NaCaB_5O_6(OH)_6 \cdot 5(H_2O)$	White, colourless	White	2.5	Uneven, brittle	{010} perfect, {110} good	Acicular, fibrous	Vitreous, silky/satiny	Triclinic	1.96

TABLE 19.18 SULPHATES

Mineral		Colour in Hand Specimen	Streak	Hardness (Mohs)	Fracture	Cleavage with Miller Indices of Planes	Habit	Lustre	Crystal System	Specific Gravity (g/cm^3)
	Gypsum $CaSO_4 \cdot 2H_2O$	White, yellow–white, green–white, brown, colourless	White	2	Fibrous	{010} perfect, {100} distinct, {011} distinct	Crystalline, massive, tabular	Pearly	Monoclinic	2.31
	Anhydrite $CaSO_4$	White, blue–white, violet, grey, colourless	White	3.5	Brittle	{010} perfect, {100} perfect, {001} good	Fibrous, massive, granular	Vitreous, pearly	Orthorhombic	2.97
	Celestine $SrSO_4$	Grey, green, brown, blue, colourless	White	3–3.5	Conchoidal, brittle	{001} perfect, {210} good	Crystalline, massive	Vitreous	Orthorhombic	3.95
	Barite $BaSO_4$	Dark brown, brown–white, grey–white, yellow–white, white	White	3–3.5	Uneven	{010} perfect, {210} perfect, {010} imperfect	Massive, tabular	Vitreous	Orthorhombic	4.48

19.5 OXIDES, HYDROXIDES, HALIDES AND SOME NATIVE ELEMENTS

TABLE 19.19 OXIDES, HYDROXIDES AND HALIDES

Group/Mineral			Colour in Hand Specimen	Streak	Hardness (Mohs)	Fracture	Cleavage with Miller Indices of Planes	Habit	Lustre	Crystal System	Specific Gravity (g/cm^3)
Oxides	Spinel Group	Spinel $MgAl_2O_4$	Red, blue, brown, green, black, colourless	Grey–white	8	Uneven	Indistinct	Euhedral, massive	Vitreous	Isometric	3.58–3.65
		Magnetite Fe_3O_4	Black, grey–black, brown–black	Black	5.5–6.5	Uneven	Indistinct	Crystalline, massive\\	Metallic	Isometric	5.15–5.18
		Chromite $FeCr_2O_4$	Brown, brown–black	Brown	5.5	Uneven	None	Granular, massive	Sub-metallic	Isometric	4.8–5.05
	Other oxides	Cuprite Cu_2O	Brown–red, purple–red, black, red	Brown–red	3.5–4	Conchoidal, brittle	{111} imperfect	Massive	Adamantine	Isometric	6.10–6.14
		Haematite Fe_2O_3	Red–grey, black–red, black	Brown–red	6–6.5	Conchoidal	none	Blocky, tabular	Metallic	Trigonal	5.3

(Continued)

TABLE 19.19 *(Continued)* OXIDES, HYDROXIDES AND HALIDES

Group/Mineral			Colour in Hand Specimen	Streak	Hardness (Mohs)	Fracture	Cleavage with Miller Indices of Planes	Habit	Lustre	Crystal System	Specific Gravity (g/cm^3)
Oxides	Other oxides	Corundum Al_2O_3	Blue, red, yellow, brown, grey	None	9		None	Euhedral, tabular	Vitreous	Trigonal	3.98–4.01
		Ilmenite $FeTiO_3$	Black	Brown–black	5–5.5	Conchoidal	None	Massive, tabular	Sub-metallic	Trigonal	4.79
		Rutile TiO_2	Red, blue, brown–yellow, brown–red, violet	Grey–black	6–6.5	Uneven	{110} good	Acicular, massive, prismatic	Adamantine	Tetragonal	4.23–4.27
Hydroxides		Goethite FeO(OH)	Yellow, red, red–brown, dark brown, black	Brown, brown–yellow	5–5.5	Uneven	{010} perfect, {100} good	Acicular	Adamantine to dull	Orthorhombic	3.3–4.29
		Gibbsite $Al(OH)_3$	Blue, green, green–white, grey, grey–white	White	2.5–3		{001} perfect	Spherical, acicular	Vitreous	Monoclinic	2.35

(Continued)

TABLE 19.19 *(Continued)* OXIDES, HYDROXIDES AND HALIDES

Group/Mineral			Colour in Hand Specimen	Streak	Hardness (Mohs)	Fracture	Cleavage with Miller Indices of Planes	Habit	Lustre	Crystal System	Specific Gravity (g/cm³)
Halides		Halite $NaCl$	Colourless, white, blue, purple, pink, orange, grey, yellow	White	2–2.5	Conchoidal	{001} perfect, {100} perfect, {010} perfect	Crystalline, euhedral	Vitreous	Cubic (isometric)	2.16–2.17
		Fluorite CaF_2	White, green, blue, red, yellow	White	4	Sub-conchoidal, uneven	{111} perfect,	Crystalline, massive	Vitreous	Cubic (isometric)	3.18 (up to 3.56 if REE high in sample)

TABLE 19.20 SOME NATIVE ELEMENTS

Mineral/ Metal	Colour in Hand Specimen	Streak	Hardness (Mohs)	Fracture	Cleavage with Miller Indices of Planes	Habit	Lustre	Crystal System	Specific Gravity (g/cm^3)
Graphite (C)	Black, dark grey, dark blue	Black	1.5–2	Even to uneven	{0001} perfect	Tabular, foliated	Metallic, sub-metallic	Hexagonal	2.09–2.281
Diamond (C)	Colourless, pink, blue, yellow, grey	None	10	Conchoidal	{111} perfect	Euhedral, granular	Adamantine	Cubic (isometric)	3.52
Gold (Au)	Golden, yellow, light yellow to white/silver	Shining yellow	2.5–3	Hackly	None	Massive, grains, dendritic, arborescent	Metallic	Isometric	19.3
Copper (Cu)	Copper, red	Copper red	2.5–3	Hackly	None	Nodular, arborescent, dendritic	Metallic	Isometric	8.94

Chapter 20

Rock Types and Occurrences

Rocks are classified in various ways, using indicators such as mineral content and chemistry, texture and grain size, to identify rock types and to infer their origins. The general idea is to fit the conclusions into a broad and simple template of three categories, igneous, sedimentary and metamorphic, with subdivisions, as shown in Figure 20.1, but this is not always obvious and easy to apply. The simple notion that igneous rocks are formed by solidification of melt, sedimentary rocks are aggregations of detritus from pre-existing rocks and metamorphics are those that have been seriously modified by heat, pressure or percolating water remains valid as a starting point. But, in many cases, more than one of these processes have contributed to rock formation, requiring judgement, or application of some rule or principle, to decide which is to be regarded as primary. The 'rock cycle' of transformations between rock types (Figure 20.2) is still simplistic but gives more scope for recognition of the complications and intermediate states that occur.

20.1 IGNEOUS ROCKS

20.1.1 Classification by Mineral and Chemical Composition

In the early 1900s, petrologist Norman L. Bowen explained why certain minerals are often found in association with one another, but others are rarely found together. He developed a series that shows both continuous and discontinuous reactions with those at the top of Figure 20.7 being the minerals that are first to crystallise from a hot magma and those at the bottom the minerals that remain in the melt longest and crystallise at the lowest temperatures. The sequence can be used to explain susceptibility to weathering. As minerals are most stable in the conditions under which they formed, the greater the departure from those conditions, the more susceptible they are to weathering. The minerals at the surface of the Earth that are least susceptible to weathering

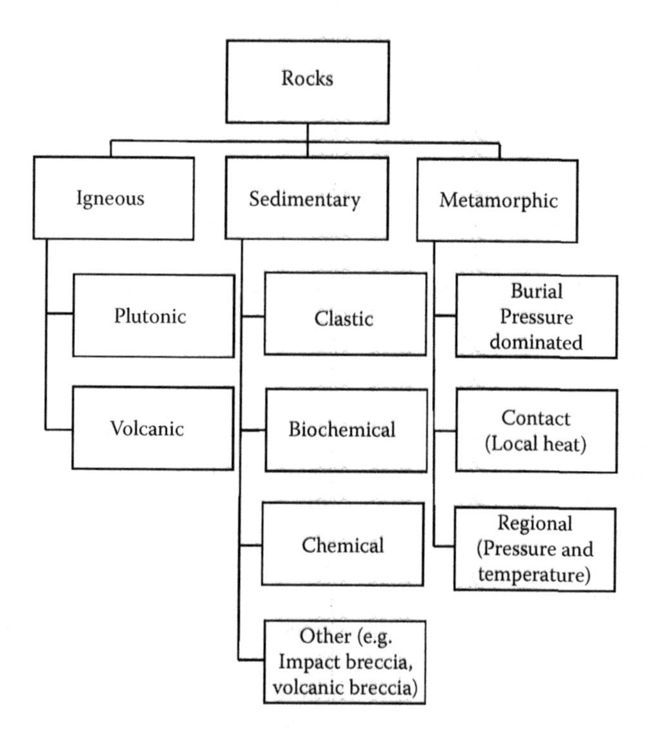

Figure 20.1 The broad scheme of rock classification.

are those formed at low temperature. Silicate stability and the relationship to polymerisation are mentioned in Section 19.3. There are several ways of classifying igneous rocks that are conveniently represented diagrammatically (Figures 20.3 through 20.9). For igneous rocks, Tables 20.1 through 20.5 give prescriptions of named types.

20.1.2 Basalt and Gabbro Classifications

These rocks are broadly similar to one another, distinguished by their emplacement conditions, as in Table 20.1. Being major components of the igneous crust, they have been subjected to detailed study, leading to subclassifications and identification of relationships that derive from the chemical processes of formation and point to evidence of the composition and evolution of the mantle. The variety of basalt mineral structures is summarised in Table 20.5, and the relationships between them and the constituent minerals are illustrated by the basalt classification tetrahedron (Figure 20.8), which applies to basalts

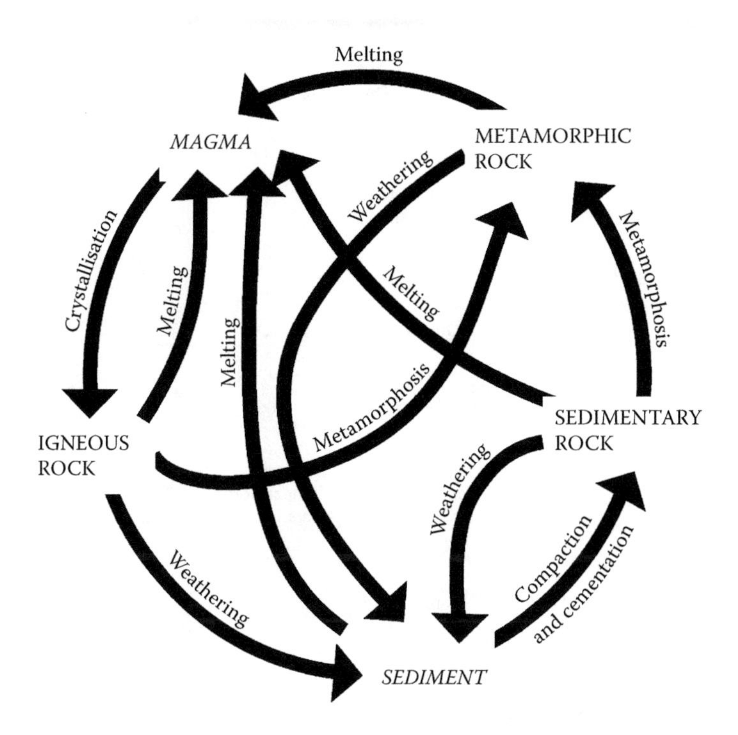

Figure 20.2 The rock cycle.

solidifying at the low pressures of extrusive rocks. The apexes of the tetrahedron and the mid-points of two sides mark the six minerals that contribute to the various basalt types. They also mark the corners of two internal triangular planes that are demarcation boundaries between compositions that are controlled by the overall silica abundance. They divide the overall tetrahedron into three component tetrahedra, identifying the mineral contents of the major basalt types, with silica concentrations increasing from left to right in the diagram. To the right of the plane of silica saturation, the rocks contain more quartz than can be accommodated in silicate minerals, leaving free silica (quartz) as one of the constituent minerals. To the left of the plane of silica undersaturation, there is insufficient silica to combine with other oxides and produce silicates with as much silica as they could have, although minerals with none are rare (nephelene, at the left tip of the figure, is basically $NaAlSiO_4$). Between these two planes is a range of silica saturation, meaning that all of the minerals are silicates, with no free quartz. The significance of this classification is that it arises from the fact that the different mineral compositions have

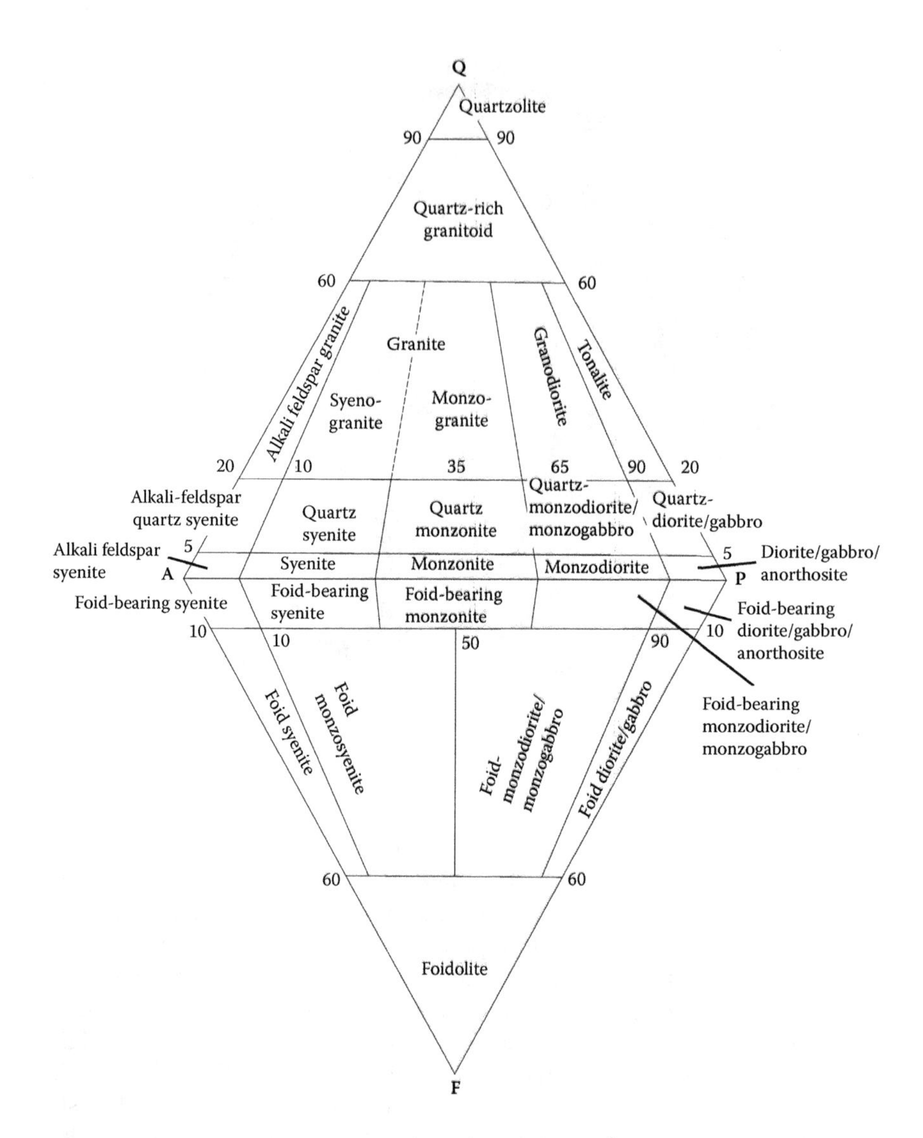

Figure 20.3 The Streckeisen (QAPF) diagram, a classification of intrusive igneous rocks by mineral composition showing the proportions of four mineral groups: quartz (Q), alkali-feldspar (A), plagioclase (P) and feldspathoid or foid (F).

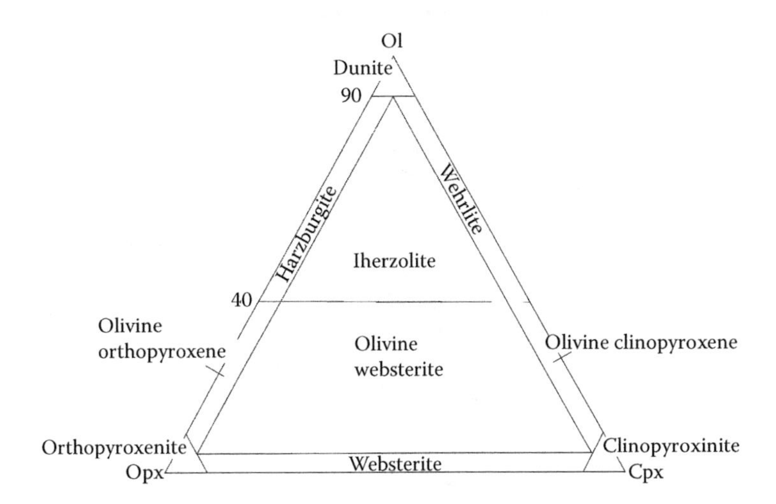

Figure 20.4 Classification of ultramafic igneous rocks by mineral composition, showing the proportions of three mineral groups: olivine (Ol), orthopyroxene (Opx) and clinopyroxene (Cpx).

different melting points, which increase from left to right in the diagram. When a silica-saturated liquid (of olivine tholeiite composition) starts to solidify, it does not do so uniformly. The minerals with high melting points separate out first and, being more silica rich than the fluid average, they drive the remaining fluid to the left in the diagram until it reaches the plane of silica undersaturation and starts to produce silica undersaturated rock. If this is remelted, it cannot produce rock with a higher melting point. For this reason, the plane of silica undersaturation is also referred to as a thermal divide, distinguishing the rock compositions according to the melting points of their minerals.

Melting points are pressure dependent and the thermal divide does not occur at the pressure range of intrusive rocks at which basalt melts produce gabbro. Figure 20.9 presents two triangular classifications for gabbroic rocks of different compositions.

20.1.3 Some Other Rock Types Classified as Igneous

Many of the economically interesting rocks and mineral deposits have complicated histories that invite doubt about their classifications. If they were produced or modified by circulation of water heated by igneous activity, as is commonly the case, they are deemed to be igneous. Especially at mid-ocean ridges, but also on land, as at Yellowstone in USA, water can flow through

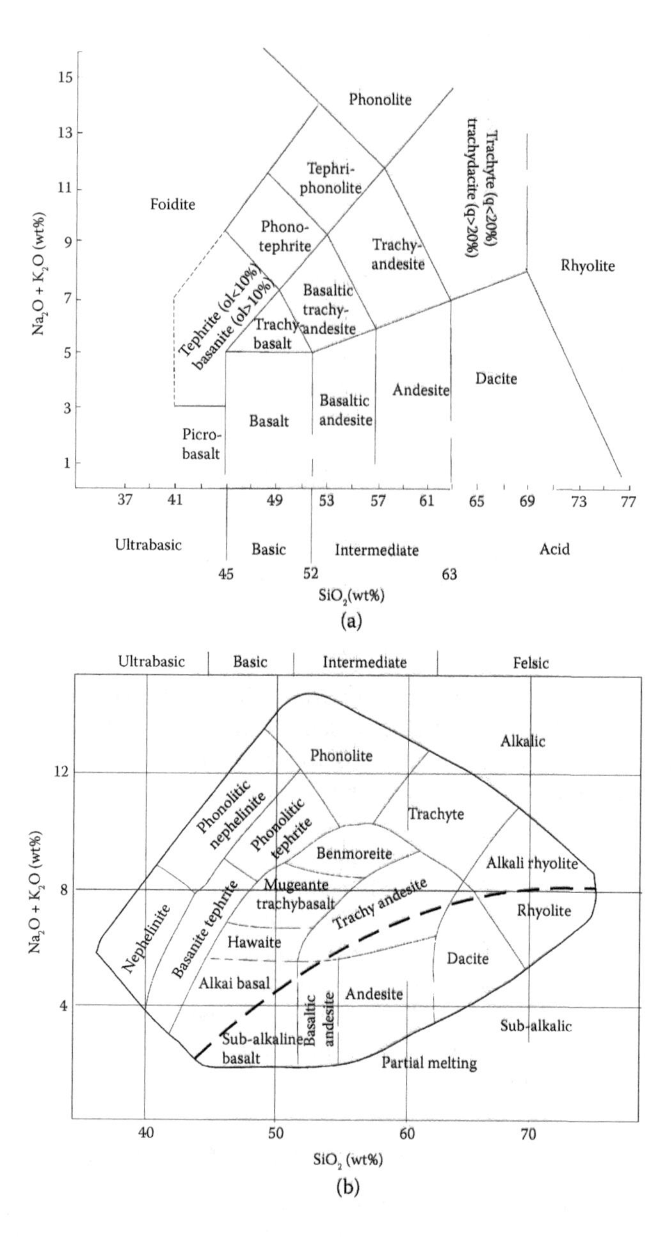

Figure 20.5 Classification of igneous rocks by chemical composition: Total Alkali Silica (TAS). Two versions (a, b) are commonly in use. The curved broken line in B separates the subalkalic and alkali rocks.

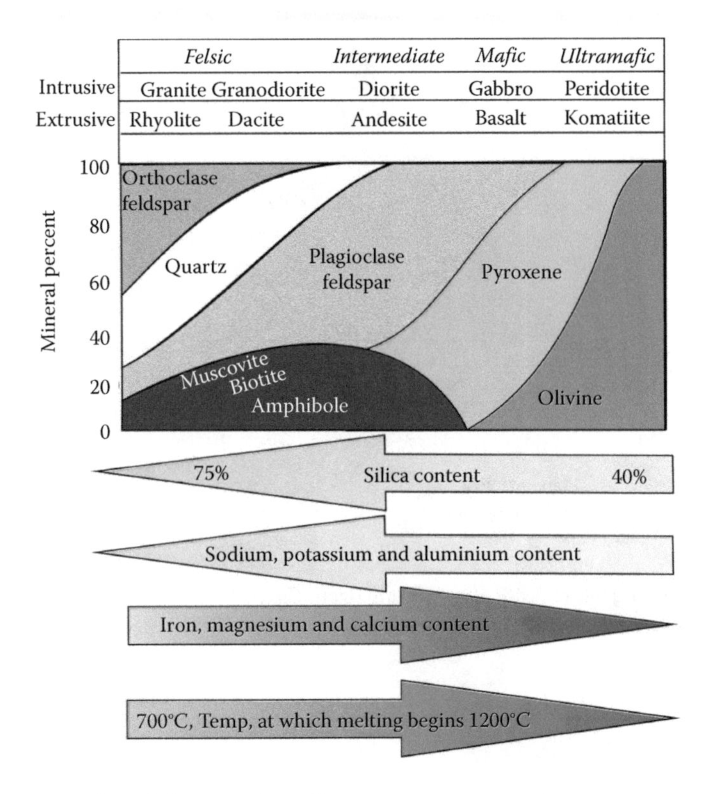

Figure 20.6 A broad classification of igneous rock compositions.

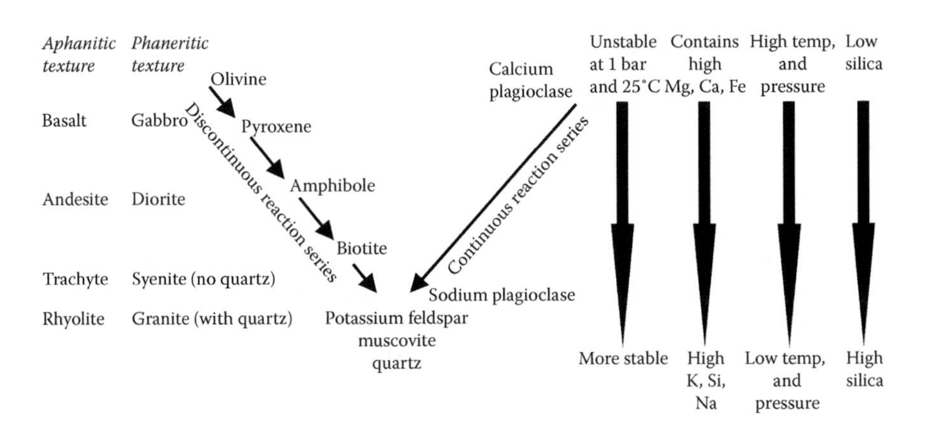

Figure 20.7 Bowen's reaction series.

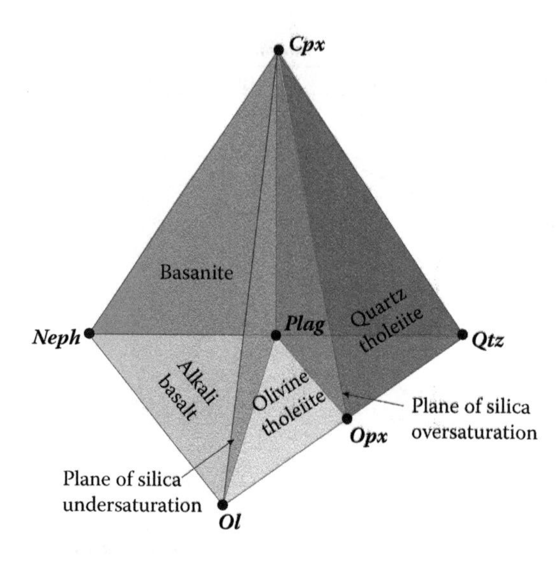

Figure 20.8 Basalt classification tetrahedron. *Cpx*, clinopyroxene; *Qtz*, quartz; *Opx*, orthopyroxene; *Ol*, olivine; *Neph*, nepheline; *Plag*, plagioclase.

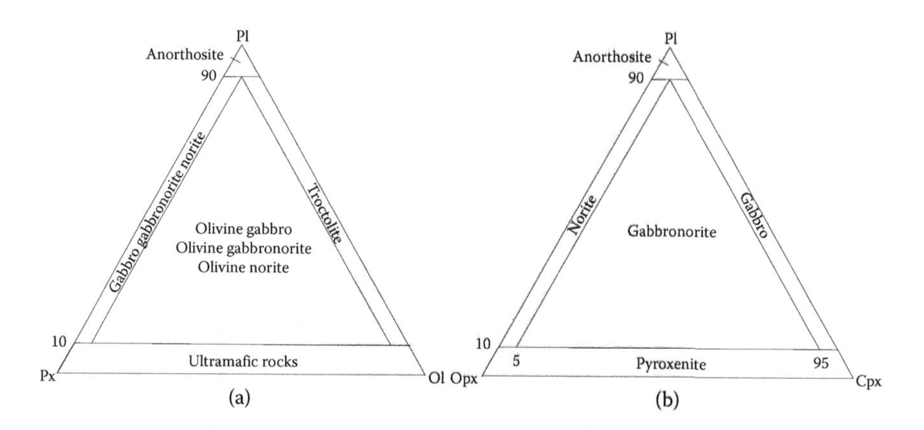

Figure 20.9 Classification diagrams for gabbro which is either olivine-rich (a), or clinopyroxene-rich (b). Gabbroic rocks bearing more than 10% olivine are typically prefixed 'olivine-'. Gabbroic rocks with little or no pyroxene are troctolite. *Pl*, plagioclase; *Px*, pyroxene; *Ol*, olivine; *Opx*, orthopyroxene; *Cpx*, clinopyroxene.

TABLE 20.1 FEATURES DISTINGUISHING PLUTONIC VERSUS VOLCANIC IGNEOUS ROCKS AND EXAMPLES OF CORRESPONDING ROCKS WITH SIMILAR CHEMICAL COMPOSITIONS

	Plutonic	Volcanic
Formation location	Intrusive	Extrusive (including sea floor)
Cooling rate	Generally slow	Rapid except for thick lava flows
Grain size	Typically medium to coarse	Generally fine, can be glassy
Formed from	Magma	Lava
Gas emission	Very little	Generally strong, may be vesicular
Examples of rock formation	Stock, laccolith, sill, dyke, batholith	Flow (sheet, ah-ah, pahoehoe, pillow), explosive (ignimbrite, tuff)
Felsic rock (high Al, low Mg, Fe)	Granite, granodiorite	Rhyolite, dacite
Intermediate rock	Diorite	Andesite
Mafic rock (high Mg, Fe, low Al)	Gabbro	Basalt
Ultramafic	Dunite (peridotite)	Komatiite

TABLE 20.2 GRAIN SIZE CLASSIFICATION FOR IGNEOUS ROCKS

Term	Grain Size
Fine grained	<1 mm
Medium grained	1–5 mm
Coarse grained	5 mm–3 cm
Very coarse grained	>3 cm

Note: For non-terrestrial rocks (meteorites), classes generally refer to smaller grain sizes.

cracks and fissures to depths exceeding a kilometre and be strongly heated by hot rock, or even magma, returning to the surface as superheated water and dissolving mineral components on the way. Meeting cool surface conditions, most obviously on the ocean floor, minerals are precipitated out, commonly as sulphides, yielding what are termed volcanic massive sulphide (VMS) deposits. These are important sources of copper, zinc, lead, gold, silver and several secondary minerals. VMS deposits on land commonly occur with felsic igneous rocks and, where the hydrothermal flow penetrates surrounding sediments,

TABLE 20.3 CLASSIFICATION OF IGNEOUS ROCKS BY % DARK MINERALS (RICH IN Fe AND Mg) SUCH AS PYROXENE, AMPHIBOLE AND BIOTITE

% of Dark Material	Type	Example
<40	Felsic	Granite
40–70	Intermediate	Diorite
70–90	Mafic	Gabbro
>90	Ultramafic	Peridotite

Note: Rocks dominated by dark-coloured minerals are described as melanocratic. Rocks poor in dark-coloured minerals are described as leucocratic.

TABLE 20.4 CLASSIFICATION OF IGNEOUS ROCKS BY A COMBINATION OF SILICA CONTENT AND GRAIN SIZE (AS DEFINED IN TABLE 20.2)

Silica (% SiO_2)	Type	Fine Grained or Glassy	Medium to Fine Grained	Coarse to Medium Grained
>66	Acid	Rhyolite	Microgranite	Granite
52–66	Intermediate	Andesite, trachyte	Microchlorite, microsyenite	Diorite, syenite
45–52	Basic	Basalt	Dolerite	Gabbro
<45	Ultrabasic	Komatiite		Peridotite

they are identified as sedimentary exhalative (SedEx) deposits, notable for zinc and lead.

Less vigorous hydrothermal flow through cracks and fissures in sediments as well as igneous rocks leaves minerals in narrow veins, known as lode deposits, which are sources of gold, silver and sulphides of other metals. Extended cracks are not necessary if the fluid can flow through porous rock, producing disseminated mineralisation, widely distributed through the host rock in low concentrations (down to 0.2%), but economically exploitable if the total volume is large.

Another group of economically interesting rocks with an uncertain origin is the pegmatites, which have the appearance of being conventionally igneous, solidifying from melt, but may be better described as metamorphic. They are formed deep in the crust, where the cooling rate is slow and the crystal size can become very large, normally several centimetres and up to 10 metres in extreme cases. There is a wide range of compositions, commonly granitic. It is not clear

TABLE 20.5 BASALT DESCRIPTIONS

Type	Typical Composition	Examples
Tholeiite	Silica rich, sodium poor, calcic, iron rich	Mountainous areas, Deccan Traps, Parana Basin S. America, Mauna Loa and Kilauea, Hawaii
Alkali basalts	Low silica, pyroxene, plagioclase, olivine, relatively sodium rich	Ocean islands (e.g. Hawaii, Madeira), continental rift locations (e.g. East Africa)
MORB (mid-ocean ridge basalts)	Low in elements with large ions, such as uranium, potassium, REE	Ocean ridges
Boninite (sometimes regarded as an andesite)	High silica and magnesium, low titanium and trace elements (back arc basins)	Fore-arc basins; upper pillow lavas at Troodos, Cyprus; Baja California, Mexico; Mariana Trench
Trachybasalt	Calcic plagioclase, alkali-feldspar, augite, olivine and minor leucite	Long Valley Caldera, California
Hawaiite	Alkali basalt and mugearite (an oligoclase-bearing basalt)	Hawaii
Tachylites	Very glassy, few crystals (basaltic obsidian)	Fragments found in Iceland, northern Britain, Stromboli, Etna

that they were ever liquid. They are primary sources of caesium, lithium and beryllium with secondary tungsten, tin, molybdenum and bismuth.

20.2 SEDIMENTARY ROCKS

Sediments, mostly deposited in water, become sedimentary rocks by processes collectively called diagenesis, which may involve some thermal, chemical or biological changes with cementation accompanying compaction, but leaving the original deposition process obvious. Four basic types are recognised (Figure 20.10). Further classifications are based on chemical composition (Table 20.6), grain size (Table 20.7), depositional texture (Table 20.8) and, for sandstones, mineral content (Figure 20.11).

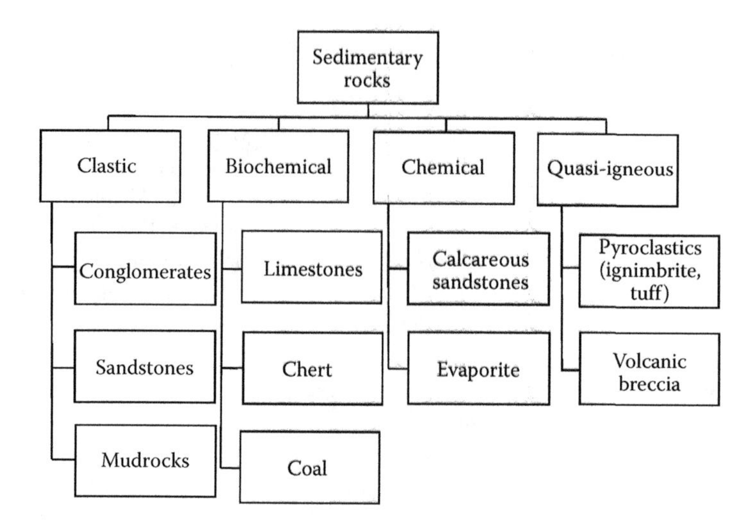

Figure 20.10 Broad categories of sedimentary rocks, expanding the sedimentary class in Figure 20.1.

TABLE 20.6 CLASSIFICATION OF SEDIMENTARY ROCKS BY COMPOSITION

Compositional Group	Defining Constituents	Process	Subdivisions and Examples
Siliciclastic	Dominantly silicate minerals, particles	Bed load, suspended load or gravity flow	Conglomerate, breccia, sandstone or mudstone
Carbonate	Calcite, aragonite, dolomite and other related minerals	Chemical and biological precipitation	Limestone, dolostone
Evaporite	Carbonates, chlorides, sulphates	Evaporation of water	Dolomite, halite, gypsum, anhydrite
Phosphatic	Contains >6.5% phosphorus	Accumulation of animal material	Bone beds, phosphate nodules
Siliceous	Dominantly micro-crystalline silica	Biological, chemical precipitation	Chert, chalcedony, opal, diatomite (fossilised diatoms)
Organic rich	High carbon (>3%)	Accumulation and burial of vegetation	Coal, oil shale
Iron rich	High iron (>15%)	Chemical precipitation	Banded iron formations (BIFs) and iron stones

TABLE 20.7 SEDIMENTARY GRAIN SIZE CLASSIFICATIONS COMMONLY IN USE: THE WENTWORTH GRADE SCALE, EQUIVALENT KRUMBEIN PHI SCALE AND THE ISO STANDARD FOR SOILS

The Wentworth Grade Scale (Wentworth 1922)			Krumbein phi Scale (Krumbein 1937)	ISO 14688-1:2002 (SOIL)
Name		**Particle Size (mm)**	**Φ**	**Particle Size (mm)**
Gravel	Boulder	>256	<−8	200–630
	Cobble	64–256	−8 to −6	63–200
	Pebble	4–64	−6 to −3	2.0–63
	Granule	2–4	−1 to −2	
Sand	Very coarse sand	1–2	0 to −1	0.63–2.0
	Coarse sand	0.5–1	1 to 0	
	Medium sand	0.25–0.5	2 to 1	0.2–0.63
	Fine sand	0.125–0.250	3 to 2	0.063–0.2
	Very fine sand	0.0625–0.125	4 to 3	
Silt	Coarse silt	0.031–0.0625	8 to 4	0.002–0.063
	Medium silt	0.0156–0.031		
	Fine silt	0.0078–0.0156		
	Very fine silt	0.0039–0.0078		
Mud	Clay	0.004–0.00006	>8	≤0.002

TABLE 20.8 DUNHAM CLASSIFICATION OF LIMESTONES ACCORDING TO DEPOSITIONAL TEXTURE

<table>
<tr><th>Mudstone</th><th>Wackestone</th><th>Packstone</th><th>Grainstone</th><th>Boundstone</th></tr>
<tr><td colspan="4">Original components not organically bound together during deposition</td><td>Organically bound during deposition</td></tr>
<tr><td colspan="3">Contains carbonate mud</td><td>No carbonate mud</td><td>With or without mud</td></tr>
<tr><td colspan="2">Mud supported</td><td colspan="2" rowspan="2">Grain supported</td><td rowspan="2">Grain or mud supported</td></tr>
<tr><td><10% grains</td><td>>10% grains</td></tr>
</table>

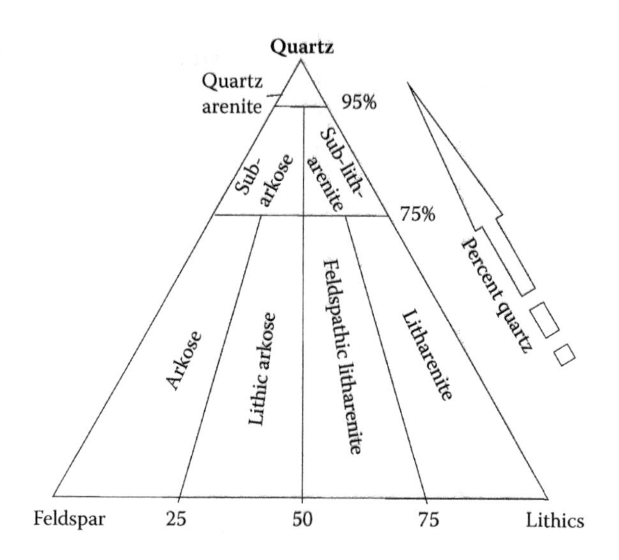

Figure 20.11 Mineral classification of sandstones. This is the Folk classification, which defines sandstones on the basis of contents of quartz, feldspar and lithics (rock fragments).

20.3 METAMORPHIC ROCKS

Rocks may be altered in various ways so drastically that they are identified as different rocks, with a general description metamorphic. The principal agents of change are heat and pressure, often in combination, and in some cases percolating water, but although the minerals are changed, the overall chemistry is almost unaffected. The changes take time, in many cases a long time, and often do not go to completion, so that some metamorphic rocks are given the identity of the original rock (protolith) prefixed by meta-, as in metagranite. If time is not a limitation, the end-points of metamorphic processes are rocks that are given names with no obvious relationships to their protoliths, even though these may be identifiable from the geological context. Figure 20.12 outlines the general scheme of end-point metamorphics resulting from heat and pressure.

The scale and cause of metamorphic events have strong influences on the resulting products. Local thermal events, such as intrusions of dykes or sills produce contact aureoles of fine-grained material ('baked margins') in the intruded country rock ('contact metamorphism'). As inferred from Figure 20.12 for such high temperature, low pressure events, often of limited duration, the mineralogy is generally dominated by hornfels compositions, graded according to depth of heat penetration ('burial metamorphism'). Thicker, well-developed

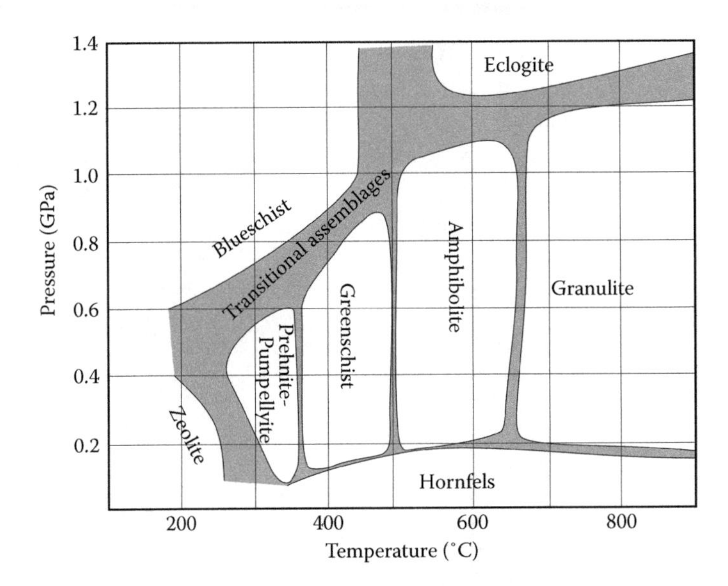

Figure 20.12 Some stable rock/mineral assemblages (facies) resulting from metamorphism at various temperatures and pressures. Based on Yardley 1989. *Note*: 1 GPa corresponds to about 30 km of burial.

aureoles occur around large granite intrusions ('contact metamorphism'). Much larger, regional scale metamorphism ('regional metamorphism') accompanies tectonic activity during orogenic (mountain building) events or large-scale intrusion. This occurs also at cool, convergent zones, yielding blueschists with glaucophane mineralisation. The formation of eclogite is a higher pressure effect, primarily in the upper mantle.

Stressed rocks in fault zones can undergo low-temperature plastic deformation and fragmentation ('cataclastic' or 'dynamic metamorphism'). Cataclastic metamorphism may also occur if rocks are impacted by extra-terrestrial objects, altering minerals by extremely high pressures at low temperatures. Invading fluids may cause metasomatic reactions, chemical alterations by fluid. This is hydrothermal metamorphism at low temperatures and low pressures.

Where mineralisation is a direct result of tectonic activity with deviatoric (non-hydrostatic) stresses, the resulting rocks develop textures as summarised in Table 20.9. Planar layering (foliation) is common, giving rocks strongly anisotropic properties, and also, but less often, compositional layering (banding). In fault zones, rocks may have accumulated shear strain (ductile deformation), resulting in recrystallisation of minerals, grain size alteration and modification of their appearances. Such rocks, classified as mylonites, can have many

TABLE 20.9 TEXTURES OF ROCKS RESULTING FROM DEVIATORIC STRESS DURING METAMORPHISM

	Foliated							Non-Foliated		
	Not banded				May be banded		Banded	Phaneritic	Aphanitic or phaneritic	Aphanitic
	Aphanitic			Phaneritic						
Description (rock name)	Slaty (slate)	Phyllitic (phyllite)	May be grainy (greenstone)	Schistose (schist)	Coarse grained, pebbles preserved (metaconglomerate)	Medium to coarse grained some large crystals (porphyroblastic schist)	Medium to coarse grained gneissose (gneiss)	Coarse grained, pebbles preserved (metaconglomerate)	Fine to coarse grained. Granoblastic (quartzite, marble)	May be grainy (greenstone)
Protolith	Shale, siltstone		Basalt	Shale, siltstone, sandstone, basalt	Conglomerate	Sandstone, granite, other igneous rock		Conglomerate	Quartz arenite, limestone, dolostone	Basalt

Note: Aphanitic = grains too small to be seen with the naked eye, Phaneritic = grains can be seen with the naked eye.

different mineral compositions. Previously metamorphosed rock that was later intruded by magma is known as migmatite.

20.4 ELEMENTAL COMPOSITIONS OF ROCKS

Abundances of elements in the core, mantle and crust are listed in Tables 6.4, 7.3 and 8.4, respectively. The crust is particularly heterogeneous and the listed global abundances reflect this only to the extent that the upper continental crust is distinguished from the crustal average. The heterogeneity is more clearly indicated by considering the diversity of elemental compositions in crustal rocks, as listed in Table 20.10.

TABLE 20.10 TYPICAL/AVERAGE ELEMENT ABUNDANCES IN REPRESENTATIVE ROCKS (PPM BY MASS)

Element (ppm)	Granite	Ultramafic	Kimberlite	Nephelinite	Alkaline olivine basalt	Diabase	Tholiite	Shale	Coal
H	400					600		6300	50,000
Li	22	2	25	16	12	15	7	66	40
Be	3	0.4	1			0.8	0.7	3	0.99
B	1.7	7				15	5.8	100	66
C	2000	100	16,200	205		100		1000–12,000	676,000
N	59	14				52		60–1000	15,000
O	455,000	429,000				449,000		495,000	96,000
F	700	97	1900			250		740	140
Na	24,600	2230	2030	25,300	23,000	16,000	17,580	9600	450
Mg	2400	247,500	160,000	71,200	45,200	39,900	36,910	15,000	270
Al	74,300	14,300	18,900	61,600	79,390	79,400	80,400	80,000	22,000
Si	339,600	203,300	147,000	188,400	226,200	246,100	239,200	273,000–280,000	37,000
P	390	220	3880	3800	2090	610	960	700	900
S	260	4000	2000	620		123		2400	1500
Cl	70	110	300	518		200		180	540
K	45,100	390	10,400	12,200	13,280	5300	6970	26,600	3400
Ca	9900	27,200	70,400	90,000	65,040	78,300	72,260	16,000–22,100	38,000
Sc	2.9	15	15	21	20	35	30	13	5.1
Ti	1500	780	11,800	16,800	14,390	6400	9710	4600	1100
V	17	50	120	221	213	264	251	130	24
Cr	20	3090	1100	344	202	114	168	90	22
Mn	195	1040	1160	1500	1472	1280	1356	850	35
Fe	13,700	64,830	71,600	91,080	90,780	77,600	85,540	47,200	7800
Co	2.4	110	77	52	43	47	48	19	3.9
Ni	1	1450	1050	291	145	76	134	68	9.4

(Continued)

TABLE 20.10 *(Continued)* TYPICAL/AVERAGE ELEMENT ABUNDANCES IN REPRESENTATIVE ROCKS (PPM BY MASS)

Element (ppm)	Granite	Ultramafic	Kimberlite	Nephelinite	Alkaline olivine basalt	Diabase	Tholiite	Shale	Coal
Cu	13	47	80	63	85	110	90	45	12
Zn	45	56	80	102	108	86	100	95	11
Ga	20	2.5	10	15	15	16	17	19	5.3
Ge	1.1	1	0.5	1.6		1.4	1.4	1.6	0.98
As	0.5	1				1.9	1.5	13	7.6
Se	0.007	0.02	0.15			0.300		0.6	1.5
Br	0.4	0.24				0.4		4–20	11
Rb	220	1.2	65	39	32	21	22	140	32
Sr	250	22	740	1350	530	190	328	300	340
Y	2.88	22	36	33	25	28	26	13	5.9
Zr	165	16	250	205	189	105	137	160	24
Nb	24	1.3	110	103	69	9.5	13	11	2.1
Mo	6.5	0.2	0.5			0.6	1	2.6	0.19
Ru			0.007						0.25
Rh			0.007						0.25
Pd	0.002	0.01	0.053			0.025		0.004	0.11
Ag	0.05	0.05				0.08	0.011	0.07	0.011
Cd	0.03	0.06	0.07	0.052	0.082	0.15	0.17	0.3	0.058
In	0.02	0.02	0.1	0.034		0.7	0.70	0.1	0.76
Sn	3.5	0.52	15			3.2	1.5	6	0.34
Sb	0.31	0.3					0.3	1.5	0.7
Te		0.001						0.08	
I	<0.03	0.13				<0.03		2.2–19	
Cs	1.5	0.006	2.3			0.9	1.1	5	2.3
Ba	1220	20	1000	1046	528	160	246	580	150
La	101	0.92	150	89	54	10	15	24–32	13
Ce	170	1.93	200	171	105	23	32.9	50–70	24
Pr	19	0.32	22	18	28	3.4	4.7	6.1–7.9	7.6

(Continued)

TABLE 20.10 *(Continued)* TYPICAL/AVERAGE ELEMENT ABUNDANCES IN REPRESENTATIVE ROCKS (PPM BY MASS)

Element (ppm)	Granite	Ultramafic	Kimberlite	Nephelinite	Alkaline olivine basalt	Diabase	Tholiite	Shale	Coal
Nd	55	1.44	85	66	49	15	18.9	24–31	8.5
Sm	8.3	0.4	13	14.5	9.1	3.6	4.9	5.7	1.9
Eu	1.3	0.16	3	4	3.5	1.1	1.5	1.2	0.47
Gd	5	0.74	8	12.1	8.1	4	5.5	5.2	1.7
Tb	0.54	0.12	1	1.7	1.8	0.65	1.2	0.85	0.33
Dy	2.4	0.57		7.3	4.7	4	4.9	4.2	2.5
Ho	0.35	0.16	0.55	1.7	1.9	0.69	1.3	1.1	0.76
Er	1.2	0.4	1.45	3.3	2.4	2.4	2.8	2.7–3.4	1.1
Tm	0.15	0.067	0.23	0.88	0.7	0.3	0.46	0.5	0.52
Yb	1.1	0.38	1.2	2.3	109	2.1	2.6	2.2–3.1	1.1
Lu	0.19	0.065	0.16	0.39	0.5	0.35	0.46	0.51	0.15
Hf	5.2	0.6	7	5		2.7	2.5	2.8	1
Ta	1.5	< 0.1	9	19		0.5	0.5	0.8	0.36
W	0.4	0.0003		11		0.5	0.7	1.8	0.6
Re		0.00023	0.001						1.1
Os		0.0031	0.005					0.00005	2.5
Ir		0.0032	0.007					0.00008	1.7
Pt	0.0019	0.06	0.19			0.0012	0.00002		0.52
Au		0.007	0.004		0.0016		0.004	0.0025	1.1
Hg	0.1	0.03	0.01	0.02	0.017	0.2	0.01	0.04	0.22
Tl	0.15	0.01	0.22	0.009	0.05	0.3	0.1	1.1	0.52
Pb	48	0.2	10	7.8	4.3	7.8	3.7	20	4.8
Bi	0.07	0.006	0.03	0.014	0.03	0.05	0.03	0.40–0.43	1.1
Th	50	0.07	16	11	3	2.4	1.8	12	3.9
U	3.4	0.025	3.1	3.2	0.7	0.6	0.5	3.7	1.0

Note: Data from: Schweinfurth, S.P. and Finkelman, R.B., Coal—A complex natural resource: an overview of factors affecting coal quality and use in the United States. USGS Circular 1143, USGS, Denver; Lodders, K. and Fegley, B., Jr., *The Planetary Scientist's Companion*, Oxford University Press, Oxford, 1998; Mason, B. and Moore, C.B., *Principles of Geochemistry*, Wiley, New York, 1982.

Chapter 21

Physical Properties of Minerals and Rocks

21.1 ELASTIC PROPERTIES

21.1.1 Isotropic Materials

Polycrystalline materials, metals and igneous rocks with randomly oriented crystals, and amorphous materials, such as glass, are elastically isotropic, with elasticity completely represented by two moduli. Although there are several ways of choosing moduli to describe various deformations and six that are commonly used, only two are independent, and all the others can be represented in terms of them. In geophysics it is convenient to select as the two independent moduli, bulk modulus (incompressibility, K or B) and rigidity (shear modulus, μ or G). There are two reasons for this. K represents pure compression, a volume change caused by hydrostatic pressure, with no change in shape. The values for constant temperature (isothermal) compression, K_T, and adiabatic compression (constant entropy, with no heat exchange), K_S, are related by a simple thermodynamic identity (Equation 21.24). The value of μ is the same for both isothermal and adiabatic deformation. The distinction between isothermal and adiabatic behaviour is much more complicated for the other moduli. The second reason is that μ and the combination $\chi = (K_S + 4\mu/3)$ are the moduli that control seismic wave speeds, observations of which provide values of K_S, and hence the variation of density within homogeneous regions of the Earth. The other moduli are:

- Young's modulus, E, is the ratio of the axial stress on a material to the strain in that direction, with the transverse directions unconstrained. This is sometimes referred to as THE elastic modulus in engineering literature.
- Poisson's ratio, ν, is not strictly a modulus but the ratio of lateral contraction to axial extension of a material subject to stress in one direction only.

- The modulus of simple longitudinal strain, χ, is the ratio of stress in one direction to the strain in that direction, with stresses in perpendicular directions self-adjusted to allow no strains in those directions. This is the modulus-controlling seismic P-waves, with wavelengths that are short compared with dimensions of the medium, allowing no lateral strains.
- Lamé parameter, λ, is one of two Lamé parameters (λ, μ), which are the two independent moduli representing the elasticity of an isotropic solid in the tensor notation. λ can also be thought of as the ratio of stress in a direction of zero strain to the areal strain in the perpendicular plane.

The seismic wave speeds are:

$$\text{Compressional (P) waves } V_P = \sqrt{(\chi/\rho)} = \sqrt{[(K_S + (4/3)\mu)/\rho]} \qquad (21.1)$$

$$\text{Shear (S) waves } V_S = \sqrt{(\mu/\rho)} \qquad (21.2)$$

The nomenclatures P and S refer to primary and secondary, that is, first (fastest) and second arrivals.

Table 21.1 gives the relationships between these moduli. For mathematical convenience, it is sometimes assumed that there is only one independent modulus, with $\lambda = \mu$ and therefore $K = 5\mu/3 = 2E/3$, $\chi = 3\mu = 9K/5$, $\nu = 1/4$. This is referred to as a Poisson solid, but it is not a good approximation for rocks and becomes increasingly unsatisfactory with increasing pressure.

21.1.2 Cubic Crystals

Elasticities of individual crystals are anisotropic and must be described by more than two moduli, the required number depending on the crystal symmetry. The moduli are identified by a numbering system based on Cartesian coordinates, with x, y and z axes numbered 1, 2 and 3, respectively. The simplest case is cubic symmetry, for which there are three independent moduli and the axes are identified as cube edges. If stress σ_1 is applied in direction 1 causing a linear strain ε_1 in that direction, with no constraints in the other directions, then the elastic response is described by modulus c_{11}. With cubic symmetry, axes 2 and 3 are equivalent to axis 1 and c_{11} applies also to them, if stressed in those directions, and there is no separate identification of moduli c_{22} or c_{33}. However, axes 2 and 3 are strained by application of σ_1 and those responses are represented by the modulus $c_{12} = \sigma_1/\varepsilon_2 = \sigma_1/\varepsilon_3$. This notation is general. Its advantage becomes apparent when applied to anisotropic materials for which a linear stress may cause a shear deformation. This is accommodated by adding numbers 4, 5, 6, with the convention $4 \equiv yz$, $5 \equiv zx$, $6 \equiv xy$, to represent shear strains and stresses. In the case of cubic symmetry, these are equivalent and introduce the third

TABLE 21.1 RELATIONSHIPS BETWEEN ELASTIC MODULI OF AN ISOTROPIC SOLID (OR POLYCRYSTALLINE AVERAGE)

	K	μ	ν	E	λ	χ
K, μ	K	μ	$\frac{3K-2\mu}{6K+2\mu}$	$\frac{9K\mu}{3K+\mu}$	$K-\frac{2}{3}\mu$	$K+\frac{4}{3}\mu$
K, ν	K	$\frac{3K(1-2\nu)}{2(1+\nu)}$	ν	$3K(1-2\nu)$	$\frac{3K\nu}{1+\nu}$	$\frac{3K(1-\nu)}{(1+\nu)}$
K, E	K	$\frac{3KE}{9K-E}$	$\frac{1}{2}-\frac{E}{6K}$	E	$\frac{3K(3K-E)}{9K-E}$	$\frac{3K(3K+E)}{9K-E}$
K, λ	K	$\frac{3}{2}(K-\lambda)$	$\frac{\lambda}{3K-\lambda}$	$\frac{9K(K-\lambda)}{3K-\lambda}$	λ	$3K-2\lambda$
K, χ	K	$\frac{3}{4}(\chi-K)$	$\frac{3K-\chi}{3K+\chi}$	$\frac{9K(\chi-K)}{3K+\chi}$	$\frac{1}{2}(3K-\chi)$	χ
μ, ν	$\frac{2\mu(1+\nu)}{3(1-2\nu)}$	μ	ν	$2\mu(1+\nu)$	$\frac{2\mu\nu}{1-2\nu}$	$\frac{2\mu(1-\nu)}{(1-2\nu)}$
μ, E	$\frac{\mu E}{3(3\mu-E)}$	μ	$\frac{E}{2\mu}-1$	E	$\frac{\mu(E-2\mu)}{3\mu-E}$	$\frac{\mu(4\mu-E)}{3\mu-E}$
μ, λ	$\lambda+\frac{2}{3}\mu$	μ	$\frac{\lambda}{2(\lambda+\mu)}$	$\frac{\mu(3\lambda+2\mu)}{\lambda+\mu}$	λ	$\lambda+2\mu$
μ, χ	$\chi-\frac{4}{3}\mu$	μ	$\frac{\chi-2\mu}{2(\chi-\mu)}$	$\frac{\mu(3\chi-4\mu)}{\chi-\mu}$	$\chi-2\mu$	χ
ν, E	$\frac{E}{3(1-2\nu)}$	$\frac{E}{2(1+\nu)}$	ν	E	$\frac{E\nu}{(1+\nu)(1-2\nu)}$	$\frac{E(1-\nu)}{(1+\nu)(1-2\nu)}$
ν, λ	$\frac{\lambda(1+\nu)}{3\nu}$	$\frac{\lambda(1-2\nu)}{2\nu}$	ν	$\frac{\lambda(1+\nu)(1-2\nu)}{\nu}$	λ	$\frac{\lambda(1-\nu)}{\nu}$
ν, χ	$\frac{\chi(1+\nu)}{3(1-\nu)}$	$\frac{\chi(1-2\nu)}{2(1-\nu)}$	ν	$\frac{\chi(1+\nu)(1-2\nu)}{1-\nu}$	$\frac{\chi\nu}{1-\nu}$	χ
E, λ	$\frac{E+3\lambda+p}{6}$	$\frac{E-3\lambda+p}{4}$	$\frac{p-E-\lambda}{4\lambda}$	E	λ	$\frac{E-\lambda+p}{2}$

(Continued)

TABLE 21.1 *(Continued)* RELATIONSHIPS BETWEEN ELASTIC MODULI OF AN ISOTROPIC SOLID (OR POLYCRYSTALLINE AVERAGE)

	K	μ	ν	E	λ	χ
E, χ	$\frac{3\chi - E + q}{6}$	$\frac{E + 3\chi - q}{8}$	$\frac{E - \chi + q}{4\chi}$	E	$\frac{\chi - E + q}{4}$	χ
λ, χ	$\frac{1}{3}(2\lambda + \chi)$	$\frac{1}{2}(\chi - \lambda)$	$\frac{\lambda}{\lambda + \chi}$	$\frac{(2\lambda + \chi)(\chi - \lambda)}{\lambda + \chi}$	λ	χ

Note: Any one modulus may be expressed in terms of any other two. $p = \sqrt{E^2 + 2E\lambda + 9\lambda^2}$; $q = \sqrt{E^2 - 10E\chi + 9\chi^2}$.

TABLE 21.2 ISOTROPIC MODULI (POLYCRYSTALLINE AVERAGES) OF COMMON MATERIALS AND SELECTED MINERALS AT LABORATORY *T* AND *P*, EXCEPT AS NOTED

Material	K_S (GPa)	μ (GPa)
Iron	170	82
Liquid iron (at MP)	130	0
Gold	172	28
Granite	55	30
Basalt	67	37
Sea water	2.05	0
Ice (–3°C)	8.7	3.4
Air	1.42×10^{-4}	0
Diamond	443	540
Graphite	161	109
α quartz	38	44
Fused silica	37	31
Corundum, Al_2O_3	254	163
Periclase, MgO	164	132
Wüstite, FeO	153	47
Magnetite, Fe_3O_4	162	91
Hematite, Fe_2O_3	207	91
Enstatite, $MgSiO_3$	108	76

(Continued)

TABLE 21.2 *(Continued)* ISOTROPIC MODULI (POLYCRYSTALLINE AVERAGES) OF COMMON MATERIALS AND SELECTED MINERALS AT LABORATORY *T* AND *P*, EXCEPT AS NOTED

Material	K_S (GPa)	μ (GPa)
Ferrosilite, $FeSiO_3$	101	52
Forsterite, Mg_2SiO_4	129	82
Fayalite, Fe_2SiO_4	136	51
Calcite, $CaCO_3$	73	32
Aragonite, $CaCO_3$	47	39
Halite, NaCl	25	15

TABLE 21.3 ELASTIC MODULI OF SOME CUBIC MATERIALS WITH AVERAGES FOR RANDOM CRYSTAL ALIGNMENTS, AS OUTLINED IN SECTION 21.1.4

	Crystal Data			Polycrystalline Averages	
Material	c_{11}	c_{12}	c_{44}	K	μ
Iron	230	135	117	167	81.5
Gold	191	162	42.1	172	27.6
Diamond	1076	125	576	442	533
Periclase, MgO	296	95	156	162	131
Wüstite, FeO	217	121	46	153	47
Magnetite, Fe_2O_3	275	104	95.5	161	91.3
Halite, NaCl	49.5	13.2	12.8	25.3	14.5

Note: Values are in GPa.

modulus, c_{44}. The cubic crystal moduli translate to the K, μ notation of the previous section with two different shear moduli for different shear orientations:

$$K = (c_{11} + 2c_{12})/3 \tag{21.3}$$

For shear across a (100) plane, normal to one of the 1, 2, 3 axes

$$\mu_{100} = c_{44} \tag{21.4}$$

For shear across a (110) plane, normal to a diagonal of a cube face

$$\mu_{110} = (c_{11} - c_{12})/2 \tag{21.5}$$

The problem of averaging the moduli for application to a polycrystalline material with random grain alignments is considered in Section 21.1.4 and the results of averaging are listed for common materials in Tables 21.2 and 21.3.

The assertion that specification of the elasticity of cubic crystals requires three independent moduli is compromised in the case of the diamond structure, for which Keating (1966) proved that there is a relationship between them.

$$2c_{44}(c_{11}+c_{12})=(c_{11}-c_{12})(c_{11}+3c_{12}) \tag{21.6}$$

This applies to silicon and germanium, as well as diamond, and presumably to α-tin, although that would be difficult to observe as it is obtained as a powder by a slow low-temperature phase transformation from metallic β-tin. This relationship does not imply that the diamond structure is isotropic. That would require $\mu_{100}=\mu_{110}$ and vanishing c_{12}.

21.1.3 More General Elastic Anisotropy

For a crystal structure completely lacking symmetry, each of the 6 components of stress (3 linear and 3 shear) would cause 6 components of strain, but the number of independent moduli would be restricted to 21 by the fact that, for $i \neq j$ ($i, j = 1$ to 6), $c_{ij}=c_{ji}$. The limiting case of 21 moduli applies to triclinic crystals, but for most minerals various structural symmetries reduce the number. Bass (1995) lists room temperature values for cubic (3 moduli), hexagonal (5), trigonal (6 or 7), tetragonal (6 or 7), orthorhombic (9) and monoclinic (13) minerals. Seismic anisotropy arises from crystal alignments (texture or preferred alignment) in the Earth, and in this connection the anisotropy of olivine, $(Mg, Fe)_2SiO_4$ (Table 21.4), which is orthorhombic with 9 moduli, is of particular interest.

The moduli listed in Table 21.4 are isothermal values. Relationships to adiabatic moduli are complicated. The differences arise only from the volume component of strain. Seismic anisotropy is caused by crystal alignments in the upper mantle, where olivine is a major mineral, and is presumed to be responsible for the higher horizontal than vertical wave speeds (both P and S). In the inner core, alignment of Fe–Ni alloy crystals (of composition in Table 6.4) with hexagonal symmetry gives higher wave speeds in the axial direction than transversely. The upper mantle anisotropy (Table 21.5), as modelled by PREM (Dziewonski and Anderson 1981), is a global average. Local horizontal anisotropies are averaged out, leaving uniaxial anisotropy, symmetric about the vertical and requiring 5 moduli, as for a hexagonal structure. Although the pattern of these moduli appears to require 6, only 5 are independent as $c_{66}=(c_{11}-c_{12})/2$.

In an anisotropic medium, S-wave speed depends on polarisation (direction of particle motion) because the shear moduli are different for different stress

TABLE 21.4 ROOM TEMPERATURE ELASTIC CONSTANTS, C_{ij}, OF OLIVINE WITH THE COMPOSITION $Mg_{0.9}Fe_{0.1}SiO_4$ (9 COEFFICIENTS), REPORTED BY ANDERSON AND ISAAK (1995)

i/*j*	1	2	3	4	5	6
1	320.6	69.8	71.2			
2	69.8	197.1	74.8			
3	71.2	74.8	234.2			
4				63.7		
5					77.6	
6						78.3

Note: Values are in GPa and axes of the table are values of i and j (1 to 6). The values are symmetrical about the diagonal because $c_{ij} = c_{ji}$.

TABLE 21.5 ELASTIC MODULI OF THE UPPERMOST MANTLE, AVERAGED OVER THE TOP 200 km OF THE PREM MODEL

i/*j*	1	2	3	4	5	6
1	224	84.8	85.4			
2	84.8	224	85.4			
3	85.4	85.4	224			
4				65.7		
5					65.7	
6						69.6

Note: Values are in GPa. A material with elastic properties horizontally averaged in this way is referred to as transversely isotropic.

orientations. This results in S-wave splitting, with a wave of initially arbitrary polarisation axis propagating as two independent waves having polarisations selected by the transmitting medium and travelling at different speeds. A common distinction is between SH and SV waves, polarised horizontally and vertically, respectively.

Anisotropy of a layered rock, with layers with different elasticities, is observed even when the individual layers are elastically isotropic. If the layers are much thinner than the wavelength of a seismic or an elastic wave propagating parallel to them, then the wave 'sees' an average modulus for the medium as a whole, but a wave travelling perpendicular to them samples the layers individually in sequence, resulting in a different average speed. Layering and crystalline alignment are commonly superimposed and are not easily distinguished.

21.1.4 Elasticity of a Composite

Most rocks are composites of minerals with different elastic properties. They often occur as grains that are randomly distributed and randomly aligned (causing no anisotropy), but averaging their individual moduli to represent whole rock properties has been a very long-standing problem to which there is no exact or rigorous solution. There are similar difficulties with thermal and electrical properties. The elasticity problem is outlined here for the case of hydrostatic compression of isotropic materials. As a preliminary comment, it is noted that properties generally depend to some extent on mechanical and thermal history. The reasons for this are illustrated by considering igneous rocks. If they are extrusions and have solidified and cooled at low pressure, even if there are no surviving gas bubbles and the grains fit together precisely when first formed, gaps and grain boundary cracks develop as they cool because of different thermal contractions. This occurs even with polycrystalline rocks of single minerals unless they have cubic crystal symmetry with isotropic expansion coefficients, and even then cracks may result from temperature gradients during cooling. Plutonic rocks formed at pressures sufficient to prevent cracking can remain uncracked as long as the pressure is maintained but crack when released from pressure because of differing elastic moduli. Virtually all rocks have some porosity, which may be a fraction of 1% or several per cent of their volumes. Consistent and readily interpreted observations of elasticity require rocks to be held under sufficient pressure to ensure pore closure, at least 1 GPa (10 kilobars). The discussion in this section assumes that condition to be satisfied. Otherwise porosity biases measured elasticities systematically low to an extent far greater than its effect on density.

Calculation of the compression of a composite from the bulk moduli of its constituents starts with two simple assumptions between which the truth must lie. The Voigt approximation assumes that all constituents are equally strained, in spite of their different moduli. It implies a composite bulk modulus

$$K_V = (V_1K_1 + V_2K_2 + V_3K_3 + \dots)/(V_1 + V_2 + V_3 + \dots) \qquad (21.7)$$

This overestimates the composite modulus because it assumes diminished compression, with the harder grains supporting a higher fraction of the ambient stress than the softer ones. The other extreme assumption is that all grains are equally stressed, giving the Reuss limiting modulus, K_R, where

$$1/K_R = (V_1/K_1 + V_2/K_2 + V_3/K_3 + \dots)/(V_1 + V_2 + V_3 + \dots) \qquad (21.8)$$

For low stresses, this is equally unrealistic, neglecting the problem of different grain responses to the same stress and resulting in grain misfitting. It underestimates the composite modulus. A simple compromise is an average of K_V and K_R, the Voigt–Reuss–Hill average

$$K_{VRH} = (K_V + K_R)/2 \tag{21.9}$$

which is sufficiently accurate for virtually all purposes. Geometric and harmonic averages have also been tried. If more restrictive bounds are wanted, an alternative has sometimes been favoured: for a binary mixture, this is an average of spheres of A in a matrix of B and spheres of B in a matrix of A, giving Hashin–Shtrikman bounds.

Seismic waves in the Earth, outside the immediate fault zones, and ultrasonic waves used in laboratory measurements of acoustic velocities, impose low stresses, to which a composite response is represented by K_{VRH}. This is termed the unrelaxed modulus, meaning that it describes a non-equilibrium response to an applied stress. If the stress is very high or very prolonged, especially at high temperature, then strains are readjusted by grain deformation to equalise the stresses and the Reuss assumption becomes valid, yielding K_R, which is the relaxed modulus. This describes the compressions resulting from the high steady pressures within the Earth. The adiabatic bulk modulus derived from seismic velocities, being the unrelaxed modulus, is slightly higher than the relaxed modulus describing the variation of compression with pressure deep in the mantle (where conditions are understood to be very close to adiabatic to be compatible with convection). This difference is a consequence of inelastic relaxation of stress, but there is another, shorter-term thermal relaxation effect. Adiabatic compression or rarefaction causes temperature changes that are unequal in grains of different thermal properties, so that diffusion of heat between them gives a delayed elastic response. This occurs with indefinitely small stresses. It is much more rapid than the inelastic relaxation and is one reason for attenuation of seismic waves and frequency dependence of their speeds. As illustrated for a mineral mix representing the upper mantle in Table 21.6, the difference between relaxed and unrelaxed moduli is quite small.

In the lower mantle, two further considerations need to be taken into account. The pressure range is such that, as the minerals are compressed, not only do the individual moduli increase, but the less compressible components become increasing volume fractions of the whole, giving a greater increase in bulk modulus with depth than inferred from variations in the individual mineral moduli. This is particularly significant in equation of state fitting, as in Section 21.1.5, involving the pressure dependence of bulk modulus, $K' = \mathrm{d}K/\mathrm{d}P$, added to the VRH calculation in Table 21.7. In all cases, 'bulk modulus' refers to the response

TABLE 21.6 A COMPARISON OF RELAXED AND UNRELAXED BULK MODULI FOR A MODELLED UPPERMOST MANTLE COMPOSITION [60% OLIVINE, 30% PYROXENE, 10% GARNET WITH Fe/(Mg+Fe) = 0.15]

	K_S (GPa)
Olivine	130.05
Pyroxene	106.95
Garnet	177
Voigt	127.81
Reuss (relaxed)	125.26
VRH (unrelaxed)	126.54

Note: The difference is almost 1% at low pressure and decreases with pressure $[(\partial K_s/\partial P)_S \approx 4.5]$.

TABLE 21.7 UNRELAXED AND RELAXED BULK MODULI FOR THE LOWER MANTLE

Moduli	Surface	Top of Lower Mantle	Bottom of Lower Mantle
P	0	23.83	135.75
K_{pv}	264	350.14	704.89
K_{mw}	162	251.65	605.69
$V_{mw}/(V_{mw}+V_{pv})$	0.2	0.1939	0.1854
K_V	243.60	331.04	686.50
K_R	234.47	325.44	684.12
K_{VRH}	239.04	328.24	685.31
K'_{pv}	3.8	3.469	2.993
K'_{mw}	4.1	3.531	2.969
K'_V	3.899	3.499	2.992
K'_R	4.166	3.582	3.003
K'_{VRH}	4.032	3.540	2.997

Note: These values are for a mix of 80% perovskite (pv) and 20% magnesiowustite (mw) by volume, approximating the lower mantle composition, at zero pressure and at pressures, P, corresponding to the top and bottom of the lower mantle. P and K values are in GPa and $K' \equiv dK/dP$. Non-significant digits are retained to emphasise small differences.

of a material to small pressure increments, that is, $K = -V\mathrm{d}P/\mathrm{d}V$, although there may be a much stronger superimposed compression.

21.1.5 Non-Linearity and Finite Strain

In descriptions of the response of a material to pressures that are not very small compared with their elastic moduli, a standard assumption of elasticity theory, that the moduli are constant, becomes invalid. Strains are no longer infinitesimal and the moduli are pressure dependent. They are, however, still valid parameters, describing small strain increments that may be superimposed on finite strains. The only finite strains considered in Earth science are volume compressions, described by bulk moduli that increase with pressure, requiring integrals for finite compressions. Rigidity moduli are also pressure dependent (Section 21.1.6), but elastic shear strains (in seismic waves) are infinitesimal superpositions on finite volume changes, so finite strain theories are theories of volume changes caused by high pressures. All such theories are empirical, with no general form. Indeed, it is now apparent that no universal form is possible, in principle, with the result that there are numerous (>40) proposed equations, some with sophisticated developments but varying plausibility and applicability. However, the understanding of fundamental (thermodynamic) constraints has improved dramatically in the last decade, and the summary presented here draws attention to them. Most applications are concerned with the variation of density with pressure, here referenced to zero pressure and denoted by $x = \rho/\rho_0$, but the slight curvature of a P–ρ plot can be fitted by almost any equation with a modest number of fitting constants. Tests of validity require examination of derivative properties. Bulk modulus $K \equiv \mathrm{d}P/\mathrm{d}\ln\rho$ is the first step in this direction and $K' \equiv \mathrm{d}K/\mathrm{d}P$ is a crucial parameter. It is dimensionless and is a positive finite quantity for all materials that remain constant in mineral and phase structure over the pressure range of interest. Extrapolations to zero and infinite pressure limits give asymptotic values, indicated by subscripts 0 and ∞. These are equations of state parameters that are sensitive to small differences between equations and strongly constrain them. The advantages of using K_0, K'_0 and K'_∞ as fitting constants are that the physical meaning is obvious and they are subject to fundamental thermodynamic constraints. It is not necessary that they be directly observable properties, particularly for K'_∞, and they commonly refer to conditions in which a material cannot exist. But the thermodynamic relationships that link these properties under the range of conditions in which they do exist remain valid outside that range. These constants can be manipulated in the same way as real physical properties and they have the important advantage that thermodynamic and algebraic relationships generally simplify in the zero and infinite pressure extrapolations.

Although it hardly features in the physics literature, the finite strain theory developed by F. Birch has been widely adopted in geophysics because, in using it to pioneer the analysis of material compression in the deep Earth, he established a methodology that others have followed. In spite of difficulties referred to below, the long record of its use will ensure continued attention and so it is outlined here. The theory originated from a mathematical requirement that, in three dimensions, strain should be invariant with interchange or rotations of axes, which was satisfied by defining it in terms of variations in the squares of the separations of material points. In the Birch formalism, this means writing volume strain in terms of $x = \rho/\rho_0$ as $\varepsilon_B = (x^{2/3} - 1)/2$, in which the subscript 'B' identifies ε_B as the 'Birch strain'. In the limit of small compressions, it reduces to the standard representation of linear strain (for compression positive). The Birch theory represents strain energy as a polynomial in ε_B, starting with $\varepsilon_B{}^2$

$$E_S = c_2\,\varepsilon_B{}^2 + c_3\,\varepsilon_B{}^3 + c_4\,\varepsilon_B{}^4 \qquad (21.10)$$

It is referred to as the second-order theory if only the first term is used, the third-order theory if two terms are used and so on, with the understanding that these are successive approximations to a convergent series. Substituting for ε_B and multiplying out, we see that E_S is a polynomial in $x^{2/3} = (a/r)^2$, where the atomic spacing is r and its zero pressure value is a. Thus, the Birch theory effectively assumes a power law atomic potential as a sum of even-powered terms. Its adoption in geophysics has been justified by the fact that the second-order version gives $K'_0 = 4$ and this is not far from the values for some common minerals, implying $c_3 << c_2$ and a convergent series. For this reason, the third-order Birch theory is generally assumed to suffice, giving

$$P = (9/8)K_0[(K'_0 - 16/3)\,x^{5/3} - (2K'_0 - 28/3)\,x^{7/3} + (K'_0 - 2)\,x^3] \qquad (21.11)$$

However, when earth model data are fitted to the fourth-order theory it gives $c_4 \approx c_2$. The series is not intrinsically convergent but must be restricted to small strains, and departures from this have led to some seriously misleading conclusions. Fitted to terrestrial data, both third- and fourth-order theories become non-physical, giving negative pressures at compressions not far beyond the terrestrial range.

To make effective use of the fundamental constraints that we now have, we turn to derivative equations, that is, equations designed to satisfy the constraints on derivative properties. The reciprocal K-primed (RKP) equation is the most successful of these equations. It is based on an approach that is opposite to that of Birch (and most others), which starts with an assumed atomic potential function and differentiates it to obtain properties such as P, K and K'.

The RKP equation starts with what is understood about a particular derivative property (K') and is integrated to obtain relationships between x, P and K. It originated from the mathematical connection between $1/K'$ and the ratio P/K (normalised pressure), which become equal in the infinite pressure limit. This is one of the fundamental constraints that can be applied by appealing to the infinite pressure extrapolation. Apart from the obvious ones, such as density increasing with pressure, there are three general conditions required for validity of an equation (and the search for others continues):

(i) $(K'P/K)_\infty = 1$ (21.12)

(ii) $K'_\infty > 5/3$ (21.13)

(iii) $[d(1/K')/d(P/K)]_\infty > 0$ (21.14)

Condition (iii) means that $[KK''/(1 - K'P/K)]_\infty$ is finite negative. The RKP equation is designed to satisfy these constraints but most of the other equations fail one or more.

$$\text{The RKP equation:} \; 1/K' = 1/K_0' + \left(1 - K_\infty'/K_0'\right) P/K \quad (21.15)$$

Fits of this equation to the lower mantle and core data are shown in Figure 21.1. The integral forms:

$$K/K_0 = \left(1 - K_\infty' P/K\right)^{-K_0'/K_\infty'} \quad (21.16)$$

$$\ln\left(\rho/\rho_0\right) = -\left(K_0'/K_\infty'^2\right)\ln\left(1 - K_\infty' P/K\right) - \left(K_0'/K_\infty' - 1\right) P/K \quad (21.17)$$

Inconvenient iterative calculations are needed to separate P and K in fitting laboratory $P(\rho)$ observations, but in dealing with Earth model data, such as PREM (Tables 6.1 and 7.1), which list both P and K, the ratio P/K can be treated as observable. The result is that four fitting constants (ρ_0, K_0, K'_0 and K'_∞) are obtained from two equations, each of which involves only three of these constants.

Figure 21.1 illustrates a problem that comes to light when derivative properties, such as K', are used. There is nothing obviously unphysical about the PREM values of ρ, K and P (Tables 6.1 and 7.1), but the variation of K' is incompatible with any plausible equation of state. It is an artefact of the manner in which PREM is parameterised, with density and the seismic wave speeds fitted to polynomials in radius, unconstrained by implications of physical properties. It is presumed that the average values of $1/K'$ over most of the lower mantle and outer

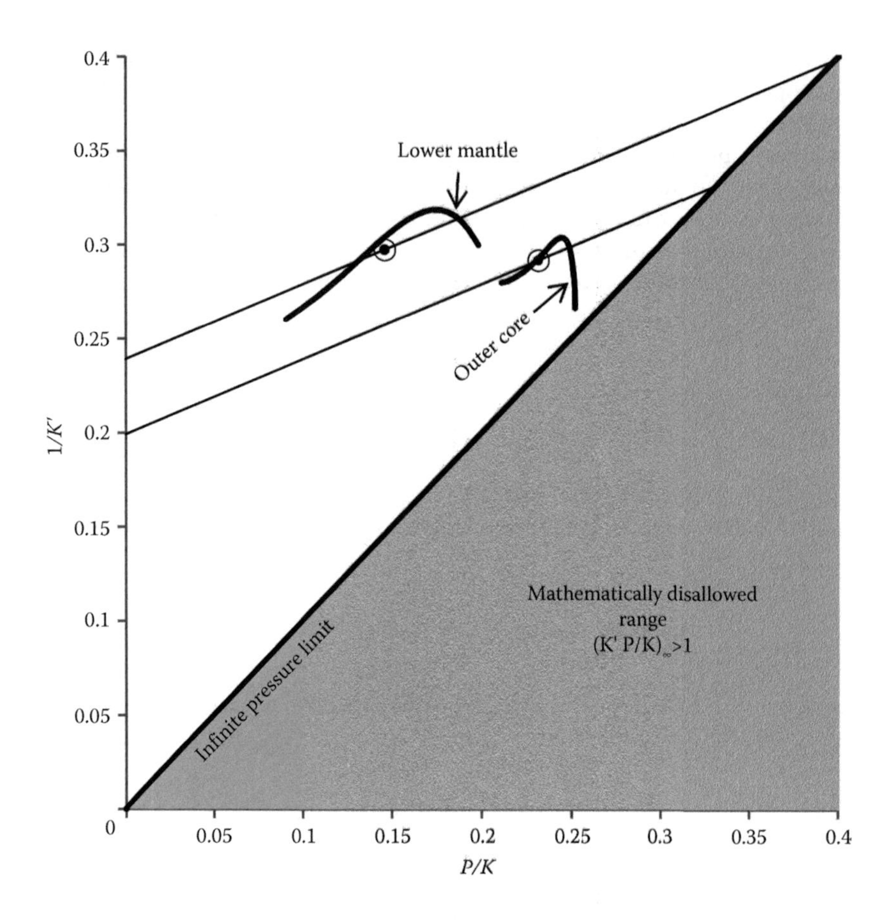

Figure 21.1 Lower mantle and outer core data as listed by PREM, excluding the lowermost 150 km and uppermost 100 km of the lower mantle which are modelled as heterogeneous, fitted to Equation 21.15, which are shown as straight lines passing through the circled points marking volume averages of P/K and $1/K'$ for the PREM data.

core ranges are much more reliable than individual values at specified depths and the graphs are drawn through them. The volume averages are:

$$\text{Lower mantle: } \langle P/K \rangle = 0.1458_0, \langle 1/K' \rangle = 0.2975_1 \quad (21.18)$$

$$\text{Outer core: } \langle P/K \rangle = 0.2316_0, \langle 1/K' \rangle = 0.2918_4 \quad (21.19)$$

It is noticeable that the two RKP graphs are parallel and require $K'_\infty / K'_0 = 3/5$ for both the lower mantle and outer core, but independent data would be needed for a conclusion that this ratio is more general.

Extrapolation of lower mantle and outer core data to laboratory pressure and temperature, using the RKP equation, gives the following density estimates, permitting inferences about composition:

Lower mantle material, with mineral and phase structure, assumed to be preserved during decompression and cooling, $\rho_0 = 3830 \pm 15$ kg m^{-3}.
Outer core material, solidified to the hcp crystal structure, presumed to be preserved during decompression and cooling, $\rho_0 = 7488 \pm 30$ kg m^{-3}. For pure iron with the same (ε) structure, $\rho_0 = 8352 \pm 23$ kg m^{-3}.

21.1.6 Pressure Variation of Rigidity Modulus

Seismological observations give measures of rigidity modulus, μ, with precision comparable to those for bulk modulus, K, so an understanding of μ is necessary to obtain maximum value from the data. Interpretation is not as straightforward as for K, partly because of a lack of comparable thermodynamic relationships, but a consideration of interatomic forces yields a relationship between μ and K that simplifies the extrapolation of μ to high pressure:

$$(\mu/K) = (\mu/K)_0 - [(\mu/K)_0 - (\mu/K)_\infty]K'_\infty P/K \tag{21.20}$$

This is the μ–K–P equation (sometimes referred to as the mu–K–P equation). It fits the PREM lower mantle tabulation of μ and K (Table 7.1), with a precision far higher than the accuracy of the data as such (Figure 21.2). It gives an even better fit to another model (ak135), although with coefficients differing from those of the PREM fit by up to 10 standard deviations. The inference is that the relationship is fundamental and applies precisely to internally consistent data sets, even if they differ from one another. Both K and μ increase with pressure, with μ increasing less than K, and the ratio μ/K decreases, as is noticeable from a progressive increase in Poisson's ratio, $\nu = (3K - 2\mu)/(6K + 2\mu)$, with depth in the Earth. This is not an indication of an approach to a liquid state (which gives $\mu = 0$), but a fundamental property of atomic forces, which gives μ/K decreasing with compression, approaching the limit $(\mu/K)_\infty$ as in Equation 21.20.

The similarity of Equations 21.15 and 21.20 is helpful in interpretation of the infinite pressure extrapolation of equations of state. For the core, a μ–K–P plot, as in Figure 21.2, confirms the observation from an RKP plot that, when considered in terms of normalised pressure (P/K), the core is quite close to the infinite pressure extrapolation. It also supports the estimate of K'_∞ because, disallowing negative μ/K, the μ-K-P core graph must terminate at $P/K <\sim 0.35$ and therefore $K'_\infty > \sim 1/0.35 = 2.86$. However, some details in the figure call for comment. The lower mantle plot extrapolates both ways from the very precisely fitted range of PREM data, indicated by the heavy line, and terminates at the

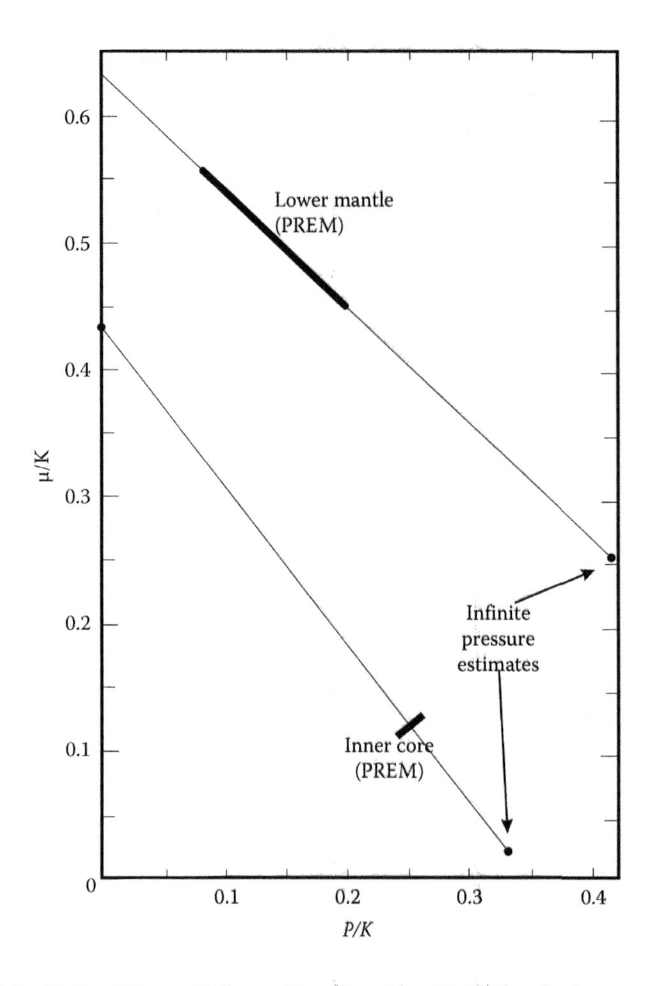

Figure 21.2 Plots of the μ–K–P equation (Equation 21.20) for the lower mantle and core, showing the lower mantle fit to PREM data and the short inner core range of reversed gradient, attributed to an artefact of PREM. [*Note*: At inner core pressure, iron is in the ε (hexagonal close packed) phase, but the zero pressure point is estimated from α phase (body-centred cubic) iron].

P/K limit in the RKP plot. For the outer core there are no μ data, but there are estimates for the inner core and the PREM range is indicated by the short heavy line. The graph assumes that the average is valid, but the opposite gradient is an artefact of PREM. The difficulty is that μ is not well observed for the inner core and its gradient is virtually unobserved. Its high PREM gradient is compensated by a value of dK/dP that is so much smaller than the outer core value as to be physically implausible, and the gradient of $\chi = (K + (4/3)\,\mu)$, which is reasonably well controlled by the observed P-wave speed, is not anomalous. Compensating adjustments of $\mathrm{d}K/\mathrm{d}P$ and $\mathrm{d}\mu/\mathrm{d}P$ give theoretically satisfactory values for both and realign the inner core gradient with the trend line in Figure 21.2.

21.1.7 Inelasticity and Anelasticity

The distinction between these departures from ideal elasticity is not always clear, and their causes are not always independent, but it is convenient to use these terms to identify effects that are subtly different. If a material is subjected to a very prolonged stress or a transient very high stress, it may not recover its original form when released from the stress. This is inelastic deformation. If the material recovers its original form, without permanent deformation, particularly for repeatedly cycled small stresses, but the strain shows a lag relative to the stress, following a stress–strain hysteresis loop instead of having the same stress–strain path on increasing and decreasing stress, this is anelasticity. Both effects are important in the Earth. Convective motion in the mantle is inelastic, and attenuation of seismic waves is anelastic.

Although the term inelasticity formally includes fracture, this is the subject of a quite different rock mechanics analysis. Here, the interest is in progressive deformation under prolonged stress, commonly referred to as creep, of material that remains coherent. This occurs at the high temperatures ($T > T_M/2$) that prevail in most of the Earth (T_M is the melting point). The expression 'steady state creep' is used when necessary to distinguish it from transient creep of material that hardens as it is deformed. The Weertman equation

$$\dot{\varepsilon} = B\left(\sigma/\mu\right)^n \exp\left(-gT_M/T\right) \tag{21.21}$$

is a general representation of the rate of steady increase in inelastic strain, $d\varepsilon/dt \equiv \dot{\varepsilon}$, under shear stress σ for material with a rigidity modulus μ and melting point T_M. Variation with temperature, T, is represented by the exponential term, with $g \approx 27$ for minerals. B is a constant and the value of the exponent n (1 to 6) depends on the atomic process(es) involved. These processes are basically the same as those involved in melting and T_M takes account of the energy of thermal activation and its pressure dependence. Creep processes involving crystalline point defects (vacancies and interstitial atoms) are represented by Newtonian viscous behaviour, with $n = 1$. In this case, viscosity, $\nu = \sigma/\dot{\varepsilon}$, is an unambiguous quantity. For simplicity, in the absence of clear contrary evidence and not because it is demonstrably valid, this is generally assumed in discussions of mantle convection and post-glacial rebound, but since these processes involve similar strain rates, a discrepancy would not be obvious. Creep with $n \approx 3$

is normally observed when it results from movements of dislocations, extended crystal defects, as in Figure 21.3, which illustrates another important effect, weakening by water.

Loss of energy by elastic waves occurs by several processes, collectively termed internal friction or anelasticity. Some are relaxation processes that cause a time delay between stress and strain and result in elliptical hysteresis loops in a strain–stress diagram. In rocks, they are most important at very low strain amplitudes ($\varepsilon < 10^{-6}$). For higher strain amplitudes, grain boundary movements become significant, and in this case the lag in strain is controlled by the state of stress, not a time delay. It is a true hysteresis phenomenon, producing pointed strain–stress loops. Relaxation phenomena result in frequency-dependent attenuation, but true hysteresis phenomena do not. However, in rocks there is generally a superposition of a wide range of relaxation times that smear out the frequency dependence. This approximates a frequency-independent value of the parameter Q (quality factor), which represents the

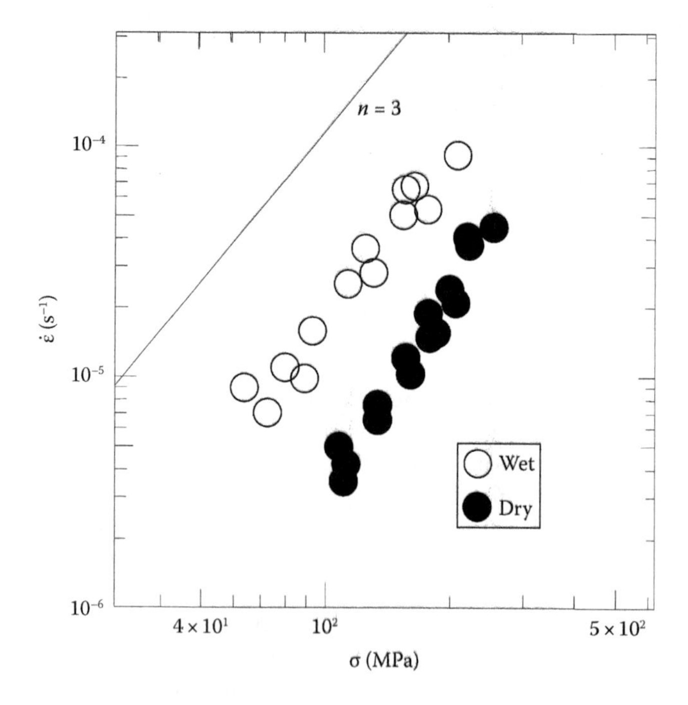

Figure 21.3 Creep of polycrystalline olivine maintained under dry or wet conditions at a confining pressure of 300 MPa and T = 1250°C. This is dislocation-mediated creep with $n \approx 3$ in Equation 21.21. Data by Mei and Kohlstedt (2000).

fractional loss of energy per cycle by a wave or oscillatory strain, $\Delta E/E = -2\pi/Q$. This is not always a good approximation. Q is systematically higher for seismic body waves, with periods of 1 to 100 seconds than for free oscillations with periods of 5 to 50 minutes, although this difference may arise partly from different distributions of stress within the Earth as well as different ratios of shear to compression. The energy loss is caused almost entirely by the shear component of strain, with very high Q for pure compression. This is evident from the global Q values for the simplest modes of free oscillation and the semidiurnal tide:

${}_0S_0$ $Q \approx 5900$; this mode is an alternating dilation and contraction of the Earth as a whole and therefore involves volume changes with no shear strains except for local effects of heterogeneity. The very high Q reflects this, but it is still lower than would be expected for perfectly homogeneous material.

${}_0S_2$ $Q \approx 800$; colloquially known as the football mode, this is an alternation between prolate and oblate deformations, involving both volume and shear deformation.

${}_0T_2$ $Q \approx 250$; the simplest torsional mode, this is an alternating twist between hemispheres with no volume change accompanying the shear strain.

Semidiurnal tide $Q \approx 280$; this is an estimate of the dissipation in the solid Earth by an ellipsoidal deformation which rotates (relative to the body of the Earth). Being a shear deformation, the Q is not clearly different from that of the ${}_0T_2$ mode.

For estimates of Q more locally, PREM separates the values for pure compression, Q_K, and pure shear, Q_μ. Q_K is uniformly too high for detailed estimates (~58,000), except in the inner core (1330). Q_μ is low in the inner core (85) and in the 'low-velocity zone' immediately below the lithosphere (80), with a very high estimate for the lithosphere (600) and intermediate values in the mantle, increasing with depth from 143 to 312. All these numbers are unavoidably approximate. There are two methods of estimating Q for individual seismic wave paths, but they are also very approximate. Estimates based on the diminution of wave amplitude with distance require corrections for refraction and wave scattering. With this proviso, the attenuation of the amplitude of a travelling wave with distance x is written as $A = A_0 \exp(-\alpha x)$, where $\alpha = \pi/\lambda Q$ is called the attenuation coefficient and λ is the wavelength.

The spectral ratio method of estimating Q records the greater attenuation of high frequencies by comparing wave spectra at different distances (or for different wave paths). Effectively, this means measuring distances in wave lengths, which depend on frequency. It is normal to assume that both Q and scattering are independent of frequency, but neither assumption is reliable. The dependence of Q on frequency, f, is represented by writing $Q \propto f^{\varepsilon}$, generally with $\varepsilon \approx 1/3$. This has an important consequence for wave dispersion. Attenuation is accompanied by a frequency variation of wave speed (or elastic modulus). This is reasonably simple if Q is independent of frequency, but if not then the ratio of wave speeds at frequencies f_1 and f_2 can be written as:

$$v(f_1) / v(f_2) = 1 + [(1 - \varepsilon)/\pi] \ln (f_1/f_2) \langle Q^{-1} \rangle \tag{21.22}$$

For $\varepsilon = 0$, this reduces to the standard expression for frequency-independent Q.

21.2 THERMAL PROPERTIES

21.2.1 Relationships between Properties and Application of the Thermodynamic Grüneisen Parameter

Thermal and elastic properties are controlled by the same atomic forces and there are relationships between them that are useful for calculating or verifying numerical values, especially for very small mineral samples or inaccessible parts of the Earth. Central to these relationships is the thermodynamic Grüneisen parameter

$$\gamma = \alpha K_T/\rho C_V = \alpha K_S/\rho C_P \tag{21.23}$$

where α is the volume expansion coefficient; ρ is the density; K_T and K_S are the isothermal and adiabatic bulk moduli, respectively; and C_V and C_P are specific heats at constant volume and pressure, respectively. Fundamental studies generally emphasise K_T and C_V, but most direct measurements are of K_S and C_P, which are also the quantities of geophysical interest, and a convenient relationship between them is the ratio:

$$K_S/K_T = C_P/C_V = 1 + \gamma\alpha T \tag{21.24}$$

This departs from unity by a few per cent, deep in the Earth (at $T \approx 2500$K) and generally about 1% in samples at laboratory temperature. γ is a dimensionless number with a value close to unity for all materials and generally 1.2 to 1.5 for terrestrial materials, with an extreme range of 1.0 to 1.8. Unlike α, C_V and C_P, its dependence on temperature is slight, even at low temperatures, and it decreases with pressure, but the variation is much less than for the other quantities in Equation 21.23. A feature that makes γ particularly useful in mineral physics and geophysics is that it can be represented in terms of elastic constants and their pressure derivatives. There are two formulae that are conveniently applied in different situations. For materials with well-determined values of elastic constants and their pressure derivatives (this includes the deep mantle), the 'acoustic γ' can be applied:

$$\gamma_A = \frac{1}{6}\frac{K_T}{K_S + \frac{4}{3}\mu}\left[\left(\frac{\partial K_S}{\partial P}\right)_T + \frac{4}{3}\left(\frac{\partial \mu}{\partial P}\right)_T\right] + \frac{1}{3}\frac{K_T}{\mu}\left(\frac{\partial \mu}{\partial P}\right)_T - \frac{1}{6} \tag{21.25}$$

Although this implicitly assumes homogeneous, isotropic material, it is satisfactory for polycrystalline mixtures of minerals and therefore for the solid parts of the Earth's interior. An approximation to γ_A is the 'Debye γ', γ_D, which assumes a single averaged modulus as in the Debye-specific heat theory:

$$\gamma_D = -(\partial \ln\theta_D/\partial \ln V)_T \tag{21.26}$$

but the Debye temperature, θ_D, is generally not well known and is usually estimated from elastic moduli anyway. The other widely used formula is:

$$\gamma_{FV} = [(\partial K_T/\partial P)_T/2 - 1/6 - f/3 - (f/9)(P/K_T)]/[1 - (2f/3)(P/K_T)] \tag{21.27}$$

This is referred to as the free volume-type formula because, with the parameter $f = 2$, it was derived from free volume theory, but there are several other derivations giving different values of f that depend on what is assumed about correlations between thermal vibrations of neighbouring atoms in a crystal. An 'observed' value, $f = 1.45$, virtually independent of pressure, is obtained by equating γ_{FV} to γ_A using lower mantle data and, since it gives values of γ agreeing with zero pressure values for other materials, it has been widely adopted. γ_{FV} has the advantage of requiring no information about μ and can be applied to liquids, including the outer core. Also, it is conveniently used with Equation 21.15, which relates the required parameters K' and P/K.

A physical significance of γ is seen in the individual values of the numerator and denominator of Equation 21.23. $\alpha K_T = (\partial P/\partial T)_V$ is the thermal pressure per degree temperature rise, the pressure required to prevent thermal expansion of a material held at constant volume. The denominator, $\rho C_V = (\partial Q/\partial T)_V/V$, is heat input per degree temperature rise per unit volume for a material held at constant volume. Thus, γ is the ratio of thermal pressure to thermal energy per unit volume. In many situations, γ simplifies the presentation of thermodynamic relationships, by allowing them to be seen as generalisations of the intuitive logic of ideal gas physics. For example, in adiabatic temperature variations, $(\partial \ln T/\partial \ln \rho)_S = \gamma$, which is familiar for an ideal gas [for which $\gamma = (C_P/C_V - 1)$]. For any material, if γ is assumed to be constant, $T_1/T_2 = (\rho_1/\rho_2)^\gamma$ on an adiabat.

Tables 21.8 through 21.11 present thermodynamic derivative relationships in a compact form convenient for geophysical applications, making use of γ and

TABLE 21.8 THERMODYNAMIC NOTATION AND DEFINITIONS

Parameter	Notation
Specific heat	
Constant P	$C_P = (T/m)(\partial S/\partial T)_P$
Constant V	$C_V = (T/m)(\partial S/\partial T)_V$
Derivatives of C_V	$C'_S = (\partial \ln C_V/\partial \ln V)_S$; $C'_T = (\partial \ln C_V/\partial \ln V)_T$
Helmholtz free energy	$F = U{-}TS$
Gibbs free energy	$G = U{-}TS{+}PV$
Enthalpy	$H = U{+}PV$
Bulk modulus, adiabatic	$K_S = -V(\partial P/\partial V)_S$
Derivatives of K_S	$K'_S = (\partial K_S/\partial P)_S$; $K''_S = (\partial K'_S/\partial P)_S$
Bulk modulus, isothermal	$K_T = -V(\partial P/\partial V)_T$
Derivatives of K_T	$K'_T = (\partial K_T/\partial P)_T$; $K''_T = (\partial K'_T/\partial P)_T$
Pressure	P
Isothermal derivative of γ	$q = (\partial \ln\gamma/\partial \ln V)_T = (\partial \ln(\gamma C_V)/\partial \ln V)_S$
Adiabatic derivative of γ	$q_S = (\partial \ln\gamma/\partial \ln V)_S = q{-}C'_S$
Heat	Q
Entropy	$S = \int dQ/T$
Temperature	T
Internal energy	U
Volume	V
Volume expansion coefficient	$\alpha = (1/V)(\partial V/\partial T)_P$
Grüneisen parameter	$\gamma = \alpha K_T/\rho C_V = \alpha K_S/\rho C_P$
Anderson–Grüneisen parameter	
Adiabatic	$\delta_S = -(1/\alpha)(\partial \ln K_S/\partial T)_P = (\partial \ln(\alpha T/C_P)/\partial \ln V)_S$
Isothermal	$\delta_T = -(1/\alpha)(\partial \ln K_T/\partial T)_P = (\partial \ln\alpha/\partial \ln V)_T$
Density	$\rho = m/V$
Second derivative of γ	$\lambda = (\partial \ln q/\partial \ln V)_T$

Note: Parameters V, S, U, H, F and G refer to arbitrary mass, m, but C_V and C_P refer to unit mass.

TABLE 21.9 FIRST-ORDER DERIVATIVES OF THERMODYNAMIC PARAMETERS

Differential Element	Constant			
	T	***P***	***V***	***S***
∂T	–	1	1	γT
∂P	$-K_T/V$	–	$\alpha K_T = \gamma\rho C_V$	K_S
∂V	1	αV	–	$-V$
∂S	$\alpha K_T = \gamma\rho C_V$	mC_P/T	mC_V/T	–
∂U	$\alpha K_T T - P$	$mC_P - \alpha VP$	mC_V	PV
∂H	$-K_T(1-\alpha T)$	mC_P	$mC_V(1+\gamma)$	$K_S V$
∂F	$-P$	$-S - \alpha VP$	$-S$	$PV - \gamma TS$
∂G	$-K_T$	$-S$	$-S + \alpha K_T V$	$K_S V - \gamma TS$
	Constant			
	U	***H***	***F***	***G***
∂T	$P - \alpha K_T T$	$1 - \alpha T$	P	1
∂P	$-\rho C_V(K_S - \gamma P)$	$-\rho C_P$	$K_T(S/V + \alpha P)$	S/V
∂V	mC_V	$\alpha V(1 + 1/\gamma)$	$-S$	$\alpha V - S/K_T$
∂S	$mC_V P/T$	mC_P/T	$mC_V(P/T - \gamma S/V)$	$mC_P/T - \alpha S$
∂U	–	$mC_P - PV\alpha(1 + 1/\gamma)$	$mC_V P - S\alpha K_T T + SP$	$mC_P - \alpha TS - P\alpha V + SP/K_T$
∂H	$mC_V[P(1+\gamma) - K_S]$	–	$SK_T(1-\alpha T) + mC_V P(1+\gamma)$	$mC_P + S(1-\alpha T)$
∂F	$\rho C_V(\gamma TS - PV) - PS$	$-S(1-\alpha T) - PV\alpha(1 + 1/\gamma)$	–	$-S(1 - P/K_T) - P\alpha V$
∂G	$mC_V(\gamma TS/V + \gamma P - K_S) - PS$	$-S(1-\alpha T) - mC_P$	$S(K_T - P) + PV\alpha K_T$	–

TABLE 21.10 THERMODYNAMIC DERIVATIVES EXTENDED TO SECOND ORDER AT CONSTANT *T, P, V* AND *S*

Differential Element	Constant			
	T	***P***	***V***	***S***
∂T	–	1	1	γT
∂P	$-K_T/V$	–	$\alpha K_T = \gamma\rho C_V$	K_S
∂V	1	αV	–	$-V$
∂S	$\alpha K_T = \gamma\rho C_V$	mC_P/T	mC_V/T	–
∂U	$\alpha K_T T - P$	$mC_P - \alpha VP$	mC_V	PV
∂H	$-K_T(1-\alpha T)$	mC_P	$mC_V(1+\gamma)$	$K_S V$
∂F	$-P$	$-S - \alpha VP$	$-S$	$PV - \gamma TS$
∂G	$-K_T$	$-S$	$-S + \alpha K_T V$	$K_S V - \gamma TS$
$\partial\alpha$	$\alpha\delta_T/V = -(\partial K_T/\partial T)_P/K_T V$	$\alpha^2(2\delta_T - K'_T + C'_T/\gamma\alpha T)$	$\alpha^2(\delta_T - K'_T + C'_T/\gamma\alpha T)$	$-\alpha[K'_S - 1 + q + \gamma\alpha T(\delta_S + q)]$
∂K_T	$-K_T K'_T/V$	$-\alpha K_T\delta_T = K_T^2(\partial\alpha/\partial P)_T$	$\alpha K_T(K'_T - \delta_T)$	$K_T[K'_T + \gamma\alpha T(K'_T - \delta_T)]$
∂K_S	$(-K_T/V)(K'_S + \gamma\alpha T\delta_S)$	$-\alpha K_S\delta_S$	$\alpha K_T(K'_S - \delta_S)$	$K_S K'_S$
∂C_V	$(C_V/V)C'_T = (C_V/V)(1 - q + \delta_T - K'_T)$	$(C_P C'_T - C_V C'_S)/\gamma T$	$(C_V/\gamma T)(C'_T - C'_S)$	$-TC_V(\partial\gamma\partial T)_V$
∂C_P	$(C_P/V)[C'_T + \gamma\alpha T(q + \delta_T)/(1 + \gamma\alpha T)]$	$(C_P/\gamma T)[C'_T - C'_S + \gamma\alpha T(\delta_T - \delta_S + C'_T)]$	$(C_P/\gamma T)\{C'_T(1 + \gamma\alpha T) + [\gamma^2\alpha T + (\gamma\alpha T)^2(q-1) - C'_S]/(1 + \gamma\alpha T)\}$	$-C_P[T(\partial\gamma/\partial T)_V + \gamma\alpha T(\delta_S + q)]$
$\partial\gamma$	$\gamma q/V$	$\gamma\alpha q + C'_S/T$	C'_S/T	$-\gamma(q - C'_S)$

TABLE 21.11 RELATIONSHIPS BETWEEN DERIVATIVES
$K_S/K_T = C_P/C_V = 1 + \gamma\alpha T$
$K'_T = K'_S(1 + \gamma\alpha T) + \gamma\alpha T[3q - 2 - \gamma + \gamma(\partial\ln C_V/\partial\ln T)_V]$
$K'_S = K'_T(1 + \gamma\alpha T) - \gamma\alpha T(\delta_S + \delta_T + q)$
$\delta_S = -(1/\alpha)(\partial\ln K_S/\partial T)_P = K'_S - 1 + q - \gamma - C'_S = (\partial\ln(\alpha T/C_P)/\partial\ln V)_S$
$\delta_T = -(1/\alpha)(\partial\ln K_T/\partial T)_P = K'_T - 1 + q + C'_T = (\partial\ln\alpha/\partial\ln V)_T$ $= (\delta_S + C'_T)(1 + \gamma\alpha T) + \gamma + C'_S + \gamma\alpha T(2q - 1)$
$C'_S = C'_T - \gamma(\partial\ln C_V/\partial\ln T)_V = (\partial\gamma/\partial\ln T)_V$
$C'_T = \gamma(\partial\ln(\gamma C_V)/\partial\ln T)_V$
$q_S = q - C'_S$
$(\partial\ln(\alpha K_T)/\partial\ln V)_T = \delta_T - K'_T = -(1/\alpha)(\partial\ln K_T/\partial T)_V$
$(\partial\ln(\alpha K_T)/\partial\ln T)_V = (\partial\ln(\gamma C_V)/\partial\ln T)_V = C'_T/\gamma$
$(\partial(\alpha K_T)/\partial T)_P = K_T(\partial\alpha/\partial T)_V$
$(\partial\ln(\alpha K_T)/\partial\ln V)_S = q - 1$
$(\partial\ln(\alpha K_S)/\partial\ln V)_S = q - 1 + \gamma\alpha T(\delta_S + q)$
$(\partial\ln(\gamma\alpha T)/\partial\ln V)_T = \delta_T + q$
$(\partial\ln(\gamma\alpha T)/\partial\ln V)_S = (1 + \gamma\alpha T)(\delta_S + q)$
$(\partial K'_T/\partial T)_P = \alpha\delta_T[\delta_T - K'_T + (\partial\ln\delta_T/\partial\ln V)_T]$
$(\partial K'_S/\partial T)_P = \alpha\delta_S[\delta_S - K'_S + (\partial\ln\delta_S/\partial\ln V)_S]$
$(\partial\delta_T/\partial\ln V)_T = -K_T K''_T + \lambda q + (\partial C'_T/\partial\ln V)_T$
$(\partial\delta_S/\partial\ln V)_S = -K_S K''_S - \gamma q_S + (\partial q_S/\partial\ln V)_S$
$(\partial C_P/\partial P)_T = -(\partial(\alpha/\rho)/\partial\ln T)_P$

other parameters defined in Table 21.8. Table 21.9 presents all possible partial derivatives of the eight primary parameters. Any one of them may be differentiated with respect to any other one with any third one held constant. Individual entries in the table have no meaning; they must be taken in pairs so that, for example, to find $(\partial T/\partial P)_S$ look down the constant S column and take the ratio of entries for ∂T and ∂P, that is, $\gamma T/K_S$. Table 21.10 extends the constant T, P, V and S columns to second derivatives, that is, to derivatives of the first derivative parameters. An arbitrary mass m of material is assumed, so that m appears in many of the entries. There are alternative forms for many of the Table 21.10 entries, summarised in Table 21.11, along with some derivatives of products and third derivatives that have been found useful.

21.2.2 Specific Heat and Expansion Coefficient

Minerals and rocks are almost all electrical insulators, with thermal properties unaffected by the conduction electrons that are important to the thermal properties of metals. This means that at high temperatures, their specific heats are well represented by the Dulong–Petit law of classical physics, $C_V = 3R$, where $R = 8.314\ldots$ J K^{-1} per mole is the gas constant. But the mole is not a very useful unit for materials that have compositions with non-integral ratios of elements, and it is more convenient to relate specific heat in J K^{-1} per kilogram to the mean atomic weight, $\bar{m}$

$$C_V = 24943/\bar{m}\mathrm{JK^{-1}kg^{-1}} \tag{21.28}$$

Adjustment to C_P by Equation 21.24 is slight, as is a correction for anharmonicity, but the important precaution in using this number is that it applies only to high temperatures. This means $T > \theta_D$, the Debye temperature, below which the high-frequency modes of thermal vibration become inactive and specific heat is reduced. For most of the minerals that are thermally interesting 500 K $< \theta_D <$ 1000 K, so that almost the entire Earth is in the high temperature, classical range, $T > \theta_D$, but measurements at laboratory temperatures are not and specific heats are lower. At $T = \theta_D$, C_V is 95% of the value in Equation 21.28, and it rapidly approaches this limiting value at higher temperatures, but at low temperatures C_V approaches zero. The reduced activity of high-frequency thermal modes affects the thermal expansion coefficients, α, similarly. The ratio α/C_V, and therefore the Grüneisen parameter, γ (Equation 21.23), are much more nearly independent of temperature. Values for selected minerals with properties well measured over a wide temperature range give an indication that these properties can be quite well estimated, using the Grüneisen parameter, where no observations are available (Table 21.12).

Materials with tetrahedral atomic bonds, especially the elements with diamond structures, diamond itself, silicon and germanium, and also quartz and silicates generally, have small expansion coefficients, particularly at low temperatures. The crystals have open structures with wide angles between bonds, allowing strong excitations of soft modes of thermal vibration, in which the atoms move relatively easily in directions with no neighbouring atoms. This makes the atomic forces that cause expansion weak in those directions and thermal expansion may even be negative over limited temperature ranges.

The conduction electrons in metals give contributions proportional to temperature to both specific heat and thermal expansion and, as with insulators, leave the Grüneisen parameter more or less independent of temperature. Iron is

TABLE 21.12 TEMPERATURE DEPENDENCES OF THERMAL PROPERTIES OF SELECTED MANTLE MINERALS

	300 K			1000 K		
	C_P ($JK^{-1}kg^{-1}$)	α ($10^{-6}K^{-1}$)	γ	C_P ($JK^{-1}kg^{-1}$)	α ($10^{-6}K^{-1}$)	γ
Periclase MgO	938	31.3	1.54	1270	44.7	1.54
Forsterite Mg_2SiO_4	840	27.4	1.20	1240	38.1	1.14
Enstatite $MgSiO_3$	820	30.0	1.23	1208	35.4	0.93[a]
Corundum Al_2O_3	1054	16.2	1.32	1666	27.3	1.37

[a] This value, obtained from component parameters in Equation 21.23, appears anomalously low. It is possible that this high-temperature state is transitional between polymorphs of enstatite with subtly different thermodynamic properties.

the only geophysically important metal, being the major constituent of the core, where the high temperature gives the conduction electrons an important role in controlling thermal (as well as electromagnetic) properties. At high temperatures, the electron contribution, C_e, to the specific heat of pure iron at temperature T and density ρ (relative to the density ρ_0 at 300 K and zero pressure) is the same for all phases, including liquid, and is given by:

$$C_e = (0.1094\rho_0/\rho - 0.0205)T \text{ J K}^{-1}\text{kg}^{-1} \tag{21.29}$$

Low temperature complications arising from magnetic alignments of electrons are irrelevant. Values of specific heat for the core alloy, listed in Table 6.3, include a component with the form of Equation 21.29. In the classical component of C_V (Equation 21.28) $\bar{m}$ is lower than for pure iron.

21.2.3 Melting

In consideration of mineral melting, there is a distinction between congruent and incongruent melting. In the simpler case of congruent melting, the liquid and solid phases are chemically identical. Some minerals melt congruently, but even for simple compounds incongruent melting is common and for rocks with mixed mineralogy the melting process involves chemical exchanges between ingredients over a range of temperatures. Examples of materials that melt congruently, with melting points at laboratory pressure, T_M, in parentheses, are:

MgO, periclase (3100 K); Al_2O_3, corundum (2330 K); Mg_2SiO_4, forsterite (2160 K); NaCl, halite (1074 K); Fe, iron (1812 K).

$MgSiO_3$, enstatite, melts incongruently, separating at 1830 K into a residuum of solid forsterite and a silica-rich liquid. SiO_2, quartz, is metastable at its melting point, 2000 K, and transforms to other phases with melting points of 1943 K and 1986 K.

Igneous rocks are composites of minerals with different individual melting points. A rock melts over a range of temperatures, from the solidus, the highest temperature at which it is entirely solid, to the liquidus, the lowest temperature at which it becomes entirely liquid. Chemical exchanges between constituents complicate the melting process, so that a rock that is melted and resolidified may have a mineral structure very different from its starting condition. The mineral processes involved are illustrated for the case of basalt in Section 20.1.2. There are wide variations in melting temperature arising from differences in mineralogy and from volatile constituents, especially water. Laboratory dried basalt has a solidus at ~1350 K and liquidus at ~1500 K, and the corresponding numbers for granite are ~1000 K and ~1150 K. The melting points of the individual minerals in an igneous rock indicate the order of their separation as the rock solidifies and cools. But the dry rock melting points are relevant only at low (nominally zero) pressure because water is ubiquitous in the Earth, and it reduces melting temperatures dramatically. Water can have no effect on melting at zero pressure because it is driven off well before the melting point is reached, but at pressures of order 1 GPa solidus temperatures are lowered by as much as 400° if the rocks are close to water saturation. This is the situation in subducted sea floor crust and sediment in which the lava for subduction zone volcanism is generated. At high pressure, regardless of the presence of water, there is a general increase in melting point with pressure, as observed with dry rocks. Other volatiles, notably carbon dioxide, also cause melting point depression, introducing a complication to plans for geo-sequestration of CO_2 under pressure. In this case, 'melting' may not be the appropriate word because there are chemical changes, making 'metasomatism' (see Section 20.3) more correctly descriptive.

Although congruent melting is simpler, it is not generally simple enough for application of the melting theories that are applied to estimates of the variations of melting points with pressure. The general thermodynamic relationship that all such theories depend on is the Clausius–Clapeyron equation

$$dT_M/dP = \Delta V/\Delta S \quad (21.30)$$

where ΔV and ΔS are the volume and entropy changes of the melting process, respectively. The problem is to calculate ΔV and ΔS, which vary widely between materials. Identical chemical compositions of solid and liquid give no evidence of the structural changes that control these parameters. The familiar example is water ice, for which melting involves a rearrangement of molecules, producing a liquid more dense than the solid, chemically the same but structurally quite different. Although, from a chemical perspective, congruent versus incongruent melting distinguishes melting types, a physical distinction is made on the basis of structural rearrangement. It is convenient to identify simple melting as a process with minimal rearrangement, in which a crystal becomes saturated with dislocations. The atoms are displaced from equilibrium positions but the short-range structural arrangement is preserved. In this situation, a liquid is a fully dislocated solid and the mobility of dislocations accounts for its fluidity. The ratio of volume change to energy, $T\Delta S$, for the introduction of dislocations is simple and applied to melting by Equation 21.30, it gives the variation of melting point with the density on the melting curve, a more general representation of what is known as Lindemann's law

$$\mathrm{d}\ln T_{\mathrm{M}}/\mathrm{d}\ln\rho = 2\gamma \tag{21.31}$$

It is crucial for the development of the Earth that this gradient is steeper than the adiabatic gradient, $(\partial \ln T/\partial \ln \rho)_S = \gamma$, although the difference is less than a factor of 2 because the density variation on an adiabat is less than on the melting curve. As is demonstrated in the core, convecting (adiabatic) planets solidify from the inside outwards and, given a sufficient understanding of the solidification of iron alloy at a pressure of 329 GPa, the inner core/outer core boundary presents a fixed point for calculations of the Earth's temperature profile. Extrapolation of laboratory measurements on ε-iron (the hexagonal high pressure phase) by Equation 21.31 gives $T_{\mathrm{M}} \approx 5750$ K at the inner core boundary pressure and, allowing for an estimated 750 K melting point depression by alloying ingredients, we arrive at a boundary temperature of 5000 ± 400 K.

21.2.4 Thermal Conduction

Heat conduction is important at four levels in the Earth, the lithosphere and crust, the upper mantle transition zone, the base of the mantle (D″) and the outer core. In the bulk of the mantle, heat transfer is by convection, with only about 3% by conduction. The thermal conductivities of silicates (the major components of most rocks) are generally low, and conduction in the rocky parts of the Earth is important only in the limited depth ranges of steep temperature gradients. The conductivities of metals, including core iron, are much higher, and thermal conduction in the core is a crucial component of its energy budget.

The conductivities of minerals have a limited range (Table 21.13), being controlled more by structural complexity than by composition. As an illustration of this point, quartz crystals have relatively high conductivity, $\kappa = 6.4$ to $10.7\ W\ m^{-1}\ K^{-1}$ in different crystal orientations and $7.8\ W\ m^{-1}\ K^{-1}$ for a polycrystalline average, but for fused silica $\kappa = 1.37\ W\ m^{-1}\ K^{-1}$, with both temperature and pressure variations opposite to those of the solid. The prime example of a nearly perfect crystalline insulator is diamond, for which reports give $\kappa \approx 545\ W\ m^{-1}\ K^{-1}$, for high-quality specimens, although this is difficult to measure with tiny samples. In all these materials, thermal energy is carried by phonons, quantised lattice vibrations (elastic wave packets), and conductivity is controlled by their mean free paths, the distances travelled between scattering by crystal irregularities. Conduction is limited by several effects: grain boundaries in fine-grained material, internal crystal irregularities such as dislocations and point defects, impurities and the multiplicity of elements more generally and even the distribution of isotopes of individual elements. Not surprisingly, measurements of conductivities of different samples of nominally the same material can give results that differ, in some cases by a factor of 2. Values are highest for high-quality crystals of the simplest minerals, such as corundum (Al_2O_3) and periclase (MgO), for which conductivity can exceed $30\ W\ m^{-1}\ K^{-1}$, but most values are much lower, in the range of 1.5 to $7\ W\ m^{-1}\ K^{-1}$.

TABLE 21.13 AVERAGE THERMAL CONDUCTIVITIES (FOR SOME MATERIALS VALUES ARE VERY VARIED)

Material	Conductivity, κ ($W\ m^{-1}\ K^{-1}$)	Diffusivity ($10^{-6}\ m^2\ s^{-1}$)
Quartz	7.8	4.1
Calcite	3.6	1.6
Forsterite	5.1	1.9
Plagioclase	1.9	0.9
Limestone	3.0	0.85
Sandstone	3.7	1.05
Granite	3.2	1.35
Basalt	1.9	1.0
Clay	2.3	0.95
Iron	75	21
Sea water	0.59	0.14
Air	0.025	20.4

Porosity, either water ($\kappa = 0.06$ W m^{-1} K^{-1}) or air ($\kappa = 0.025$ W m^{-1} K^{-1}) saturated, lowers the value in all cases, and crustal mineralogy is so diverse that a global average is best estimated from borehole data. A mean crustal value is 2.5 W m^{-1} K^{-1}. For the mantle, there are reliable estimates for prominent minerals, such as olivine, $\kappa \approx 6.0$ to 3.0 W m^{-1} K^{-1}, for the composition range of forsterite to fayalite. Others are generally in this range, allowing a global average estimate. For the uppermost mantle, $\kappa \approx 4.0$ W m^{-1} K^{-1}, but there is a general increase with depth and for the lowermost mantle, $\kappa \approx 7.0$ W m^{-1} K^{-1}.

Phonon conduction also occurs in metals but is obscured by a much stronger contribution by conduction electrons. The electron component of thermal conductivity, κ_e, is related to electrical conductivity, σ_e, by the Wiedemann–Franz rule:

$$\kappa_e = L\,\sigma_e T \qquad (21.32)$$

where the Lorenz number, $L = 2.443 \times 10^{-8}$ W Ω K^{-2}, is a universal coefficient and T is absolute temperature. Estimates of σ_e are easier than estimates of κ_e, so the thermal conductivity of the core is deduced from estimates of its electrical conductivity (Section 21.3). The contribution by lattice conduction (~ 2 W m^{-1} K^{-1}) is lost in the uncertainty of the total, 29 W m^{-1} K^{-1}.

There is no rigorous general solution to the problem of estimating rock conductivity from values for constituent minerals, and several empirical methods are used. The one with the most sophisticated mathematical pedigree is adopted from the analogous electrical problem, originating from the work of J. C. Maxwell in the 1800s. For the simplest case of a two component mix (A and B), this treats either spheres of A in a matrix of B or spheres of B in a matrix of A, according to which is more abundant. But this physical model is too far removed from the real structures of rocks to justify its mathematical complexity. There are straightforward solutions for plane layers of homogeneous materials with volume fractions (total thicknesses) V_1, V_2, V_3... and conductivities κ_1, κ_2, κ_3... analogous to electrical conduction in parallel or series resistors. Parallel to the layers, conductivities are added

$$\kappa_{/\!/} = (V_1\kappa_1 + V_2\kappa_2 + V_3\kappa_3 + \ldots)/(V_1 + V_2 + V_3 + \ldots) \qquad (21.33)$$

and perpendicular to the layers, their reciprocals (thermal resistivities) are added

$$1/\kappa_{\perp} = (V_1/\kappa_1 + V_2/\kappa_2 + V_3/\kappa_3 + \ldots)/(V_1 + V_2 + V_3 + \ldots) \qquad (21.34)$$

An obvious possibility is to take a suitable average of these special cases for a random mix of mineral grains, such as a weighted geometric average:

$$\kappa = \kappa_{/\!/}^{2/3} \times \kappa_{\perp}^{1/3} \qquad (21.35)$$

A favoured alternative is to take a weighted average of the square roots of the conductivities

$$\kappa = [(V_1\kappa_1^{1/2} + V_2\kappa_2^{1/2} + V_3\kappa_3^{1/2} + \ldots)/(V_1 + V_2 + V_3 + \ldots)]^2 \quad (21.36)$$

but, as with Equation 21.35, this gives values that are biased high, which is especially obvious if metal grains are included. For porous rocks, these equations are applied with pore volume, either water or air, as one of the constituents.

For time-dependent thermal conduction problems, it is often more convenient to use thermal diffusivity, η, than conductivity, κ. They are related by density, ρ, and specific heat, C_V:

$$\eta = \kappa/(\rho\, C_V)\ \mathrm{m^2\ s^{-1}} \quad (21.37)$$

Values of κ and η are listed in Table 21.13 for selected minerals, rocks and common materials.

21.3 ELECTRICAL PROPERTIES

21.3.1 Dielectric Constants and Molecular Polarity

When an insulating material, such as a mineral crystal or rock, is subjected to an electric field, its constituent electric charges (electrons and atomic nuclei) are slightly displaced, modifying the field. The material is said to be polarised, with the flux density of the field within it (termed the displacement field) enhanced by the factor k, the dielectric constant of the material. A particular reason for geophysical interest in k is that the speed of electromagnetic waves is proportional to $1/\sqrt{k}$ so that refraction and reflection of the microwaves of ground-penetrating radar are controlled by it. The dielectric constants of most pure, dry minerals are in the range of 2 to 10 (Table 21.14), but they are sensitive to impurities and, for polycrystals, to porosity. Moisture causes a sharp increase because for water, $k = 80$. Reported values of k for what are nominally the same geological materials often differ by 50% or more. Measurements are normally made at high frequencies (MHz to GHz). k decreases with frequency and is virtually impossible to measure with DC because even slight conductivity obscures the dielectric response.

In some materials, individual molecules have mutually displaced ionic charges, even in the absence of a field, but with their dipole moments randomly oriented, causing no bulk polarisation. These are referred to as polar materials. An imposed field causes partial alignment of the molecular dipoles, usually resulting in a higher value of k than the induced polarisation of non-polar materials. Water is a particular example, systematically biasing high the values of k in wet material, becoming dominant in very porous samples, such as sediments.

TABLE 21.14 TYPICAL VALUES OF DIELECTRIC CONSTANTS FOR REPRESENTATIVE MATERIALS

Material	Dielectric Constant
Diamond	4.6
Quartz	5
Corundum	9
Periclase	9
Olivine	7
Enstatite	8
Calcite	8
Halite	7
Dry granite	8
Dry basalt	12
Ice	4
Liquid water	80

There are two technically important special cases, ferroelectricity and piezoelectricity. A few polar materials, notably $BaTiO_3$, are ferroelectric, with molecular dipole moments that are aligned by an electric field and remain aligned when the field is removed. This produces electrets, electrical analogues of permanent magnets, with several laboratory and industrial applications. The 'ferro-' in ferroelectricity refers to this analogy to ferromagnetism and not to the presence of iron. Piezoelectricity has a very wide application in the quartz crystals of digital clocks and watches. The internal asymmetry of a quartz crystal causes electrical polarisation in a particular crystal direction when the crystal is stressed. Conversely, polarisation by an electric field causes deformation of the crystal. Quartz crystals are cut to have resonant frequencies of mechanical oscillation that are independent of temperature and, fitted with electrodes, act as stable frequency control elements of oscillatory electrical circuits. Manufactured discs, or stacked discs, of piezoelectric materials are used for electrically driven displacements in small instruments and in acoustic transmitters and receivers.

Searches for naturally occurring piezoelectric effects, such as electrical signals of seismic waves in quartz veins, have had little success because bulk quartz is not significantly piezoelectric. Quartz crystals occur as twins, with subtly different structures that have opposite piezoelectric polarities. The two types are equally common and, even in a hand sample of pure quartzite, they are so interwoven that any piezoelectric effect of the whole sample depends on the statistical chance of unequal abundances. Piezoelectricity of quartz is useful only with single crystals.

21.3.2 Electrical Conduction in Minerals, Rocks and the Mantle

In their pure forms almost all simple minerals are good insulators and conduction arises from impurities. Conductivities are very variable between samples and depend on treatment, increasing with temperature in the manner of semiconductors. Of the impurities, water (effectively hydrogen) has the biggest effect, but iron is also significant because it readily changes valency between Fe^{2+} and Fe^{3+}, making conductivity sensitive to the oxidation state. The strong temperature dependence of conductivity is represented by equations of the form

$$\sigma = A\exp(-E/kT) \qquad (21.38)$$

but data fits are uncertain and different values of parameters A and activation energy, E, may be fitted to different temperature ranges, or multiple terms of this form added. The very low conductivities at ambient temperature are too variable to be of interest and more useful is an indication of the temperature at which conductivity is high enough to be regarded as an intrinsic property of a mineral and not an artefact of its impurities (Table 21.15).

Being composed of grains of minerals with different properties and imperfectly fitting boundaries, rocks inevitably have some porosity, with electrical conductivity enhanced by moisture. Even for hard igneous rocks, prolonged vacuum drying is needed to reduce moisture to the level that conductivity is a property of the mineral mix and not a moisture effect. Very dry granite may have a conductivity below 10^{-10} S m^{-1} at laboratory temperature, but increasing with time when exposed to air. For basalt it is difficult to get below 10^{-8} S m^{-1}, probably because of the abundance of iron. Sediments are generally much more conducting.

Electromagnetic methods of probing the crust and mantle use a very wide range of frequencies. The highest frequencies, extending into the GHz range used in ground-penetrating radar, penetrate only a few metres, even in favourable situations, and have both dielectric and conducting responses. At lower frequencies the dielectric response is not significant and at the lowest frequencies the observations are magnetic rather than electrical and rely on currents induced in the Earth by magnetic variations. The lowest frequency that has been used is that of the 11-year cycle of solar activity, but those data are very

TABLE 21.15 A BRIEF LEAGUE TABLE OF THE TEMPERATURES AT WHICH THE CONDUCTIVITIES OF MINERALS ARE 1 mS m^{-1}

Mineral	T(K) for $\sigma = 10^{-6}$ S m^{-1}
Quartz, SiO_2	480
Corundum, Al_2O_3	1300
Periclase, MgO	1340
Forsterite, Mg_2SiO_4	1100
Orthopyroxene, $MgSiO_3$	1000
Calcite, $CaCO_3$	570
Halite, NaCl	700

Note: The numbers are a rough guide to scattered data.

uncertain. However, detailed work on the diurnal variation of the magnetic field has outlined the conductivity deep in the mantle. There is a shallow (crustal) layer of locally variable conductivity averaging nearly 0.1 S m^{-1}, where ground water has a strong influence, falling to about 0.01 S m^{-1} in the upper mantle but rising quite sharply at ~660 km depth to about 1 S m^{-1} in the lower mantle, without a further significant increase. Observations of the geomagnetic secular variation have suggested that there are small patches of higher conductivity at the base of the mantle but not approaching the core conductivity.

21.3.3 Electrical Conductivities of Iron and the Core

By the standards of metals, iron is not a particularly good conductor because it has overlapping bands of electron states with different properties. The highly mobile (4s) electrons are outnumbered by much less mobile (3d) electrons, and the 3d states scatter the 4s electrons. The conductivity is less than a tenth of that of copper, which has no free 3d states. This difference is important to the core and the geomagnetic field because a high electrical conductivity means a high thermal conductivity (the Wiedemann–Franz law, Equation 21.32) and a high conductive heat loss in the core. If the core had the conductivity of copper, its energy would be drained too completely to allow magnetic dynamo action. This is a trade-off in the operation of planetary dynamos: conductivity must be high enough for dynamo action but not high enough to kill that action by a conductive heat loss.

At laboratory pressure, the conductivity, σ, of liquid iron at its melting point is 7.4×10^5 S m^{-1}, with resistivity, $\rho_e = 1/\sigma$, increasing approximately linearly with temperature, but this does not lead directly to an estimate of core conductivity. For simple metals, the variation with pressure can be understood by recognising that resistivity is caused by scattering of conduction electrons by thermal vibrations, in effect by instantaneous thermal disorder of the crystal structure, and that this is closely similar to the thermal disorder of melting, so that resistivity is roughly constant on the melting curve. But this simple picture fails for iron because pressure affects the 4s and 3d states differently, reducing the number of active 4s electrons more strongly that it reduces the number of 3d states into which they can be scattered and decreasing conductivity relative to the model of a simple metal. There is an additional problem of impurity resistivity, electron scattering by the lattice distortions caused by impurity atoms, which increases with pressure. These complications have led to widely disparate estimates of core conductivity but an important constraint is imposed by the related thermal conductivity and consequent core heat loss. This is small enough to leave the core hotter than the mantle, a constraint not accommodated by some of the higher conductivity estimates. Values consistent with available data give a slight decrease with depth in the outer core from 2.7×10^5 S m^{-1} to 2.1×10^5 S m^{-1} and 2.7×10^5 S m^{-1} in the inner core.

21.4 MAGNETIC PROPERTIES

21.4.1 Spontaneous Magnetisation

Although a few pure minerals are either diamagnetic or paramagnetic, in rocks these properties are generally masked by the much stronger ferromagnetism or ferrimagnetism of iron-bearing minerals or impurities. The underlying physical difference is that in ferromagnetic materials, the intrinsic magnetic moments of electrons do not respond individually to a magnetic field, but act collectively, because they are coupled by what are known as exchange interactions. These interactions take several forms, illustrated in Figure 21.4. Except at very low temperatures, the energies of individual atomic moments in a field are small compared with ambient thermal energy, kT, and there is only a weak average alignment. In ferromagnetic materials, large groups of atomic moments are coupled into domains and the energies of the domains in a field are not small compared with kT, allowing strong alignment. The magnetic alignment within the domains, referred to as spontaneous magnetisation because it occurs without any external field, varies with temperature in the manner of Figure 21.5, disappearing at a Curie temperature, θ_C, characteristic of each material, above which thermal activation disrupts it. Spontaneous magnetisations at room temperature and Curie points of magnetic minerals are listed in Table 21.16.

Interaction type	Ferromagnetic	Antiferromagnetic	Ferrimagnetic	Canted antiferromagnetic
Examples	Fe, Co, Ni	NiO, MnO	Magnetite (Fe_3O_4)	Hematite (Fe_2O_3)
Atomic magnetic moments	↑↑↑↑↑↑	↓↑↓↑↓↑↓	↑↓↑↓↑↓	↗↘↗↘↗
Net spontaneous magnetisation	↑	Zero	↑	→

Figure 21.4 Alternative types of spontaneous alignment of magnetic moments of neighbouring atoms in crystals. Strictly, ferromagnetism is restricted to a few metals, but the word is commonly taken to include ferrimagnetism and canted antiferromagnetism, which are the important phenomena in minerals.

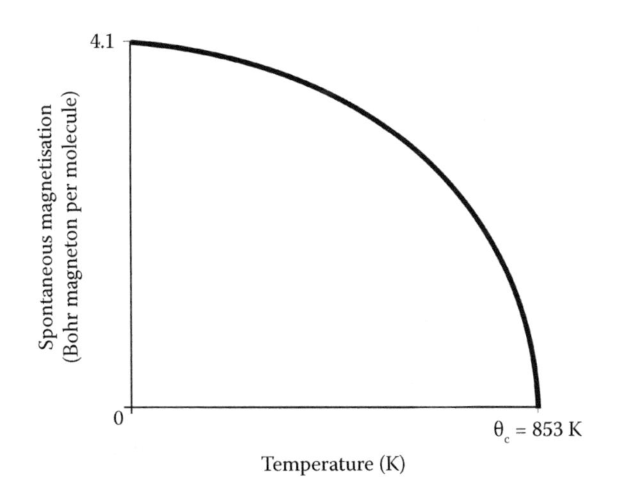

Figure 21.5 Temperature variation of spontaneous magnetisation of magnetite.

21.4.2 Susceptibility and Remanent Magnetism

The spontaneous magnetisations of minerals, as listed in Table 21.16, are their saturation magnetisations when exposed to strong magnetic fields. Although the magnetic domains are individually magnetised to saturation, except for the smallest mineral grains, in the absence of any field exposure the domains are self-arranged within the grains to minimise magnetic field energy. The magnetic flux follows closed loops within each grain, giving it zero magnetic moment. Magnetisation is caused by the application of a field which rearranges

TABLE 21.16 SPONTANEOUS MAGNETISATIONS AND CURIE TEMPERATURES OF THE MAJOR MAGNETIC MINERALS

Mineral	Formula	Density ($kg\ m^{-3}$)	Spontaneous Magnetisation at 0°C		Curie Point (°C)
			kA/m = emu/cm³	Bohr magnetons[a] per formula unit	
Magnetite	$Fe^{2+}Fe^{3+}{}_2O_4$	5200	448	3.8	580
Titanomagnetite	$Fe_{3-x}Ti_xO_4$	5200–0.430x	448(1–x)	3.8(1–x)	580–520x–240x^2
Maghemite	γ-$Fe^{3+}{}_2O_3$	5070	355	2.2	~640
Hematite	α-$Fe^{3+}{}_2O_3$	5270	2.1	0.024	675
Goethite	α-FeOOH	4260	~1	0.009	120
Pyrrhotite	Fe_7S_8 (often less magnetic $Fe_{1-x}S$)	4660	~60	~0.47 per Fe atom	320

[a] The Bohr magneton is the intrinsic magnetic moment of an electron (see Section 1.3).

the domains. If the field is weak the process is reversible. The weak magnetisation disappears when the field is removed and the susceptibility of a material is the ratio of magnetisation to field strength in this situation. The geomagnetic field is a weak field in the sense considered here. Its surface strength averages 24 A m^{-1} (0.3 oersted) at the equator and is twice as strong at the poles, but the corresponding mean field intensity, 41,000 nanotesla, is more often quoted (as in Table 5.5). Saturation of the magnetisation of a magnetite grain by the parallel alignment of all of its domains requires the application of a field several thousand times as strong. After removal from a saturating field, some magnetisation remains. This is the saturation remanence. Cancellation of that requires application of a reversed field, the coercive force. The characteristic properties of the dispersed magnetic grains in rocks, susceptibility and coercive force, are very dependent on grain size, impurities and various crystal imperfections. For a volume fraction f of magnetic mineral, typical values of susceptibility are $\chi = 0.5f$ for magnetite-bearing igneous rocks or $10^{-2}f$ for hematite-bearing sediments, increasing with grain size, and corresponding coercive forces H_C = 12,000 or 120,000 A m^{-1} (150 to 1500 oersted), decreasing with grain size.

Although, in the absence of other effects, domain structures are negligibly affected by fields comparable to that of the Earth, that is not so when heat or chemical changes are involved. At temperatures close to the Curie points, the domains adjust to equilibrium in whatever ambient field prevails. They are effectively paramagnetic (often called superparamagnetic), with magnetic alignment disappearing if the field is withdrawn. But when they are cooled in the field, they reach a temperature at which thermal activation becomes ineffective and domain adjustments are blocked by energy barriers caused by crystal heterogeneities and domain interactions. With further cooling the domain structure and grain magnetisation at the blocking temperature are retained, whatever field changes occur. The resulting magnetisation is referred to as TRM (thermoremanent magnetisation). It can be very stable and maintain indefinitely a record of the field at the time of cooling. This is basic to palaeomagnetic studies of the history of the geomagnetic field. In spite of complications, such as domain interactions and a range of blocking temperatures, elaborated by numerous authors (e.g. Dunlop and Özdemir 1997), TRM gives a reliable indication of the direction of the field. Estimates of its strength require more care but are also routinely obtained. For rocks in which the dominant mineral is magnetite, the magnetisation is normally TRM and the finest grains are preferred for maximum reliability of palaeomagnetic inferences.

In many cases, magnetic minerals are chemically altered after formation and cooling or are even formed in a field. This is common in the case of hematite, which results from oxidation of magnetite, via maghemite or via hydrated minerals such as goethite (Table 21.16). Then the resulting magnetisation is termed CRM (chemical remanent magnetisation), which is generally slightly weaker than TRM but is just as stable and can be used in the same way. A third mechanism by which rocks acquire natural remanence is in the sedimentary deposition of magnetic grains from eroded igneous rocks, retaining a slight magnetic alignment by the field as they are deposited. The resulting DRM (depositional, or detrital, remanent magnetisation) may give an inclination error in its indication of the field direction, caused by the rolling of grains into their final positions, but this is often not observed. A post-depositional alignment has been postulated, but probably more important are chemical changes, so that the magnetisation becomes CRM. The intensity of magnetisation is also a clue. Typically, in a rock with a fraction f of magnetic mineral, CRM $\approx 2f\,\mu\text{T}$, but DRM is weaker by a factor of 10 or so. This compares with TRM, which is commonly $\sim 5f\,\mu\text{T}$, but may be substantially stronger for the finest grains.

21.4.3 Self-reversing Remanence

Recognition that rocks with remanence opposite to the direction of the geomagnetic field were common preceded wide acceptance of field reversals, prompting a search for mechanisms by which remanence could be opposite to the field that induced it. That this was possible was demonstrated by the discovery of a Japanese rock (a dacite from Mt. Haruna) which acquired reverse thermoremanence in laboratory experiments. It was shown to be a property of a limited range of ilmeno-haematite minerals in which exsolution produces disordered intergrowths of phases with electron spins coupled negatively to one another, so that a developing phase is magnetised oppositely to the mineral in which it is growing and may become magnetically stronger. In the case of Haruna dacite, the reversed remanence was a property of an intermediate partially ordered state of the mineral and not either the original or final fully ordered states. Similar behaviour has been seen in a few specially selected materials, but is certainly rare in rocks, leaving no doubt that the field reverses. The possibility remains that partial reversals of components in rocks with complex mineralogy occur and may need to be considered in palaeointensity studies.

21.5 OPTICAL PROPERTIES

Only a few minerals, the strongly magnetic ones such as magnetite and those with significant electrical conduction, especially sulphides, are optically opaque. Most are sufficiently transparent to visible light to permit identification by their refractive behaviour, which depends on crystal structure. In the simplest cases, minerals are optically isotropic, with light propagation independent of orientation relative to crystal axes. In these cases, refraction is described quantitatively by a single number, the refractive index, μ (Table 21.17). This is defined as the ratio c/v [the speed of light in vacuum, c, to the speed in the medium, v, which is the same in all directions although it is slightly dependent on the colour (wavelength, λ) of the light]. Cubic minerals are optically (although not elastically) isotropic, as are amorphous materials such as glass, including fused silica.

Refractive index is a measure of the inertia imposed on light propagation by the electrons in a material. The oscillating electric field of a light wave causes the electrons to oscillate with it, slowing it down by an amount that depends on how the electrons respond. In symmetrical structures, such as cubic crystals, they respond in the same way for all orientations of the field. In asymmetrical structures, the electrons are bound to the atomic nuclei in ways that depend on the atomic

TABLE 21.17 ISOTROPIC REFRACTIVE INDICES OF SELECTED MINERALS IN SODIUM LIGHT, AN APPROXIMATE AVERAGE FOR THE VISIBLE SPECTRUM

Material	Refractive Index, μ
Diamond, C	2.419
Periclase, MgO	1.74
Halite, NaCl	1.54

arrangement and their response to an alternating field varies with its orientation. This causes birefringence, or double refraction, in which light travels at different speeds according to the orientation of its electric field (polarisation direction). Most sources produce light that is unpolarised, with electric fields in all directions, but polarised, or plane polarised, light with an electric field in one direction only is used for optical examinations of rocks and minerals.

Many minerals have uniaxial optical anisotropy, with properties in one direction (optic axis) different from those in the perpendicular plane. For all polarisations of light travelling parallel to the optic axis, the speed is the same because they all have electric fields in the perpendicular plane and the electrons respond in the same way. For light travelling perpendicular to the axis, there are two polarisations with different speeds. For one, the electric field is also perpendicular to the axis, as for light in the direction of the axis, so its speed and corresponding refractive index are the same. This is termed the ordinary wave and is assigned a refractive index n_0. The other polarisation, with an electric field parallel to the optic axis, travels at a different speed and is termed the extraordinary wave, assigned refractive index n_e, which may be either bigger or smaller than n_0. The ordinary wave travels at the same speed in all directions and follows the familiar refractive laws for isotropic materials, but the extraordinary wave does not. n_0 and n_e are the principal refractive indices and the difference $(n_e - n_0)$ is referred to as the birefringence (positive or negative) of a material. In general, with light travelling in an arbitrary direction with respect to the optic axis, there are two waves, with perpendicular polarisations: the ordinary wave, with refractive index n_0, and the extraordinary wave, for which the index is between n_0 and n_e, according to its direction with respect to the axis. Table 21.18 gives examples of minerals in the three symmetry classes that give uniaxial birefringence: hexagonal, tetragonal and trigonal, with 6-fold, 4-fold and 3-fold symmetry, respectively, in the perpendicular plane.

The carbonates are notable for strong birefringence, calcite being of particular interest. It is available in high-quality, clear crystals (Iceland spar) and is widely used in optical instruments, cut in a way that refracts one of the polarisations of incident light out of the optical system, producing the most nearly perfect polarised light. A cheaper option is Polaroid sheeting, which makes use of dichroism, a property of certain birefringent crystals that absorb one of the polarisations. Microcrystallites of dichroic material are embedded in plastic with aligned optic axes.

Some minerals are more asymmetrical than the uniaxial ones in Table 21.18 and are referred to as biaxial, because double refraction is not observed for light travelling in the directions of two different crystal axes (as for the single optic axes of uniaxial crystals). Table 21.19 gives examples of minerals in the three symmetry classes with biaxial optical properties, which could be called

TABLE 21.18 PRINCIPAL REFRACTIVE INDICES OF SOME MINERALS WITH UNIAXIAL BIREFRINGENCE

Symmetry Class	Material	n_o	n_e	Birefringence, $(n_e - n_o)$
Hexagonal	Ice, H_2O	1.309	1.313	+0.004
	Wustite, ZnS	2.356	2.378	+0.022
Tetragonal	Rutile, TiO_2	2.616	2.903	+0.287
	Zircon, $ZnSiO_4$	1.943	1.997	+0.054
Trigonal	Quartz, SiO_2	1.544	1.553	+0.009
	Calcite, $CaCO_3$	1.658	1.487	−0.172
	Corundum, Al_2O_3	1.769	1.761	−0.008

TABLE 21.19 EXAMPLES OF MINERALS IN THE THREE OPTICALLY BIAXIAL SYMMETRY CLASSES, WITH TYPICAL OR AVERAGE REFRACTIVE INDICES

Symmetry Class	Mineral	n_α	n_β	n_γ	Birefringence, $\delta = (n_\gamma - n_\alpha)$
Orthorhombic	Enstatite, $MgSiO_3$	1.659	1.662	1.669	0.010
	Aragonite, $CaCO_3$	1.529	1.681	1.685	0.156
Monoclinic	Gypsum, $CaSO_4.2H_2O$	1.520	1.522	1.529	0.010
Triclinic	Albite, $NaAlSi_3O_8$	1.530	1.534	1.540	0.010

trirefringent because they have three different refractive indices. The symmetry classes are orthorhombic (with three unequal axes, all mutually perpendicular), monoclinic (two unequal mutually perpendicular axes with the third one misaligned) and triclinic (three unequal axes, none mutually perpendicular).

21.5.1 Thin Sections

The birefringent properties of minerals are used to identify them in rock samples prepared as thin sections on microscope slides, embedded in a medium of standard refractive index, 1.54, and ground to a nominal thickness of 30 μm

(0.03 mm). The significance of this thickness is that it is approximately 100 wavelengths of light in a mineral of typical refractive index 1.7. (Light with 0.5 μm vacuum wavelength, in the middle of the visible spectrum, has a wavelength of 0.5/1.7 = 0.29 μm in the mineral). For material of birefringence $|n_e - n_o| = 0.01$, the 30-μm thickness gives a maximum (according to crystal orientation) of one wavelength difference between the o and e rays.

The simplest observation of mineral birefringence uses a microscope with a polarising filter but no analyser to observe colour changes when a specimen is rotated in plane polarised light. This is pleochroism, which depends on mild, colour-dependent dichroism to preferentially absorb one polarisation. An interesting pleochroic effect is observed with specimens containing grains of radioactive material, which cause radiation damage in surrounding crystals, modifying the optical properties in the damaged zones. The effect is observed as pleochroic halos, coloured rings around the radioactive grains.

For most thin section work, a petrographic microscope is fitted with two polarisers, one below the specimen slide and the other above the objective lens, that are normally kept crossed, that is, transmitting only light polarised in directions perpendicular to one another. With nothing, or only isotropic material, between them, no light passes through and the field of view is black. Colours appear when birefringent materials are inserted. On specimen slides, the minerals occur at all angles and the strength of the colours arising from interference of o and e waves depends on crystal orientations. But intensities of individual colours change as specimens are rotated. The angular difference between maximum and minimum intensities is characteristic of each mineral, referred to as the extinction angle and used for mineral identification. Four extinctions are observed per 360° specimen rotation because there are four directions perpendicular to the axes of either the polariser or the analyser.

These techniques are irrelevant to examinations of opaque minerals, for which optical reflection methods are employed. A particular application to coal is considered in the footnote to Table 24.4.

Section VII

Resources

Chapter 22

Elements as Resources

22.1 A SURVEY OF THE ELEMENTS: THEIR OCCURRENCE, PROPERTIES AND USES

Naturally occurring elements are listed by atomic number, with important properties, mineral sources, uses and biological roles. Intermediate radioactive daughters that occur only in the decay chains of thorium and uranium are not included. A note on common features of rare earth elements follows element 56.

1. Hydrogen (H). By far the most abundant element in the universe (and the Sun), hydrogen constitutes only 0.05% of the mass of the Earth, where it occurs primarily as water (H_2O), and in hydrocarbons, especially methane (CH_4). It is one of the essential constituents of biological materials. Although water, as a crucial resource, comes under the heading of hydrogen in this chapter, it features separately as freshwater in Chapter 11, and as saline water in the oceans and salt lakes (Chapters 9 and 10). Hydrogen is a major component of the fossil fuels, which are categorised here as carbon, and discussed in Chapter 24.

If purity is not critical, then extraction of hydrogen from natural gas (methane and the like) or oil is the most economic method and accounts for 80% of the total global production (120 million tonnes/year), with extraction from coal providing 16%. The major uses are in oil refineries and in the production of ammonia for manufacturing fertiliser. More specialised uses require pure hydrogen, which is extracted from water by electrolysis but accounts for only 4% of total production. The energy requirement is the only limitation in this process. Hydrogen cannot be regarded as an energy source; it yields, or stores, less energy than is used to produce it by any method.

2. Helium (He). The second most abundant element in the universe (and Sun), but rare in the Earth, where it is almost entirely a decay product of uranium, thorium and their radioactive daughters. It is a very volatile and diffusive, chemically inert gas that accreted only as a very minor trace in the Earth.

It leaks to the atmosphere from its radioactive sources but escapes to space, leaving an equilibrium atmospheric concentration of 5.2 parts per million. It is non-toxic with no role in biology.

Commercial production of helium, originally in USA but now also in several other countries, is from natural gas, in which it has been trapped after diffusing from deeper sources. It is also found in interesting concentrations, but limited volumes, in volcanic hot springs, notably in the East Africa Rift Valley. Global production, about 29,000 tonnes/year, satisfies the present demand, although it was in short supply only a few years ago, and erratic prices can be problematic. The principal uses are cryogenic. The boiling point, below which it is liquid at atmospheric pressure, is 4.2 K, low enough to maintain superconductivity in the magnet coils of magnetic resonance imaging (MRI) machines and other specialised instruments. Being an electrical insulator with a high thermal conductivity, it is used in high power electrical machinery and as a controlled atmosphere for particular welding applications. It is used with oxygen in deep diving breathing apparatus, and is favoured for party balloons.

3. Lithium (Li). As an element, Li is a soft, silvery metal of density 534 kg m^{-3} and has a melting point of 180.5°C. It is highly reactive and is stored under light oil, although it floats and containers must be sealed. Although widely distributed in a range of minerals, it is not abundant because nuclear synthetic processes favour more tightly bound light nuclei, especially ^{4}He and ^{12}C. It is found only in low concentrations (e.g. 6 ppm of sea water solutes), and resource exhaustion will become a significant threat if the demand for lithium batteries increases as rapidly as appears possible. Also, extraction imposes a high energy demand. Commercial production, mainly from brines and salt pans in Chile, Argentina and Australia, is about 35,000 tonnes per year. Other major uses are as oxide in the manufacture of glass and glazes and in lubricating grease. It has no obvious biological role, although it appears to be botanically active and has a minor medical application in the treatment of depression.

4. Beryllium (Be). A light, structurally strong and dimensionally stable hard, grey metal of density 1850 kg m^{-3} and a melting point of 1287°C. It is the rarest of the light elements and exists only as the isotope ^{9}Be. Nuclear processes that might have been expected to yield ^{8}Be, produce pairs of ^{4}He nuclei, explaining the rarity of Be. It is widely dispersed in a variety of minerals, but exploitable concentrations are restricted to few minerals and locations. The element name is derived from the most widely recognised mineral source, beryl, which occurs in a few places as gem-quality emerald. Exploitable ores are estimated to be a few hundred thousand tonnes, but extraction is limited to few locations (USA, China, Kazakhstan). Both extraction and manufacturing processes using beryllium are difficult (expensive), partly because Be is seriously toxic, especially as inhaled dust, requiring facilities with appropriate precautions.

Properties of beryllium that give it special uses are its combination of lightness and strength, high rigidity (very low Poisson's ratio), low thermal expansion coefficient (at ambient temperature), diamagnetic (virtually non-magnetic) susceptibility and low atomic number (transparency to X-rays and particle radiation). It is the universal window material for X-ray machines and is used for critical components of nuclear facilities and aerospace applications. Even as a minor alloying component, Be gives enhanced mechanical properties to other metals, in particular aluminium and copper. Beryllium copper alloys are used for small springs.

5. Boron (B). Like the other elements between helium and carbon, boron is cosmically rare and in the Earth as a whole it is no more than 0.4 ppm of the total mass. It is more concentrated in the crust, where the development of exploitable deposits has resulted from the solubility of compounds, particularly borax, in water. Sea water contains 4.4 ppm of boron and the only common rock type with a significant concentration is andesite, which is a product of volcanism of subducted sea-floor basalt with an infusion of sea water. As an element, boron is a hard, black crystalline solid of density 2080 kg m^{-3}, with a low thermal expansion coefficient, and a melting point of 2076°C. It is not an electrical conductor but is not a good insulator either. Its chemical bonding behaviour has some similarity to that of carbon, allowing multiple covalent bonds between networks of B atoms, with links to various other elements. It is difficult to isolate elemental boron, but most uses are of compounds, for which separation is not needed. Boron is found as a minor ingredient of numerous minerals but not generally in useful concentrations. The most important is borax, sodium tetraborate, which is found in hydrated forms, especially the decahydrate, $Na_2B_4O_7.10H_2O$. It is extracted from evaporite deposits, the major sources being in eastern Turkey and the Mojave Desert of California (near the town of Boron).

Borax or the oxide, B_2O_3, is used in ceramics, the fibres in fibreglass and in mechanically and thermally robust borosilicate glass. Boron carbide and nitride are abrasives. Boric acid, H_3BO_3, is used as an insecticide but is not noticeably toxic to mammals. Boron is needed as a trace element in soil for plant health and may be applied as a fertiliser where a deficiency is identified.

6. Carbon (C). Cosmically abundant and the fourth most abundant element in the Sun (after H, He and O), but not a major component of the Earth because the volatility of its compounds with other common elements, CO, CO_2 and CH_4, inhibited accretion from the nebula. Nevertheless, it is abundant enough to be a crucial component of the environment and it is the central element in biological materials. The total terrestrial abundance may be much greater than is obvious because processes of terrestrial accretion and evolution have not concentrated it in the crust, in which it has accumulated from atmospheric

CO_2 as carbonate rocks (7×10^{19} kg) and reduced (organic) carbon (1.6×10^{19} kg). The mantle content may be five times as much and Table 6.4 assigns more than 100 times as much to the core. The crustal organic carbon is assessed as a resource in Chapter 24.

Elemental carbon has diverse forms with remarkably different properties. The two most familiar forms are diamond and graphite. Diamond is a transparent crystal of density 3515 kg m^{-3}, colourless in its pure form, an excellent electrical insulator and a very good thermal conductor. It has a cubic crystal structure, with a less common hexagonal alternative, and in both cases each atom is tetrahedrally bonded to four nearest neighbours. It is the hardest known material but with a rumoured potential challenge from a form of boron nitride. Graphite could hardly be more different. It is opaque black, with a density of 2267 kg m^{-3}, consisting of layers of atoms bonded in a hexagonal pattern, making it very anisotropic, being an electrical conductor in the plane of layering, and with resistance to shearing between layers so weak that it is soft enough to be used for lubrication. Amorphous carbon is essentially very finely divided graphite. Graphite is the structure in thermodynamic equilibrium under ambient conditions, and diamond can be produced from it at high pressure (>5 GPa) and high temperature, but both are stable and resistant to chemical attack, over a wide range of conditions. Diamond can also be produced by condensation from vapour and is occasionally found in meteorites, which could never have been subjected to the pressure needed for thermodynamic equilibrium. There are some variants of the graphite structure, all basically graphene, a one-atom-thick hexagonally bonded layer of carbon atoms, with interesting properties and potential applications, now under intensive investigation. The most closely studied are nanotubes, graphene sheets rolled into tubes of various sizes, that may be arranged concentrically, with properties that differ between them. One interesting property is mechanical strength, apparently a consequence of the fact that an atomic monolayer cannot accommodate the crystal defects (dislocations) that limit the strength of three-dimensional crystals. Fullerenes (buckyballs) are spherical or ellipsoidal graphene structures, resembling the geodesic domes pioneered by Buckminster Fuller. Carbon does not melt, but sublimates at 3642°C, surviving as a solid to much higher temperatures than any other element.

Both diamond and graphite occur naturally in exploitable deposits, although in very small quantities relative to the carbon accessible as hydrocarbons in coal, oil and natural gas. Diamonds are the rare form because they have been brought to the surface from deep sources in localised eruptions of kimberlite or komatiite magma that occurred too quickly for the diamonds to degrade to graphite. Diamonds are also found in placer deposits, washed out of eroded kimberlite and komatiite. The global total mass of diamonds may be as much as

1000 tonnes, 98% being of 'industrial' quality, suitable only for specialised cutting and drilling tools. Some deposits include gem-quality diamonds, which are used not only in jewellery but also in diamond anvil cells for extreme pressure experiments. They are uniquely suited to this role not just for their strength but also transparency to both light and X-rays, electrical insulation and high thermal conductivity. The industrial-grade diamonds are now dominated by 'artificial' diamonds, manufactured by exposing graphite to high pressure and temperature. Higher quality diamonds are also produced industrially, but by the vapour deposition process.

Graphite occurs naturally as the highest grade of organic carbon, from which other elements, especially hydrogen, have been expelled by heat and pressure. Anthracite, the highest grade of coal used as fuel, is more than 90% carbon and comes close to being described as rough graphite. Flakes of graphite are commonly found in metamorphic rocks. Exploitable deposits of high-quality graphite are not common but occur in several places, particularly China. Its use in the manufacture of pencils is decreasing but as an electrical conductor it is the preferred material for contact 'brushes' in electric motors and electrodes in several types of battery. In fine flakes, it is a lubricant and a component of vehicle brake linings.

Compounds of carbon are numerous and diverse. It is the central element of biological materials and the whole field of organic chemistry. Many useful materials are derived directly from plants or trees, and many more from coal, gas and, especially, oil. All these materials can be deemed products of atmospheric CO_2, fixed by photosynthesis and processed in myriad ways. CO_2 dominates the atmospheres of Venus and Mars and the total amount in the Earth's early atmosphere must have been comparable to that now on Venus (90 times our atmospheric mass) to account for the organic carbon and carbonate rocks in the crust. Although our atmosphere now contains only 400 ppm of CO_2 (increasing by 2.2 ppm/year), this much is cycled through the biosphere in less than 20 years. It is the ultimate parent material for everything organic.

7. Nitrogen (N). Cosmically, this is a common element and in the Sun it is more abundant than silicon, but its volatility limited its accretion in the inner planets. It is 78% of the atmosphere but is otherwise rare in the Earth. Nitrogen is trivalent and occurs as di-nitrogen, N_2, with a very strong triple bond between the component atoms. This bond must be broken to allow nitrogen to combine with any other element, and if the bonds to N atoms in a compound are weaker than the N_2 bond, the compound is likely to be unstable and break up, allowing the establishment of N_2 bonds. This is the situation for familiar explosive materials, nitroglycerin (the essential constituent of dynamite), trinitrotoluene and ammonium nitrate, with explosive power a consequence of the fact that the nitrogen is produced as gas. The energy of the N_2 bond presents a difficulty for the production of nitrogen compounds, both by natural processes and industrially, so that, in many situations, including planetary accretion, nitrogen has the character of an inert gas. Although its terrestrial abundance is modest, it is a crucial feature of the environment as it is an essential ingredient of all forms of biological life.

The nitrogen in the Earth's crust (~10^{18} kg, a quarter of the atmospheric content) is almost entirely of organic origin. Plants incorporate it in their structures but release it to the atmosphere when they decay so that continuous regeneration is necessary, but a small fraction survives and is buried along with fossil carbon. The nitrogen in the soil of any area exerts an important control on its biological productivity. Nitrogen fixation is the term applied to processes that convert it from a gas to other chemical forms. Several natural processes do this. Some nitrate is washed into the crust from the atmosphere by production of nitric oxide (NO) in lightning strikes and by UV radiation in the stratospheric ozone layer, and this was possibly a trigger for the beginning of biological activity. The development of nitrogen-fixing bacteria several billion years ago increased nitrate production and consequent soil fertility and continues to do so. Some cyanobacteria (blue–green algae), which are notable for photosynthesis, also fix nitrogen and live in colonies evident as stromatolites, which are found very early in the geological record. Several forms still exist but their total effect remains small and more significant now are the nitrogen-fixing bacteria in the root nodules of leguminous plants. Soil nitrogen must be continuously maintained because decay of vegetation releases it back to the atmosphere. In many parts of the world, where soil fertility is limited by low nitrogen, crop production is increased by the use of nitrogen fertilisers. Originally, these were potassium and sodium nitrates extracted from evaporite deposits, but nitrates are now produced industrially. The first step is the production of ammonia (NH_3) by heating a compressed mixture of atmospheric nitrogen and hydrogen from natural gas (the Haber process). Global production of nitrogen fertiliser is about 500 million tonnes/year. Iron and molybdenum are used as catalysts in the Haber process, and it appears not to be incidental that these elements are found in the root nodules of legumes. The exhausts of vehicles and aircraft are an incidental modern source of fixed nitrogen, with oxides of nitrogen regarded as a pollutant. There is no comparable industrial solution to another major limitation on soil fertility: phosphorous deficiency.

Nitrogen is extracted from air by fractional distillation and, being the largest fraction of air, it is less valuable than the oxygen and argon. It is readily available and widely distributed as liquid nitrogen, which can be stored and transported in vacuum flasks at its boiling point (−196°C, 77 K). It is a convenient, safely inert refrigerant used in laboratories and medical surgeries where it is routinely used to freeze skin abnormalities. As a gas, the inertness makes nitrogen suitable as a controlled atmosphere for large electrical machinery that may involve sparking.

8. Oxygen (O). This is the third most abundant element cosmically and in the Sun (after H, He), and it is the most abundant in the Earth (Table 4.7). It occurs in all parts of the Earth: by mass nearly 50% of the mantle and crust (Tables 7.3 and 8.3), an important component of the outer core (Table 6.4), almost 90% of the oceans and 23% of the atmosphere. It is a major constituent of living material and is rapidly exchanged with the atmosphere (or with gas dissolved in oceans or fresh water) by plants as well as animals that need to breathe it continuously. As with nitrogen (and hydrogen, fluorine, chlorine), gaseous oxygen is diatomic, O_2, but, unlike nitrogen, it is very reactive. If the atmospheric content were not maintained by photosynthesis and by the loss to space of hydrogen from water, it would disappear in about 10 million years. The Earth has had an oxygen-rich atmosphere only for about 500 million years, the most recent 10% of its existence. This is a very special situation, unique in the solar system and probably cosmically rare. Oxygen readily combines with most other elements and most of the Earth's oxygen occurs as mixed oxides, which constitute the mantle and crust. But, in spite of the overall abundance of oxygen, almost all of the Earth is in a reduced (oxygen-deprived) state, and geological processes consume oxygen by the weathering of igneous rocks and oxidation of volcanic gases.

The most widely used method of producing oxygen is fractional distillation of liquefied air. Although a large fraction of the cost arises from the energy required to cool it, much of the distribution is of liquid oxygen at its boiling point, –183°C, and this applies also to nitrogen, which is a product of the same industrial process. Uses are very diverse with steel production dominant. Oxy-acetylene torches for welding and steel cutting and fabrication and numerous products of the petrochemical industry depend on oxygen, but a more familiar and widespread use is in medical facilities, where it is kept on hand for managing impaired breathing.

9. Fluorine (F). The lightest of the halogens, as an element this is a diatomic gas with a boiling point of 85 K, but it is extremely reactive and combines with almost anything within range, making it dangerous to handle. Although not abundant, cosmically or terrestrially, it is not rare, but its extreme reactivity has made it a minor component of a wide range of minerals, from which it is difficult to extract. Unlike the other halogens, its compounds are not water soluble and it is not found in evaporites. The only current economic source is fluorite, CaF_2, from China and Mexico. A much larger source could be fluoroapatite, $Ca_5(PO_4)_3F$, but the fraction of fluorine makes that uneconomic, and it is used instead as a phosphate fertiliser. A third mineral, cryolite, Na_3AlF_6, is of interest for its own sake, but it is not abundant enough to be used to produce fluorine.

The most widely known fluorine product is polytetrafluoroethylene, $(C_2F_4)_n$, a remarkably stable white plastic, commonly referred to by the trade name Teflon. The properties that give it a range of uses include resistance to all water-based and other materials and a low frictional coefficient in solid–solid contacts, as in 'non-stick' cookware and bearing surfaces. Also, it has a high electrical resistivity and the excellent dielectric properties needed for cable insulation, especially at high frequencies. Two hexafluorides with industrial uses are dangerous and require very careful handling: SF_6 is a liquid insulator used in high-voltage transformers and UF_6 is used in the separation of uranium isotopes. Fluorine compounds are used in some metallurgical processing of steel and aluminium. Fluorocarbons, referred to by the collective name freons, are standard refrigerant gases that are being progressively phased out because of their damage to stratospheric ozone. Those incorporating chlorine, being the worst offenders, are already disallowed, but fluorine appears to be a component of most potential replacements. Fluorine has no known major biological role and is toxic in large doses, but is added to toothpaste and, in many places, to drinking water, because it has been found to reduce the incidence of tooth decay.

10. Neon (Ne). A cosmically abundant, completely inert 'noble' gas with a boiling point of 27 K. Being extremely volatile, it accreted only as a trace in the Earth and is a very minor component of the atmosphere, from which it is extracted for use. Unlike helium and argon, it is not a product of any radioactive decay in the Earth and is consequently much rarer. The characteristic strong red glow from neon excited by an electric discharge in the neon-filled tubes of advertising signs and in He–Ne lasers provides its only uses. Neon has no biological role.

11. Sodium (Na). This is a common element, widely distributed in the crust in a range of minerals and effectively mobile because of the solubility of its salts, especially common salt, NaCl. It is a major solute in the oceans and inland salt lakes and has accumulated in evaporite deposits. As an isolated element, sodium is a very reactive, soft and silvery alkali metal, of density 968 kg m^{-3} and a melting point of 97.8°C, which must be kept under oil or inert gas as it oxidises quickly in air and reacts vigorously with water. The metal is produced by electrolysis of liquid sodium chloride, but that is not necessary for production of most of the useful materials that incorporate it. A basic industrial starting point for production of commercial compounds of sodium, known as the chloralkali process, electrolyses brine (NaCl solution) to yield the hydroxide, NaOH. This is familiar as caustic soda and also known as lye. Chlorine and hydrogen are additional products of the same process. Caustic soda is correctly named and must be handled with care but has many uses, including the cleaning of things affected by persistent grease (ovens and drains) and the manufacture of soap. Compounds produced directly from it include the bicarbonate, $NaHCO_3$ (baking soda), and oxide used in glass manufacture.

Although the production and use of its compounds greatly exceeds the use of sodium metal, there are a few uses of the element. An obvious one to many city dwellers is in street lighting, because sodium vapour lamps, producing the characteristic orange light associated with sodium, are particularly efficient in converting electrical energy to light. A specialised use is of the alloy with potassium, colloquially referred to as NaK, which is liquid at room temperature for compositions close to the eutectic (77% K, for which the melting point is −12.8°C). As a cooling fluid in facilities such as nuclear reactors, it has the advantage of remaining fluid when cold, but the disadvantage of being dangerous to handle.

Sodium is one of the essential biological elements, usually ingested as salt where this is available. Many herbivores actively seek salt deposits. A balance with potassium is required.

12. Magnesium (Mg). This is one of the very abundant elements in the Earth, but much less concentrated in the crust than in the mantle (see Tables 8.3 and 7.3). It is nevertheless common in the crust as a constituent of numerous minerals and as a solute in the oceans, in which its abundance is 12% of that of sodium. In the Dead Sea, its abundance exceeds that of sodium. The favoured mineral for extraction of magnesium is magnesite, the carbonate $MgCO_3$, which when 'burned' yields the oxide, MgO. This is familiar in crystalline form as the mineral periclase. It has several uses, including furnace lining, and is the starting point for production of metal. China currently dominates world production.

As an element, magnesium is a grey–white metal of low density, 1740 kg m^{-3}, and a melting point of 650°C. It is very reactive, but in air it rapidly forms a thin but hard protective layer of oxide. Once ignited, it burns with an intense white light, useful in flares, and is difficult to extinguish, continuing to burn even in nitrogen or carbon dioxide. The pure metal is moderately soft and its resistance to deformation is enhanced by adding traces of other elements such as scandium or gadolinium, but these are rare and expensive and most uses of magnesium as metal are in alloys with aluminium.

Magnesium has a biological role for animals as well as plants, for which it is a central element in chlorophyll. Magnesium deficiency is sometimes identified in agriculture and horticulture and corrected by application of Mg-rich fertiliser. The hydrated sulphate, Epsom salts, $MgSO_4 \cdot 7H_2O$, has several medical applications in the treatment of cardiac arrest, asthma and as a laxative/purgative.

13. Aluminium (Al). Although it is much less abundant globally than magnesium, with which it shares both physical properties and uses, aluminium has been relatively concentrated in the crust, where it is the third most abundant element. It is very reactive and extraction makes a high energy demand. In spite of its wide distribution in numerous minerals, the only one used for extraction is bauxite, a mixed oxide/hydroxide of somewhat variable composition conventionally represented by $AlO_X(OH)_{3-2X}$. This occurs in laterites, weathering products of several rock types from which more soluble minerals and readily dislodged grains have been removed by prolonged exposure to tropical rainfall. Several countries have substantial bauxite deposits, Australia being a major exporter. An alternative representation of the bauxite formula, $\frac{1}{2}[Al_2O_3.(H_2O)_{3-2X}]$, shows that it is reduced to the oxide, alumina, by driving off water. Alumina is the starting point for manufacture of all aluminium products, including metal, and has several uses in itself. It occurs naturally as the mineral corundum, which is very hard and is used as an abrasive. Gem-quality stones, coloured by other elements, are ruby and sapphire.

Aluminium metal is produced by electrolysis of molten alumina and, for the highest purity, as required for electrical cables, a second stage of electrolysis is used, so that energy represents a large part of the cost. It is a soft, silver–white metal with a density of 2700 kg m^{-3} and a melting point of 660°C. Physical properties that make the high purity form of interest are an electrical conductivity of 36.5×10^6 S m^{-1} (resistivity 0.028 $\mu\Omega$ m) and thermal conductivity of 237 W $m^{-1}K^{-1}$. Aluminium is very reactive and will burn fiercely, but in air it is protected by a very hard, thin layer of oxide, making it corrosion resistant. Alloying with magnesium gives it the mechanical strength needed for many applications, such as aircraft construction, for which the combination of lightness and corrosion resistance gives it a particular advantage. The cables of overhead transmission lines use pure aluminium, with a steel core for strength. Although not as good a conductor as copper, it is nevertheless still a very good conductor and much lighter than copper.

Aluminium has no known biological function, but in acid soils it inhibits the growth of plants that are not specifically tolerant to it. It is widely used for cooking utensils and food containers, being understood to have negligible toxicity to humans, although some reports have indicated that it is a risk factor for Alzheimer's. The recycling of drink cans stimulated the recycling of aluminium more generally so that about half of the production is now from recycled material, a major energy saving factor.

14. Silicon (Si). Apart from oxygen, this is the most abundant element in the crust. Many of the common minerals are silicates, that is, composites of silica, SiO_2, and oxides of other elements. Silica itself is an abundant mineral, well known as quartz. It is a constituent of granite and found widely in placer deposits, such as beach sands derived from eroded granite. Silicon bonds very strongly to oxygen, from which complete separation is a difficult, energy-intensive process, but most of the bulk uses of minerals containing silicon (rock, gravel, sand, kaolin) require little or no processing. Si-O bonds have a strong preference for a tetrahedral structure, but this is slow to establish when silica is cooled through its melting point from the disordered, liquid arrangement of atoms, with the result that it is readily supercooled to a liquid-like glass structure. This has made silica the major component of most kinds of glass and ceramics, and in its pure form fused silica is a glass with properties that give it a range of special uses: a very low thermal expansion coefficient and almost lossless elastic cycling.

Elemental silicon is a brittle, grey semiconductor of density 2330 kg m^{-3} and a melting point of 1414°C. It is tetravalent, with tetrahedral bonds to four neighbouring atoms (the diamond structure). For some uses in alloying, 99% purity suffices. Alloying the steel of transformer cores with silicon strongly increases its electrical resistivity, reducing eddy current losses without seriously compromising the magnetic properties. Alloys of Si with Al are used in vehicle components. A volumetrically small but expanding high-value fraction of elemental silicon production requires extreme purity for use in the electronics industry. This was made possible by the technique of zone refining, by which a furnace melts a small part of the length of a silicon ingot, with the melt zone starting at one end, moving to the other end and repeated as often as necessary. Almost all impurities are more soluble in the liquid and are concentrated in it and swept to one end, giving increasing purity with successive passes. Precisely controlled additions of elements that give desirable semiconducting properties can be introduced in this way.

Plants incorporate some silicon in their structures, and it contributes to strength and growth. Herbivores obviously ingest significant amounts, and some studies have suggested roles in maintaining health in various ways, but they are not well understood and biochemical materials incorporating silicon must be involved because silica itself appears not to be digested. For a few marine species, silica is a major structural component, forming the skeletons of diatoms and sponges, and the trace of silica dissolved in the oceans is recognised as necessary to the marine ecosystem.

15. Phosphorus (P). Although this is not rare cosmically or in the Earth as a whole, much of it has probably been sequestered in the core (Table 6.4). The modest remaining fraction is relatively concentrated in the crust but is rated as a minor element (50 times less abundant than potassium) and its availability for exploitation is diminished by its wide distribution as a minor constituent of numerous minerals. The only mineral from which it is extracted is apatite, with the general form $Ca_5(PO_4)_3X$, where X may be F, Cl or OH. Although this is found in several places, the deposits are small and very unevenly distributed geographically. By far the largest known reserve is in North Africa, in the Western

Desert area claimed by Morocco. An easily exploited source is the guano on certain islands, accumulated over millennia from sea bird droppings, but this is now more or less exhausted, leaving some islands almost uninhabitable. Since phosphorus, as phosphate, is widely used as fertiliser in agricultural areas with soils that are deficient in it, this is a major resource concern. Soil nitrogen and phosphorus have been limiting factors in the world food supply, but atmospheric nitrogen is now industrially converted to fertiliser. There is no prospect of finding a similar source of phosphorus, leaving it as the probable primary limitation on food production. This leads to the question of recycling. The basic problem is that, like nitrogen, a continuous supply of phosphorus is required because it is a necessary constituent of food but is excreted by animals, including humans, as a waste product. It ends up as a pollutant of various waterways, causing development of excessive algal growth, and this problem is exacerbated by run-off from over-fertilised agricultural land. The challenge is to develop a recycling system that solves both problems.

As an element, phosphorus has several allotropic forms with very different properties, densities 1823 kg m^{-3} to 2090 kg m^{-3} and a melting point of 44.1°C. The allotropes are identified by their colours. Most familiar is white phosphorus, which is the form in which it is produced but requires careful handling, being not just toxic and highly inflammable but also spontaneously self-igniting. This has been used in incendiary weapons that have had devastating effects in several wars. The unpleasant reputation of phosphorus is reinforced by dangerous compounds, including nerve gas. Maintained in an inert atmosphere, white phosphorus slowly transforms to red phosphorus, but appears as yellow phosphorus with partial transformation. The denser black phosphorus is produced only by the application of pressure, although it appears to be the thermodynamically stable form under ambient conditions.

There are minor metallurgical uses of phosphorus, the best known being phosphor bronze, in which a very small addition of P (<1%) has a big effect on the elasticity of Cu–Sn bronze, yielding corrosion-resistant spring material.

Phosphorus is an essential biological constituent at several levels, including animal skeletons.

16. Sulphur (S). This is an element that can be described as geochemically gregarious and consequently mobile in the Earth, having a range of alternative valencies that allow it to combine with other elements in various ways. It is chemically similar to oxygen and may act as either an oxidising or reducing agent, appearing in volcanic gases as either H_2S or, more usually, SO_2, and is deposited as a native element on volcanic rims, hot springs and salt domes. In a reducing environment, it forms sulphides, which become ores of numerous metals (Fe, Co, Ni, Cu, Zn, Mo, Sb, Hg and Pb), but in oxidised form it appears as sulphates, especially $MgSiO_4$, which is a constituent of the oceans, salt lakes and evaporites. Sulphur combines readily with iron and has a dramatic effect on its melting point, so that a large fraction of the total Earth content must have accompanied the liquid iron that sank into the core (Table 6.4), leaving moderate abundances in the mantle and crust (Tables 7.3 and 8.3). It is a constituent of all fossil fuels and is removed from gas and oil before use to minimise the release of SO_2. Sulphur obtained in this way is the source of all industrial sulphur, making the mining of native sulphur unnecessary, simple as that is. Sulphur in coal is more problematic, requiring the identification of sulphur contents for selective use. High S contents (lower economic values) are identified with coals formed in shallow marine environments.

Elemental sulphur is a low melting point (115°C) yellow solid with three allotropic forms with densities 1920 kg m^{-3} to 2070 kg m^{-3}, all based on an S_8 structure and with properties that are not dramatically different. It is an excellent electrical insulator. It is a basic feedstock for the chemical industry, most being used to produce sulphuric acid, H_2SO_4, which is the starting point for much industrial chemistry. The range of products incorporating sulphur includes (as CS_2) cellophane and rayon and the hardened rubber in vehicle tyres, in which S is a hardening agent.

Sulphur is an essential ingredient of biological life and, as gypsum, $CaSO_4 \cdot H_2O$, a natural mineral found in evaporites, is used as a fertiliser. As an element, it is not hazardous, but many of its compounds are poisonous, obnoxious or environmentally damaging. Powdered elemental sulphur is used in horticulture as a fungicide. The well-known bad egg smell of H_2S is sufficiently offensive to be avoided, but mustard gas has been a poison with military applications. SO_2 from sulphurous coal oxidises in the atmosphere, becoming sulphuric acid and falling as acid rain, destroying vegetation in regions of heavy industry. Large quantities of SO_2 from major volcanic eruptions reach the stratosphere and remain there for months or years as sulphuric acid droplets, cooling the Earth by screening sunlight.

17. Chlorine (Cl). This is the most familiar of the halogens. In its elemental form, it is a diatomic gas, Cl_2, liquid below −34°C, but it is very reactive and occurs mainly as chlorides, especially common salt, NaCl. This is the major solute in sea water and is extracted from brines and salt pans. Chlorine production is by electrolysis of brine, a process with other useful products, H_2 and caustic soda, NaOH (or KOH if KCl is used). Cl^- ions are essential to all life forms but the element itself is seriously poisonous in more than very low concentrations and has been used as such in warfare. It is a powerful oxidant, used as a disinfectant, and the basic ingredient of common bleach, with a characteristic smell in the low concentrations of bleach and swimming pools. Chlorine readily forms organic compounds, the simplest of which is carbon tetrachloride, CCl_4, dry cleaning fluid. Polymerised vinyl chloride is a widely used plastic.

Chlorinated organic pesticides are stable compounds, very persistent in the environment, and chlorinated fluorocarbons, once in general use as refrigerants and spray propellants, have been embargoed because they reach the stratosphere and destroy UV-absorbing ozone.

18. Argon (Ar). This inert noble gas is the third most abundant constituent of the atmosphere (1.25% by mass), but the argon in the solar wind, and by inference in the Sun, is isotopically quite different. All of the noble gases accreted in the Earth in the very small abundances corresponding to their volatilities, but terrestrial argon has been supplemented over geological time by argon from radioactive decay of the rare potassium isotope ^{40}K. The decay product, ^{40}Ar, dominates atmospheric argon (99.6%), whereas this isotope is only 10.5% of the solar wind argon, which is mostly ^{36}Ar. Without convection of the mantle, which has caused much of the terrestrial argon to be released to the atmosphere, it would be quite rare. It is completely non-toxic and has no biological role.

Argon is extracted from air by fractional distillation in the same industrial process that separates oxygen and nitrogen. The boiling point is 87 K, 10 K higher than that of nitrogen. It is used as an inert atmosphere in situations where the more abundant (cheaper) nitrogen is less suitable, as in some welding applications, very high temperature graphite furnaces and food preservation. It glows blue in an electrical discharge and in argon lasers, used in surgery.

19. Potassium (K). A soft alkali metal similar to and slightly more reactive than sodium, with a density of 862 kg m^{-3} and a melting point of 63.5°C. It very rapidly oxidises in air and reacts violently with water. As the free element, it is stored under oil or kerosene (in which it just sinks). It occurs naturally as ions, mostly in salts that are soluble so that it is a solute in sea water and salt lakes, although in the oceans its concentration is much less than that of sodium. It is extracted from brine pools and evaporites that are often compositionally graded. The most useful form is nitrate, but this has an organic origin and is less common than the inorganic salts, which are derived from the minerals in eroding rocks. Of the common rocks, granite has the highest concentration, typically 3% to 3.5%. Although some potassium compounds have specialised applications, more than 90% of global production is used as fertiliser. All vegetation, including wood, contains potassium, and ash from fires (potash from which the element derives its name), has been used since ancient times as a fertiliser. Potassium is essential to all plants, which must derive it from soil. It is also essential to all animal life and a balanced intake is recognised as vital to human health. A wide range of foods include sufficient potassium to ensure a balance and bananas are notably potassium rich.

There are three naturally occurring isotopes of potassium, of masses 39, 40 and 41. ^{40}K is relatively rare (0.01167% of the element), but it is radioactive and the abundance of potassium makes it an important source of radioactivity in the environment as well as a heat source in the Earth as a whole. It is unique in having three alternative modes of decay (Table 3.2), with a combined half-life of 1.25 billion years. Around 89.28% of the decays are by β emission to ^{40}Ca, which is the common isotope of calcium. The other modes both yield ^{40}Ar, a component of the atmosphere, as mentioned above. Since all biological materials incorporate potassium, they are unavoidably radioactive, far more so than from the very much rarer radioactive isotope of carbon, ^{14}C, but that may be a more critical component of genetic material.

20. Calcium (Ca). A soft grey alkali metal of density 1540 kg m^{-3} with a melting point of 842°C. It is strongly reactive, oxidising in air, burning vigorously once ignited and also forming nitride. It reacts with water, releasing hydrogen. The element is produced by electrolysis of molten $CaCl_2$ but has no widespread use as metal. It is one of the common elements in both the mantle and crust and occurs in numerous minerals. The most concentrated form is familiar as carbonate rocks, limestone and chalk, some of which are used with little processing. It is dissolved from eroding rock and has a concentration of 415 ppm in sea water. About 2% of the crust is composed of calcium carbonate, $CaCO_3$ (or calcium–magnesium carbonate) precipitated, either inorganically or biologically as shells of marine creatures, by the combination of dissolved calcium and carbon dioxide. This is the major sink of atmospheric/oceanic carbon dioxide and, on a geological time scale, controls its abundance. The carbonate rocks are quarried as building stones of various qualities. Portland stone is particularly well known and marble is a highly prized variety. Limestones, chalks, oolites and corals have a range of uses, the major one being the manufacture of cement and lime, CaO or $Ca(OH)_2$. Calcium sulphate, $CaSO_4$, is used as blackboard chalk (in spite of the name) and plaster of Paris is a hydrated form. A conveniently controlled small-scale method of producing acetylene, C_2H_2, is immersion of calcium carbide, CaC_2, an apparently inert granular solid, in water.

Calcium is an essential biological constituent in cell physiology, and particularly (as phosphate) in bones and teeth. It is, therefore, a correspondingly important dietary ingredient, especially for children (with developing skeletons). It occurs in a sufficiently wide range of foods to ensure an adequate intake with a reasonably balanced diet, dairy products, nuts and beans being particularly rich sources.

21. Scandium (Sc). Classed as a rare earth element (see note following element 56), but not really very similar to the lanthanides. As elemental metal, it has a density of 2985 kg m^{-3} and a melting point of 1541°C. It is uncommon but is used for special light but strong alloys. It is not mined as a primary target but is extracted in very small quantities as a by-product of uranium production. It has no known biological role and appears to be moderately toxic if ingested as chloride.

22. Titanium (Ti). A light, strong, corrosion-resistant metal, with density of 4506 kg m^{-3} and a melting point of 1668°C. The corrosion resistance is due to the formation of a thin oxide layer on any surface exposed to air and extends to moderately high temperatures and a wide range of acids and other solvents, but excluding chlorine, with which it reacts. In common with transition metals generally, it has low electrical and thermal conductivities. It is reasonably common (~0.4% of the crust), but widely distributed in numerous minerals. It is normally tetravalent and the dioxide, TiO_2, is found as several mineral forms, rutile being of particular commercial interest, because it is readily extracted from beach sands that are derived from eroded igneous rocks. This is also the case for another economically important mineral, ilmenite, $FeTiO_3$. The metal is not easily extracted from its ores; probably the best method, but not yet widely used, is electrolysis of liquid chloride.

The combination of strength and lightness led to use of the metal and high Ti alloys in aircraft components and this accounts for more than 50% of its production. It is biologically completely inert and non-corroding, with no known chemical role for any living organism, and no allergenic effects, so that it has medical applications in implants, such as hip and knee replacements. The oxide can be refined much more easily and has a range of uses, especially as a white pigment of high opacity in paint and paper. An interesting specialised use is in barium titanate, $BaTiO_3$. This is strongly piezoelectric and is used in high-power audio systems, converting electrical signals to sound.

23. Vanadium (V). This is a hard but malleable metal characterised by a high tensile strength. The density is 6000 kg m^{-3} and the melting point is 1910°C. The formation of a thin oxide layer makes it corrosion resistant. It occurs in several minerals but is produced mainly as a by-product in the extraction of other metals, including uranium, and also in processing fossil fuels. Most of the production (85%) is not of isolated metal but of ferrovanadium that is used in alloying. High-quality tool steels commonly include a few per cent of vanadium, often in combination with chromium. A feature of vanadium is its multiplicity of alternative valencies and this has made the pentoxide, P_2O_5, a widely used catalyst in the manufacture of sulphuric acid and the reduction of nitrogen oxides in power station emissions. The vanadium–gallium alloy, V_3Ga, is a superconductor similar to the widely used Nb_3Ti.

Some compounds of vanadium are mildly toxic, but it possibly has a wide minor biological role. Some marine creatures, notably sea squirts, have strong concentrations in selected organs, inviting the inference that it acts as a toxic deterrent to predators, but that must be doubted. The common occurrence in fossil fuels suggests a wide distribution at low concentrations in the biological materials that formed the fuels and possibly in the biosphere generally, but its role is not known.

24. Chromium (Cr). This is a hard metal notable for its corrosion resistance due to a very thin oxide layer that is not obvious to visual examination. Its maintenance of a high sheen under normal atmospheric conditions led to one of its uses, chrome plating. It has a density of 7190 kg m^{-3} and a high melting point (1907°C). The most widely used ore is chromite ($FeCr_2O_4$), of which there are deposits in several places, with production dominated by South Africa, India and Kazakhstan. The major use is in stainless steel, of which there are numerous variants, a common one having ~18% Cr and ~6% Ni. For this purpose, it is not necessary to separate it from iron but to use ferrochrome for alloying. Very hard tool steels are alloyed with V as well as Cr. As with some other transition elements, alternative valencies led to the use of oxides, including CrO_2, as catalysts in industrial chemistry. Depending on preparation and condition, this oxide may be ferromagnetic. A biological role for chromium is marginal/doubtful and observations on its trivalent form indicate that it is benign, but there are tentative reports of a role in the human pancreas and control of glucose tolerance.

25. Manganese (Mn). A hard, brittle metal resembling iron, with a density of 7470 kg m^{-3} and a melting point of 1246°C. Globally, it is reasonably common, being about 0.1% of the mantle, slightly less in the crust but almost certainly more in the core (Table 6.4). Economically, the most important ore is pyrolusite, basically MnO_2, with the major producers being in South Africa, Gabon (West Africa) and Australia. It is not used as an elemental metal (and, like iron, it corrodes), but has wide use as an alloying element in structural steel and with aluminium. The steel industry uses as much as 90% of the global production, with Mn steel as the favoured material for things like railway lines. It is a small component (~1%) of the aluminium used for soft drink cans. Manganese phosphate is applied as a rust inhibiter to the steel of car bodies. There is a range of alloys with copper, typically $MnCu_2$ plus various other elements, known as Heusler alloys, that are ferromagnetic, although not necessarily including any ferromagnetic elements, and are a subject of interest to the electronics industry. A simple bimetallic compound with bismuth, MnBi, is ferromagnetic with extremely strong anisotropy of magnetic properties and consequent coercive force, and is of interest as potential permanent magnet material, but still has stability problems to be resolved. Several manganese compounds are strong oxidising agents and MnO_2 in non-rechargeable dry cells prevents emission of hydrogen.

Manganese has several essential roles in both plants and animals. It is a necessary dietary constituent, although in excess it is toxic. It is involved in photosynthesis and is a component of a wide range of enzymes that mediate physiological processes in animals, including humans.

26. Iron (Fe). This is a cosmically abundant element, ranked sixth by mass in the Sun and a close second to oxygen in the Earth, with most of it sequestered as metal in the core (Table 6.4). The abundance can be attributed to the fact that the dominant isotope, ^{56}Fe, is the lowest energy state of nuclear matter. Pure iron is a relatively soft metal, with a density of 7874 kg m^{-3} and a melting point of 1539°C, but lower values are reported for impure forms. It is rarely encountered in pure form because impurities, especially carbon, harden it and the familiar alloy is steel, of which there are numerous variants. Iron is a constituent of many minerals and occurs in the crust in concentrated forms as oxides, especially hematite, Fe_2O_3, and also magnetite, Fe_3O_4, which, like iron, are strongly magnetic. Major ores are the banded iron formations, found on all continents, so-called because they consist of alternating thin layers of iron oxide and silicate, mostly deposited from sea water more than 1.8 billion years ago by oxidation of soluble divalent iron, Fe^{2+}, to the insoluble trivalent form, Fe^{3+}. Iron as metal occurs in meteorites, alloyed with nickel which inhibits rusting, and this was probably the earliest form available for human use. Discovery of the method of extraction from ore by smelting began in the iron-age about 3000 years ago, and this was the crucial first step towards the iron and steel technology which is the backbone of our technically developed society. By any standards, iron is cheap and the global production of steel exceeds a billion tonnes annually, dwarfing all other metallurgical activity.

The physical properties of iron have led to a wide diversity of uses. It is strongly magnetic, a feature referred to as ferromagnetism when applied to other materials, recognising it as characteristic of iron. This has made it a key component of the electricity industry. In common with transition elements generally, by the standard of metals, its electrical and thermal conductivities are low and in transformer cores this is an advantage, enhanced by alloying with silicon. Alloys of iron with a wide range of other elements yield materials that are fine-tuned to diverse requirements. A significant limitation on its use is rusting. Exposure of iron or steel to moist air, especially in a salty marine situation, causes a coating of hydrated oxide to form and peel off, exposing fresh metal to corrosion. Unlike the situation with some other metals, notably chromium, there is no hard, protective oxide layer. The problem is addressed by introducing alloying elements, particularly Cr and Ni, to produce stainless steel, but this is too expensive for many uses.

Iron is a necessary minor constituent of all biological materials. It occurs in proteins, various enzymes and is particularly well recognised in the haemoglobin of red blood cells, the carriers of oxygen (and carbon dioxide) circulating in blood. There are standard treatments for iron deficiency (anaemia), but they are needed only in special situations as many foods, such as dark green vegetables and red meat contain enough iron for normal health.

27. Cobalt (Co). A hard, silver–grey metal with a density of 8900 kg m^{-3} and a melting point of 1495°C. It is one of the three ferromagnetic elements

(Fe, Co and Ni) in the first transition series and has the highest Curie point of any element (1115°C), but its saturation magnetisation is only 1.7 Bohr magnetons/atom, compared with 2.2 for iron. It is widely distributed in numerous minerals, especially sulphides and arsenides (linnaeite, Co_3S_4, smallite, $CoAs_2$ and cobaltite, CoAsS) but is mostly obtained as a by-product in the extraction of nickel and copper, with which it is closely associated geochemically. World production is concentrated in the Central African copper belt of Congo and Zambia. Its major uses are as an alloying element in special steels that remain strong and wear resistant at high temperatures. Its role as a constituent of magnetic materials, such as the permanent magnet alloy alnico (aluminium–nickel–cobalt), diminished as the use of rare earths developed. The addition of silicate or aluminate of cobalt to glass gives it a characteristic blue colour.

Cobalt is classed as a necessary ultra-trace element for numerous (perhaps all) life forms. It appears to have a particularly strong role for many simple life forms, algae, fungi and bacteria. It is a constituent of vitamin B_{12}, and also proteins. A balanced human diet provides an adequate intake, but diseases of cattle and sheep grazing on cobalt-deficient pastures have been identified (and corrected). As also with chromium and nickel, dermatitis-like reactions to prolonged contact with the metal have been reported, and it is possible to be poisoned by a cobalt overdose.

28. Nickel (Ni). A hard, ductile metal with a density of 8908 kg m^{-3} and a melting point of 1455°C. It is ferromagnetic with a Curie point of 355°C and saturation magnetisation of 0.6 Bohr magnetons/atom, a quarter of the value for iron. This is a cosmically abundant element closely associated with iron, so that most of the terrestrial content is in the core. It is less abundant in the crust than in the mantle and even less abundant in the upper crust, but is still rated as one of the reasonably common crustal metals. There are two main ore types, laterites (limonite, Ni-bearing Fe_2O_3 and garnierite, a mix with Ni–Mg silicates) and sulphides [pentlandite, $(Ni,Fe)_9S_8$], normally involving iron in all cases. Separation from iron is not needed if the product is used in steel alloying, in particular for stainless steel (~6% Ni), which accounts for ~70% of nickel production. Another major use is nickel plating, with or without an overlay of chromium, and nickel has featured prominently in coinage, particularly in USA. The use of nickel in rechargeable batteries is decreasing as lithium takes an increasing share. The global total production exceeds 2 million tonnes/year, the principal sources being Philippines, Indonesia, Russia, Australia and Canada.

Enzymes in plants and micro-organisms have been found to contain nickel but it has no known role in animals. Although some nickel compounds are seriously toxic, exposure to them is improbable under normal conditions. Ingestion of nickel leached from stainless steel cookware by acid foods is well below the recommended tolerance. Reports of an allergic dermatitis-like reaction to prolonged exposure to Ni metal in earrings prompted its replacement in that application.

29. Copper (Cu). A soft, malleable metal of density 8960 kg m^{-3} and a melting point of 1085°C. It has a characteristic orange colour, which slowly tarnishes to a deeper red, distinctly different from the grey of almost all metals. Brass and bronze, familiar alloys of copper, are also coloured, inviting ornamental use. Although uncommon, native metallic copper was known and used in ancient times. The most important ores are sulphides, especially chalcopyrite, $CuFeS_2$, with a hydrated carbonate, malachite, $Cu_2CO_3.(OH)_2$, and oxide, cuprite, CuO_2, less common. Major producers are Chile, USA, Peru, Australia and Congo/Zambia. Although moderately abundant in the crust, it is widely distributed, with limited opportunity for expansion of production; recycling is well established. As an electrical conductor, copper ranks a close second to silver and the principal use is for electrical wires and cables, which require HC (high conductivity) copper that has had an electrolytic refinement stage. It is also a good thermal conductor, used for heat sinks in electronic circuitry. It is regarded as decorative roofing material, developing the green colour of carbonate (verdigris) with age, but is generally corrosion resistant, notably in a marine situation. Alloys with zinc (brass) are widely used, being stronger than copper and readily worked mechanically. Bronze, an alloy with tin, is also a harder material, with a long history (the bronze age). Copper is used in coinage in many countries, generally alloyed with nickel (cupronickel).

Copper is an essential trace element in both plants and animals. It is a component of proteins and has a particular role in controlling respiration. The blood of crustaceans and crabs is copper based, an alternative to the iron in mammalian haemoglobin. The necessary dietary trace is found in a range of foods and is most abundant in oysters, liver and chocolate. Copper poisoning would not occur easily because mammalian physiology disposes of any excess, but elemental copper acts as a bactericide and fungicide and is used in wood preservation.

30. Zinc (Zn). At ambient temperatures, this is a hard, brittle metal, dull grey in commercial grades, but appearing white when very pure. The density is 7140 kg m^{-3} and the low melting point, 420°C, means that it is readily cast. The major ores are sulphide, sphalerite ZnS, and carbonate, smithsonite $ZnCO_3$. Recognised reserves identified in Iran, Australia, USA and Canada are seriously depleted and recycling is important to maintain supply. The largest single use is for zinc plating (galvanising) of steel to protect it from rusting. It is familiar as electrode material in batteries (dry cells) and is used for sacrificial anodes bolted to the hulls of ships and connected to a low voltage supply to cause a current through sea water, dissolving the zinc but protecting the steel of the hulls. Zinc readily alloys with many other metals; its alloys with copper, in various proportions, are widely used as brass. The oxide, ZnO, is used as a white pigment in paint.

Zinc is an essential biological ingredient, assimilated from soil by plants and sufficiently widely distributed in food to ensure an adequate intake in a balanced diet, although effects of a deficiency are noted in the development of children in areas that are economically depressed or have low zinc contents in agricultural soils. Dietary supplements are available for individuals with particular concern about eyesight, but beans, nuts, pumpkin and sunflower seeds, oysters, lobsters and red meat are good sources, but an excess of zinc introduces the possibility of deficiencies in other minor elements.

31. Gallium (Ga). A soft, silvery metal with a melting point close to ambient temperature (29.76°C). The solid density is 5910 kg m^{-3}, and this is one of the very few elements that contract on melting, with a 3.1% density increment. It is chemically and crystallographically complicated, with a very high Poisson's ratio and a superficially glassy appearance. It is not rare in the crust but is widely distributed with no obvious concentrations and is not mined anywhere for its own sake. It is found as a trace in bauxite and is extracted as a by-product of aluminium refinement but with global production of only a few hundred tonnes/year. The major use is in the semiconductor industry as gallium arsenide ($GaAs$), being particularly useful at microwave frequencies, at which other semiconducting materials fail. It is also a constituent of most light-emitting diodes (LEDs). The low melting point is further depressed by alloying, and an alloy with indium and tin (galinstan) is liquid down to −19°C and has been used as a replacement for mercury in thermometers, although thermal radiation meters have mostly taken over for medical use, avoiding accidental mercury poisoning. Gallium is not known to be toxic or to have any biological function, although it is under investigation as a possible anti-cancer agent. That potential role depends on Ga^{3+} ions mimicking Fe^{3+} ions and preventing the growth of cancer cells, as well as bacteria that require Fe^{3+} for reproduction and growth.

32. Germanium (Ge). This is a hard, grey semiconductor with a diamond crystal structure, density of 5323 kg m^{-3} and a melting point of 938°C. It is a rare element, widely distributed in numerous rock types in concentrations that are too low for exploitation. It is found as a minor element in the zinc sulphide ore sphalerite and some production is from the flue dust in smelters used to process zinc ore. Smaller amounts are obtained in processing other metal ores, but most of the production is from fly ash from power stations burning selected coals in North China and Far East Russia. The first transistors were made of germanium. This use continues in a minor way now that silicon has taken over as the major semiconducting material, but Si–Ge alloy offers the advantage of operating at higher frequencies and may supersede the use of $GaAs$ chips in this range.

Optical properties of germanium have now become the basis for most of its use. Although optically opaque, it is transparent to infrared light and is used in equipment such as thermal imagers, fibre optic cables and solar cells. In the optical range and extending into the infra-red, the oxide, GeO_2, has a high refractive index with low dispersion, making it useful for specialist photographic lenses as well as optical fibres. Germanium has no known biological role and the insolubility of its compounds limits the possibility of observing any toxicity.

33. Arsenic (As). This is a semimetal, dull grey in its most familiar elemental form, with a density of 5727 kg m^{-3}, but with other allotropic forms. It does not liquefy with heating but sublimates at 615°C. It is widely distributed and has even been found as the free element. It occurs as sulphide and oxide ores, but these are not used. Extraction is as a by-product of the treatment of metal ores, Fe, Cu, Co and Ni. The commercial form for production and distribution is the oxide, As_2O_3, mainly in China, Chile and Morocco. The element is used in alloying, particularly with lead in car batteries and with gallium in microwave semiconductors and lasers. Other uses, as a green dye, insecticide and wood preservative, have declined, or have been discontinued, with concern about its acute toxicity. In trace amounts, it is recognised as a necessary dietary constituent, and has even been fed as a food supplement to chickens, but it is lethal in excess and very effective as a rat poison. Accidental poisoning can occur in several ways, inhalation of smoke from fires using treated wood, evaporation from arsenic-dyed material and (in Bangladesh, for example) from use of arsenic-contaminated ground water.

34. Selenium (Se). An element that is chemically similar to sulphur, with three allotropic forms. Of greatest interest is grey selenium, which is the densest (4810 kg m^{-3}) and the most thermodynamically stable form, and is a semiconductor. Red selenium, generally produced as an amorphous powder, and black selenium, a brittle vitreous solid, are both insulators. The melting point is 221°C. There are no useful concentrations of selenium itself, but it is found as selenides of several metals, especially Cu, Pb and Ni, incorporated as minor constituents of the sulphides of these metals. Production is mostly a by-product of copper refining, principally in USA, China, Japan and Germany. Although a large fraction of the total selenium produced is used in high-quality, clear glass making, the more significant uses depend on the semiconducting, photoconductive and photovoltaic properties that have applications in light detectors, photocopiers, solar cells and laser printers.

Much has been written about the biological role of selenium. It is recognised as an essential trace nutrient and deficiencies have been noted in areas of low soil content, but an excess is poisonous and the boundary between too little and too much appears from some reports to be narrow and to be species dependent. Some plants are rich in selenium, while others have (and appear to need) none at all, inviting the inference that this is an effect on evolution of local availability. In animals, selenium has a role in cellular functions and in several enzymes, but the abundance appears to affect other trace elements; it is systematically richer in marine species than in land animals. As in at least most such situations, a balanced diet normally satisfies the dietary requirement, Brazil nuts and sea food generally being rich sources.

35. Bromine (Br). Under ambient conditions, this is a fuming deep red liquid with a boiling point of 58.8°C and density of 3103 kg m^{-3}. If left exposed, it quickly evaporates, producing toxic red fumes. As an element, it occurs as diatomic molecules, Br_2, but the natural occurrence is as ions, mostly in soluble salts, being 67 ppm of sea water, almost 100 times its average crustal concentration. It is extracted from brine pools, mainly in USA, Israel and China. Its uses are limited. As bromomethane, CBr_4, it has been used as a flame retardant in fire extinguishers, but its release destroys stratospheric ozone. It is regarded as a biological trace element, with a possible role in tissue development, and has been used medically in some countries in controlling epilepsy.

36. Krypton (Kr). A colourless, odourless inert gas with a boiling point of –154.3°C. It comprises 1.14 ppm of the atmosphere, from which it is extracted by partial distillation. Compounds with other elements are known but are generally unstable, the most stable being KrF_2. Krypton discharge tubes give white light as a consequence of spectral lines spread over a wide range and tuned lasers give high power in red light. There is no biological role and the gas is not toxic.

37. Rubidium (Rb). This is a very soft, silvery alkali metal with a melting point of 39.3°C and density of 1532 kg m^{-3}. It is highly reactive, oxidising rapidly in air and even catching fire spontaneously. It can be kept under dry mineral oil but is generally stored as one of its salts. Its uses are limited, with little motivation for extraction as a commodity of prime interest, and it is obtained as a by-product in the production of lithium from the mineral lepidolite and of caesium from pollucite. Principal sources are in Manitoba, Canada and the Mediterranean island of Elba. Sea water contains 0.12 ppm of Rb^+. Rubidium has a high vapour pressure under ambient conditions, making it convenient to use in vapour lamps, in particular those used as frequency standards based on a hyperfine line in its spectrum, although caesium lamps are now favoured for the highest precision. Rubidium is used to impart a red colour to glass, ceramics and fireworks. One of its two isotopes, ^{87}Rb, is radioactive, decaying to ^{87}Sr with a half-life of 4.92×10^{10} years and this is the basis of one of the methods of radiogenic dating of geological materials. Rubidium is not known to have a biological role but, due to its close chemical similarity to potassium, it is incorporated in both plants and animals, substituting for potassium, apparently passively in low concentrations but possibly as a poison in artificially high concentrations.

38. Strontium (Sr). A soft, silvery alkaline earth metal with a density of 2640 kg m^{-3} and a melting point of 777°C, but with two high-temperature allotropic forms before reaching the melting point. It is very reactive and rapidly develops yellow oxide when exposed to air. The most abundant exploitable ore is celestite, basically the sulphate, celestine, $SrSO_4$, but also mined is the carbonate, strontianite, $SrCO_3$. The major producer is now China, with useful deposits in several other countries and 7.8 ppm in sea water. Uses include alloying, particularly as a minor addition to Si–Al, reducing brittleness. It gives a strong red colour to pyrotechnic displays, and strontium aluminate is phosphorescent and is used in advertising signs. Demand for strontium has decreased as two of its major uses declined, processing of beet sugar and the display glass of television tubes. Strontium is chemically similar to calcium and biologically it is found with calcium, typically as ~0.1% of the calcium in the bones of humans and other animals. Evidence has accumulated that it is beneficial, and dietary supplements have been prescribed for the treatment of osteoporosis.

39. Yttrium (Y). Recognised as a rare earth element, with properties similar to the lanthanides and mined with them. It is extracted with heavy lanthanides from the mineral xenotime. It is a silvery metal of density 4472 kg m^{-3} and a melting point of 1526°C. It has some metallurgical uses but is more notable for special properties that include low-temperature ferromagnetism and high-temperature (>90 K) superconductivity when alloyed with Ba and Cu in an alloy with the acronym YBCO. It forms garnets with a wide range of fluorescent colours when combined with other rare earths used in LEDs. Yttrium aluminium garnet is the critical component of high-energy pulsed lasers as well as white light LEDs with added cerium. Being a reasonably common element, yttrium is found as a trace in both plants and animals, although it has no known role. In more than trace abundance it is toxic in most chemical forms.

40. Zirconium (Zr). A light grey, strong transition metal of density 6520 kg m^{-3} and a melting point of 1855°C. It is corrosion resistant, even at high temperatures, except in the presence of water, and is transparent to neutrons, so that it is used for fuel cell cladding in nuclear reactors. It occurs in numerous minerals, the most important one economically being zircon, $ZrSiO_4$, which is found in beach sands, along with rutile and ilmenite, from which titanium is extracted. These minerals survive as small grains from the erosion of the igneous rocks in which they formed, especially granite, and are available as easily processed placer deposits. The major producers are in Australia and South Africa. Zircon itself is used for grinding wheels. As individual mineral crystals, it is of particular interest in radiometric dating of rocks and minerals by lead–uranium methods, because uranium is accepted as a substitute for zirconium in the crystal structure, but lead is rejected and the lead isotopes that are found in zircons are almost entirely radiogenic. Zirconium dioxide, zirconia ZrO_2, is a cubic crystal closely resembling diamond in both hardness and optical properties, for which it is in no way inferior to diamond, and is widely used in cheaper jewellery. Traces of zirconium are found in living material but it has no known biological role and is not toxic.

41. Niobium (Nb). A soft, grey metal with density of 8570 kg m^{-3} and a high melting point (2477°C) with several specialised uses. It occurs in a number of minerals of which pyrochlore, a mineral of variable composition generally represented by $Nb_2O_6(Na,Ca)_2(OH,F)$, found in carbonatites, is the only one currently extracted commercially. Deposits have been identified in several places, with Brazil as the largest producer and Canada the only other substantial one. Niobium is chemically very similar to tantalum and they are separated from mixed ores. In production, it is commonly mixed with iron and the product, ferroniobium, is conveniently used directly as a minor ingredient (<0.1%) in special stainless steels because it melts more readily than niobium. Alloys with niobium as the major ingredient are very stable, resisting corrosion to high temperatures and are used for nozzles of rockets and jet engines.

Pure niobium is a superconductor below a critical temperature of 8.2 K and the superconducting properties are enhanced by alloying with some other metals. An alloy with titanium has a critical temperature of 9.2 K and is particularly favoured for the magnet coils in magnetic resonance imaging (MRI) machines and particle accelerators. Strands of Nb–Ti alloy are encased within copper or aluminium wire. A minor but perhaps expanding use of niobium nitride at low temperatures is as a detector of microwaves. Niobium has found a place in jewellery for two reasons: it is readily anodised to a range of interesting colours, and it is physiologically inert. It has neither a known biological role nor any toxicity.

42. Molybdenum (Mo). A silvery–grey metal of density 10,280 kg m^{-3} and a high melting point (2623°C). It maintains its mechanical properties to high temperatures but oxidises in air above about 500°C. The principal ore is molybdenite, MoS_2, which is commonly found with copper ores and extracted in several places (China, USA, Chile, Peru, Mexico and Canada). The major use is as a constituent of steels of high strength that are readily welded, are heat and corrosion resistant and are favoured for uses such as railway lines. The disulphide, MoS_2, is a lubricant that withstands both high stresses and high temperatures, and it is mixed with grease for more general use in situations in which reliability is important. Biologically, molybdenum is an essential trace element, and is noted as being central to the action of nitrogen-fixing bacteria. It is a constituent of numerous enzymes, but it is toxic in excess, at least partly because that leads to a calcium deficiency. A balanced diet ensures adequate intake, with liver, eggs and green beans prominent among many food sources.

44. Ruthenium (Ru). This is a very rare transition metal with a density of 12,450 kg m^{-3} and a melting point of 2334°C. Its properties are similar to those of the platinum group metals, and it is extracted with them from Ni and Cu ores. It is used as a constituent of special alloys for applications such as high-performance turbine blades, and for very corrosion-resistant plating, giving a dark finish to jewellery and high reliability in electrical contacts. There is no evidence of either a biological role or any toxicity.

45. Rhodium (Rh). A rare, hard and inert, bright silvery transition metal with a density of 12,410 kg m^{-3} and a melting point of 1964°C. It is occasionally found as a free metal but more generally with Pt group metals in Cu and Ni ores. Its major use is as one of the catalytic converters in car exhausts. It is very hard and corrosion resistant and a very thin layer applied as plating to jewellery imparts a very shiny finish. With Pt and Ir, it is used in high-temperature applications such as furnace windings and thermocouples. There is no known biological role or toxicity.

46. Palladium (Pd). This is a rare element, a silvery-white member of the platinum group of metals, and extracted with them from sulphide ores of Cu, Ni, Pb as well as from chromite. It has a density of 12,023 kg m^{-3} and a melting point of 1555°C. It is soft when freshly prepared or annealed but rapidly work-hardens if deformed. The major producers are Russia, South Africa, Canada and USA. Although further potentially exploitable deposits are known, this is one of the elements for which adequacy of the supply is under threat. As a pure metal, palladium is notable for absorbing hydrogen without the application of pressure. This is a reversible process, the hydrogen being released as readily as it is absorbed, although it appears that it is taken in as H^+ ions and not as H_2 molecules, because it progressively cancels the strong paramagnetism of the pure metal, making it diamagnetic for H/Pd atomic ratios exceeding 62%, which is close to the absorption limit. Palladium screens are used to separate hydrogen from mixtures such as coal gas. But the principal use is as one of the elements in the catalytic converters in car exhaust systems. In jewellery, it resembles silver and can be applied as thin plating or beaten into very thin foil. It is corrosion resistant and, unlike silver, does not tarnish, although it dissolves in strong acids. It has no known biological role, but there is some evidence of toxicity and an adverse effect on plant development.

47. Silver (Ag). A soft, white metal with a density of 10,490 kg m^{-3} and a melting point of 961.8°C. It has been identified since ancient times as a measure of wealth, second to gold, and used in coinage. Although occasionally found as native metal and as sulphide, argentite Ag_2S, it is extracted as a by-product in the processing of Cu, Au, Pb and Zn ores. Major producers are in Mexico, Peru and China. Significant physical properties that lead to special applications include the highest electrical conductivity of any material (under ambient conditions) and the highest optical reflectivity if protected from tarnishing. In spite of the fact that it tarnishes, it is widely used in jewellery and prestige cutlery. Silver alloys are used for musical instruments and silver is a generally minor constituent of some high-temperature solders, referred to as silver solder. Silver halides, especially the iodide, are light sensitive and are used in photography. Silver has no known biological role but is not regarded as toxic, although some of its salts are, and it is cumulative in the body. A suggested health benefit of silver foil in food is disputed.

48. Cadmium (Cd). This is a soft, white metal of density 8650 kg m^{-3} and a low melting point (321°C). It is rare, averaging about 90 ppb in the continental crust but made exploitable by its concentration as greenockite, CdS, in zinc ores and retrieved as a by-product of the processing of sphalerite, ZnS. The principal producers are in China, South Korea and Japan. The elemental metal is corrosion resistant and used for plating steel in situations requiring better protection than galvanising with zinc, but it does dissolve in acids. The major use is in rechargeable (nickel–cadmium or Nicad) batteries, but that is decreasing as nickel–metal hydride and lithium ion types take increasing market shares. Its compounds are used as pigments in paint, especially a strong yellow colour. It has a specialised use as a neutron absorber in control rods of nuclear reactors. The chemical similarity to zinc makes cadmium a minor trace element in biological systems and in a range of foods, particularly sea food, although, with one exception, it is not known to have a useful biological role and is a cumulative poison in more than trace abundance. The exception is that some marine diatoms in zinc-deficient environments appear to use cadmium as a substitute for it. There is some concern that cadmium is carcinogenic and that its concentration in tobacco smoke may be a contributory cause of lung cancer.

49. Indium (In). A soft, white metal of density 7310 kg m^{-3} with a very low melting point (156.6°C). Like cadmium, it is rare (60 ppb in the continental crust) but is found with the slag and dust from processing zinc ores, mainly in China, Japan, South Korea and Canada. It is one of the elements for which the long-term supply is particularly insecure. Indium is a metallic conductor, a superconductor below 3.4 K, and forms superconducting alloys with other metals. It is a component of low-melting-point solders. As oxide with tin (sometimes antimony), it is used in the screens of TV and computer monitors and in liquid crystal displays. Indium has neither a known biological role nor toxicity but with little evidence to go on.

50. Tin (Sn). There are two allotropic forms of tin that are very different. It is usually encountered as a white, malleable metal of density 7365 kg m^{-3} and a melting point of 231.9°C. This is referred to as β tin. Below 13.2°C, it slowly transforms to α tin, a brittle powdery grey insulator with a diamond crystal structure and a density of 5769 kg m^{-3}. The metallic form is stabilised by alloying. There are 10 stable isotopes, more than for any other element. It is more abundant than its immediate neighbours in the periodic table (5 ppm of the continental crust), but the prospect of a shortage within a few decades arises from the present rate of its use The only exploited mineral is cassiterite, SnO_2, a constituent of igneous rocks, found as placer deposits in several places and extracted mainly in China, Indonesia, Peru, Bolivia and Brazil. Tin is a superconductor below 3.72 K. An alloy with niobium, Nb_3Sn is a superconductor with a transition temperature of 18 K and tolerance to very high magnetic fields. Tin is corrosion resistant and, being non-toxic, is widely used for plating of steel used as food containers, commonly referred to as 'tins'. It is used in several alloys, with bronze, an alloy with copper, being known for almost 5000 years (the bronze age). A major use is in varieties of soft solder. Although tin is accepted as non-toxic, some of its organic compounds are very toxic, an indication that it is not biologically useful.

51. Antimony (Sb). A lustrous, grey semimetal with a density of 6697 kg m^{-3} and a melting point of 630.6°C. It is not common but is not considered rare (0.4 ppm in the upper continental crust). The most important mineral is stibnite, Sb_2S_3, but much of its extraction is as a by-product of lead production from galena, PbS. It is also found in copper ores. The major producer is China, with smaller outputs by South Africa, Bolivia and Russia. It is one of the elements in risk of a supply shortage. The metallic form is the stable one of four allotropes. It resists acid attack and corrosion in air under ambient conditions but oxidises if heated. A major use (with tin) is as an alloying component that strengthens the lead plates in lead–acid batteries, and it is a hardening ingredient of lead bullets. A lead–tin–antimony alloy that expands slightly on solidification and accurately reproduces fine details in casting has been used for printer's type for hundreds of years but is now largely superseded by computer-controlled laser printing. Minor uses of antimony are in colouring glass, pottery and paint and in flame-proofing clothing. It has no known biological involvement or toxicity, except possibly from inhaling dust, but this is probably not a problem specific to antimony.

52. Tellurium (Te). This is a rare element in the Earth (1 ppb in the crust), although more abundant cosmically. It is a silvery, brittle semiconductor of density 6240 kg m^{-3} and a melting point of 449.5°C. It is occasionally found as native element, alloyed with gold and in minor minerals that are not used for extraction. Commercial production is as a by-product of the electrolytic refining of copper and the smelting of lead ores, principally in USA, Japan, Canada and Peru. A major use is as an alloying ingredient which improves the machinability of stainless steel and copper. Other important uses depend on particular physical properties of both the element and its compounds. It is a semiconductor with conductivity dependent on crystallographic direction and increased by exposure to light (photoconductivity). It is alloyed with cadmium as sensor material in solar panels, with lead in infra-red detectors, with bismuth in thermoelectric generators and as an ingredient of optical fibres. It has no known biological function, although it can substitute for S and Se in fungi, and it is a mild cumulative poison.

53. Iodine (I). Under ambient conditions, this is a purple–black solid of density 4933 kg m^{-3} which melts at 113.7°C, but it sublimates to a violet vapour of I_2 molecules even below that temperature. It is not an abundant element but is very water soluble as I^- ions and constitutes 0.17% of sea water solutes. It is obtained from brines, mainly in Japan and USA, and is extracted as iodates, especially lauterite $Ca(IO_2)_2$, from naturally concreted limestone in Chile. It is a biologically essential trace element, with particular significance in thyroid hormones, and is crucial to the development of young children. Much of its use is as a dietary supplement, and it also has antiseptic properties, being used to control infections, although newer antibiotics have largely taken over that role. The combination of its rarity and solubility has led to low contents in soils from which it has been leached in many areas, resulting in widespread deficiency in land animals generally, including humans. A dietary supplement in the form of iodised table salt is used widely, but not universally, around the world and in at least some areas iodine is included in nutritional supplements for cattle. Natural dietary sources include sea food, dairy products and eggs. Medical applications include the use of tincture of iodine, a solution in ethanol with a concentration of a few per cent, for topical application to infections, but prolonged application can be damaging to skin. Although iodine is toxic in excess, temporary bowel insertion is used in X-ray examinations; being a heavy element, it is X-ray opaque.

54. Xenon (Xe). This is the heaviest of the inert gases (discounting radon, an intermediate decay product of uranium and thorium), but it is not quite as inert as was once believed. It liquefies at −108°C. There are nine naturally occurring isotopes, one of them, ^{136}Xe, is slightly radioactive (half-life $>10^{21}$ years), and they are found in different proportions in the atmosphere and in various meteorites. The differences are attributed to varying contributions by fission of extinct ^{244}Pu and decay of ^{129}I very early in solar system development. The radioactive isotopes ^{133}Xe and ^{135}Xe are identified as fission products of nuclear weapons and are used as indicators of weapons tests. An increasing number of compounds are known, none of them naturally occurring, mostly involving fluorine and only marginally stable or even explosive. As with the other inert gases (but excluding helium), it is extracted by fractional distillation of air, but being very rare (87 ppb of the atmosphere, by volume), it is expensive. Its uses are almost all in various light sources. It gives a violet glow in discharge tubes and intense white light in high-power short arc tubes. It is used as a starter gas in sodium discharge lamps and allows incandescent bulbs to run hotter (more efficiently) than normal, by virtue of low thermal conductivity. The high atomic mass makes it a favoured propellant in ion propulsion engines for spacecraft. A fairly recent development, but one that is likely to expand with improvements in extraction methodology and availability (cost), is the use of xenon as a general anaesthetic. Biologically it is not completely inert. It dissolves in blood and is carried to the brain, causing anaesthesia, without complications that can occur with other anaesthetics.

55. Caesium (Cs). The alkali metal of highest atomic weight, with a solid density of only 1930 kg m^{-3}. It is a silvery solid, extremely soft and close to melting even under cool ambient conditions (melting point 28.4°C). It is extremely reactive, spontaneously burning in air and explosive with water but can be stored under mineral oil. Globally, it is rare (7 ppb in the mantle) but is strongly concentrated in the upper continental crust (5 ppm). There is only one economically useful mineral source, pollucite, $(Cs,Na)AlSi_2O_6 . H_2O$, which is mined on the largest scale in Manitoba, Canada, with smaller outputs in Namibia and Zimbabwe. Its major use is as caesium formate, CsCOOH, in drilling fluids that are dense enough to bring up gravel and rock fragments. It has a highly specialised use in 'atomic' clocks, based on the frequency of radiation from a transition between hyperfine electron levels in nominally isolated Cs atoms in vapour. This frequency is used internationally as the reference standard of time and has been adopted as such in the SI system of physical units. Caesium and its alloys are used in photocells, being photoelectric over a wide wavelength range. The element is dangerous to handle but compounds are only mildly toxic, with no known biological role.

56. Barium (Ba). In pure form, this is a soft, silvery alkaline earth metal of density 3510 kg m^{-3} and melting point of 727°C. As an element, it is very reactive and is stored under mineral oil or in argon. It forms a grey-black oxide layer when exposed to air and readily catches fire. It is widely distributed as the mineral barite, $BaSO_4$, which is mined in several places, notably China, India, Morocco and USA, and was once supposed to be magical because of its phosphorescence. This is a very stable, non-reactive and insoluble compound, and is the form in which most barium is used. The major use is as a heavy additive to drilling fluid in the oil industry. The first stage of processing to produce other compounds is carbon reduction in a furnace to obtain the sulphide, BaS, which is soluble in water. The element has several minor metallurgical applications, one being as a component of solder. An alloy with yttrium and copper, known by its acronym YBCO and mentioned in the entry for yttrium, is a superconductor with a transition temperature of 93 K, clearly above the temperature of liquid nitrogen (77 K). Barium compounds are generally white or colourless, useful as filler in paint and giving a high refractive index when added to clear glass, but a green colour in fireworks. Barium has no known biological function and its water-soluble compounds are poisonous and affect several bodily functions, but are non-cumulative. The carbonate, $BaCO_3$, which occurs naturally as the mineral witherite, is used as rat poison. However, the inertness (and indigestibility) of the sulphate (barite) and its X-ray opacity, arising from the high atomic number of barium, have given it a medical application in imaging the digestive tracts of patients who have ingested it.

Rare earth elements: an overview. Rare earth elements are not all rare. They are identified as the lanthanides, elements 57–71 (but with 61 not naturally occurring), plus yttrium (39) and scandium (21), which have similar properties and are difficult to separate. The relationship between yttrium, scandium and the lanthanides is seen in the periodic table (Figure 3.1). This note summarises common features of the rare earths, to minimise repetition in the individual entries. It indicates the fundamental reason for the wide range of specialised applications of various rare earths.

The reason for the long series of similar elements is that the lanthanides are normally trivalent, with electrons in 6s and 5d states responsible for chemical bonding, and 14 available 4f electron states that are partly occupied but are not normally chemically active. The number of 4f electrons progressively increases through the lanthanide series, from zero for lanthanum (element 57) to 14 for lutetium (element 71). They give the optical and magnetic properties on which most of the specialised uses of rare earths are based. Transitions between excited states of 4f electrons occur at optical frequencies and cause colouring or selective wavelength absorption or emission of light in glass, LEDs and similar devices. The magnetic properties arise from alignment of magnetic moments (spins) of the 4f electrons. Interactions between them in isolated atoms cause the moments of all of the first seven to be aligned parallel, and the next seven antiparallel to them, although magnetic properties are modified by interaction with electron orbits. This is Hund's rule of quantum interactions, which is familiar for the common ferromagnetic elements, Fe, Co, Ni, of the first transition series. If the atoms are close together, as in a metal, then the discrete electron states are merged into spread energy bands, so that identification of specific electron states and Hund's rule are compromised, but, as solid metals, several lanthanides are ferromagnetic at low temperatures. As alloying elements, they are used to adjust the properties of those permanent magnets for which the expense of rare ingredients is justified. Some rare earth compounds or alloys are superconductors with transition temperatures up to 90 K.

None of the rare earths has a recognised biological role, but they are mostly not seriously toxic.

Similarities in chemistry have allowed rare earth relative abundance variations to be used as tracers of geochemical processes. As generally observed through the periodic table, there is a systematic trend for elements with even atomic numbers to be more abundant than those with odd atomic numbers, by a factor typically 3 to 6. All of the rare earths are more concentrated in the crust than in the mantle (compare Tables 8.3 and 7.3), particularly so for the lighter lanthanides which are the only ones specifically identified as 'crustophile' in the geochemical classification used in the periodic table (Figure 3.1). Although normally all trivalent, under controlled oxidising conditions, some of the lanthanides become divalent or tetravalent, with properties that can be used in separation processes. As an example, in its divalent state, ytterbium has a complete complement of 4f electrons and is non-magnetic.

Three minerals dominate the exploitable sources of rare earths, all of them having igneous origins, with element segregations resulting from partial solidification of magmas. The most abundant is bastnäsite, a fluoro-carbonate $(RE)FCO_3$, (RE) being rare earths dominated by the lighter lanthanides. It is found in igneous carbonatites, with the richest deposit at Bayan Obo in Inner Mongolia and the next most important, and the first to be developed, at Mountain Pass in California. The other two major mineral sources are phosphates. Heavier lanthanides are found in monazite, $(RE)PO_4$, which occurs in low concentrations in granites and related metamorphics and is extracted from placer deposits derived from erosion of these rocks and found in Australia, Brazil, India and South Africa. Xenotime, also a phosphate, includes thorium and heavier lanthanides but is mined particularly for yttrium (as YPO_4), originally in Norway and now also Brazil and USA.

57. Lanthanum (La). One of the more abundant rare earths and the first in the lanthanide series, to which its name is given. Element density is 6162 kg m^{-3} and the melting point is 920°C. It is used to produce clear glass with a high refractive index and in rechargeable batteries. For many uses, it is not isolated from other lanthanides, which have similar properties. As fine flakes, it catches fire and is the active component in lighter flints.

58. Cerium (Ce). The most abundant lanthanide, with a range of uses, especially in its tetravalent state as oxide. The element density is 6770 kg m^{-3} and the melting point is 795°C. It is used in high-quality UV-transparent glass and as an oxidising catalyst in vehicle exhausts and self-cleaning oven linings.

59. Praseodymium (Pr). Element density of 6770 kg m^{-3} and a melting point of 935°C. It imparts a yellow colour to ceramics and to glass, which it makes opaque to UV. Pr is ferromagnetic below 85 K and is used as a component, with neodymium, in permanent magnets. An alloy with neodymium, didymium, is a constituent of the glass in welders' goggles, serving to reduce excessive glare by absorbing yellow light.

60. Neodymium (Nd). Element density of 7010 kg m^{-3} and a melting point of 1024°C. This is an essential component of strong permanent magnets (Nd–Fe–B compositions), for which small size is important. Gives blue light in solid-state lasers.

62. Samarium (Sm). Element density of 7520 kg m^{-3} and a melting point of 1072°C. This is used as an alloy with cobalt to produce permanent magnets that resist thermal demagnetisation in high-temperature applications.

63. Europium (Eu). Element density of 5264 kg m^{-3} and a melting point of 826°C. This is a rare element notable for red phosphorescence, used in liquid crystal displays and television screens.

64. Gadolinium (Gd). Element density of 7900 kg m^{-3} and a melting point of 1312°C. As the seventh element of the lanthanide series, gadolinium atoms or ions that are isolated or sufficiently widely separated in insulating compounds to avoid significant interaction between the 4f electrons of one another, each have seven parallel (uncompensated) electron spins, giving the 4f electrons atomic magnetic moments of seven Bohr magnetons ($7\mu_B$). This is possible also for europium in its divalent state. The distinction of having the strongest magnetic moment of any atom has been attributed to both dysprosium and holmium, but this depends on crystalline environments that allow alignment of electron orbital magnetic moments. Below its Curie temperature of 19°C, gadolinium is ferromagnetic, the only element apart from Fe, Co and Ni that is ferromagnetic under ambient conditions, but in the metallic state, the 4f electrons are spread into energy bands with much weaker spontaneous magnetisation than suggested by the $7\mu_B$ per isolated atom. The very strong paramagnetism of gadolinium ethyl sulphate (in which the maximum atomic moments are active) is used for magnetic cooling in cryogenic work (the magnetocaloric effect). Although toxic in soluble salts, gadolinium in injected chelated organic compounds is not absorbed and is used to enhance contrasts in magnetic resonance imaging (MRI). Gadolinium is rated as rare and expensive.

65. Terbium (Tb). Element density of 8230 kg m^{-3} and a melting point of 1356°C. This is very rare. It is used mainly for its green phosphorescence, but other colours in different materials. An alloy with iron and dysprosium, Terfenol-D (~$Tb_{0.3}Dy_{0.7}Fe_2$) displays extremely strong magnetostriction, changing dimension by 0.2% when magnetised to saturation, giving it applications in instrumentation and high-power acoustic transmission.

66. Dysprosium (Dy). Element density of 8540 kg m^{-3} and a melting point of 1407°C. This is an infra-red phosphor and, as a minor additive, improves the thermal stability of Nd–Fe–B permanent magnets.

67. Holmium (Ho). Element density of 8790 kg m^{-3} and a melting point of 1461°C. This imparts a yellow colour to glass and is also used in infra-red lasers. As an alloying element, it increases the saturation magnetisation of some ferromagnetic materials, which are used in the pole tips of permanent magnets to increase field strength by concentrating the magnetic flux. Although strongly paramagnetic, holmium is ferromagnetic only below 19 K.

68. Erbium (Er). Element density of 9066 kg m^{-3} and a melting point of 1529°C. A component of optical fibres, especially as an optical amplifier, and used also in infra-red lasers.

69. Thulium (Tm). Element density of 9320 kg m^{-3} and a melting point of 1545°C. This is rare. It is a minor component of infra-red lasers and has green optical emission lines.

70. Ytterbium (Yb). Element density 6900 kg m^{-3} and a melting point 824°C. This is a component of ceramic capacitors and specialised steels.

71. Lutetium (Lu). Element density of 9841 kg m^{-3} and a melting point of 1652°C. This is very rare, with limited applications. It has been used as a catalyst in the petrochemical industry and as a positron detector in nuclear physics. It is the heaviest of the lanthanides.

72. Hafnium (Hf). A silvery grey metal of density 13,310 kg m^{-3} and a melting point of 2233°C. It is generally made corrosion resistant by formation of a thin oxide layer, but will burn in air, especially if powdered. It is quite rare, and supply is limited, but it is chemically similar to zirconium and occurs as a minor constituent of zirconium ores, notably zircon, which is extracted particularly in Australia and South Africa. Separation of pure hafnium from zirconium was stimulated by requirements of the nuclear industry, because both are used in reactors in incompatible ways. Zirconium is not a significant absorber of neutrons and is used as fuel cladding. Hafnium is a strong neutron absorber, used in control rods, but would seriously degrade reactor performance as an impurity in the zirconium. A notable application of hafnium arises from its low-surface work function, which means that it emits electrons much more readily than most other metals. Small buttons of hafnium are inserted into the tips of plasma torches used for cutting metal. Hafnium has no known biological role and is not known to be toxic, but exposure to it is, in any case, rare.

73. Tantalum (Ta). A rare, hard, corrosion-resistant metal of density 16,690 kg m^{-3} and very high melting point (3017°C). It is chemically similar to, and occurs with, niobium, from which it is not easily separated. It occurs as tantalite, $(Fe,Mn)(Ta,Nb)_2O_6$, with various proportions of Ta and Nb and consequently different emphases in processing ores. A major producer is Australia, with separation as a by-product from niobium in Brazil and Canada, and it features in resource conflicts in Central Africa. An important use is in electrolytic capacitors for use in electronics. These have tantalum as the anode, with a very thin coat of pentoxide, Ta_2O_5, as the dielectric and manganese dioxide as the other electrode. The thinness of the dielectric layer, and its high dielectric constant, result in physically small capacitors, but, as electrolytic capacitors, they operate with only one electrical polarity, as in smoothing circuits, and are destroyed by reversed voltages. Tantalum is used as coating on other metals, such as steel, but some coating methods result in a metastable layer of a less familiar allotrope (β-Ta), which is soft and ductile, very different from the hard α form. Tantalum has no known biological role, is non-toxic and corrosion resistant, notably so to body fluids, making it a favoured choice for surgical instruments.

74. Tungsten (W). Alternatively known as wolfram, from which the chemical symbol derives, this is a hard, grey metal of high density, 19,250 kg m^{-3}, and the highest melting point of any element, 3422°C. Although carbon survives as solid to a higher temperature, it sublimes and does not have a melting point. The element has two phases; the normally encountered, stable body–centred cubic α phase and a metastable β phase with a more complicated crystal structure that tends to be stabilised by impurities. Tungsten is one of the siderophile (iron-loving) metals that have accompanied iron into the Earth's core, making them rare in the silicate, although concentrated in the crust relative to the mantle. Tungsten is not as rare as most of the others. There are two basic ore types, wolframite, $(Fe,Mn)WO_4$, which is found in granitic intrusions, and scheelite, $CaWO_4$, a metamorphic product found in hydrothermal veins, although they can coexist. They are mined in several countries, including the Central African conflict area, with production dominated by China. A major use is as tungsten carbide, WC, a very hard and tough material applied to the cutting edges of industrial tools, generally bound with metallic cobalt. The high melting point has made metallic tungsten the material of choice for the filaments of incandescent light bulbs, although that use is decreasing as more efficient fluorescent and LED lighting takes over, and for special applications such as high-temperature furnace windings. Tungsten is used as target material in X-ray tubes (as well as for the filaments) and for welding electrodes and has some metallurgical use in high-strength alloy steels. It is one of the platinum group metals that is used as a catalyst in industrial processes and is the only one with any known biological role, being found in a few micro-organisms, although it is injurious to animal life, in which it interferes with molybdenum and copper metabolism.

75. Rhenium (Re). This is a very rare element, chemically related to molybdenum, but with physical properties close to those of tungsten. Its density is 21,020 kg m^{-3}, exceeded only by close neighbours in the platinum group, and the melting point, 3186°C, is second to that of tungsten. It occurs as a minor constituent in molybdenite, the molybdenum ore that is itself a minor component of copper sulphide ores and is extracted from flue condensate in molybdenite processing. Chile and USA are the principal sources. The main use is as a constituent (a few per cent) of alloys used for military hardware (turbine blades and nozzles of jet engines), cost arising from its rarity being a restriction on wider use, although it has had application as a catalyst in combination with other Pt group metals. It has no known biological role but appears to be only mildly toxic, although with very limited testing.

76. Osmium (Os). This is a very hard, brittle, silvery metal with the highest density of all the elements (22,590 kg m^{-3}) and a melting point of 3033°C. It is very rare, occurring as a trace element in platinum ores and as native metal in alloys with iridium (osmiridium). As an element, it is difficult to work with, not just because it is brittle and arguably the hardest element, but because in air, particularly if powdered, it forms the tetroxide OsO_4, which is volatile and dangerously toxic. It is obtained as a by-product, with other Pt group metals, in the processing of Pt and Ni ores, in South Africa, Russia and Canada. It has some use in wear-resistant alloys and, as the oxide, in staining biological materials subjected to transmission electron microscopy, for which its high electron density enhances contrast. It has a very high UV reflectivity, useful in space missions, for which oxidation is not a problem. There is no biological role.

77. Iridium (Ir). A rare, hard and brittle white metal of density 22,560 kg m^{-3} (a close second to osmium) and a melting point of 2446°C. As a member of the platinum group, most of its uses are in alloys with Pt. As with other members of the group, its rarity in the crust arises largely from the fact that it is strongly siderophile and dissolved in the iron that forms the core. For the same reason, it occurs in meteoritic iron. Widespread sediments of age 65 million years, that are enriched in Ir, along with other siderophile elements consistent with iron meteorite compositions, presented evidence of a major meteorite impact at that time, although the argument that the impact was responsible for a mass extinction of species is not sustainable. Iridium is extracted as a minor trace element in the processing of sulphide ores of Ni and Cu, mainly in South Africa, Russia, Canada and also from placer sediments in Colombia. In addition to maintaining its strength at high temperatures, iridium is recognised as the most corrosion-resistant element, with special uses in crucibles used for producing high-purity single crystals for the semiconductor industry, in electrodes for electrolysis of brine in the production of chlorine (the chloralkali process) and in spark plug tips. Much of its use is in alloys with platinum as a catalyst. It has no known biological role or evidence of toxicity.

78. Platinum (Pt). A silver-white, malleable and ductile metal with a density of 21,450 kg m^{-3} and a melting point of 1768°C. It belongs to a group of dense, corrosion-resistant metals collectively referred to as the platinum group (Pd, Ru, Rh, W, Re, Os, Ir and Pt). It is rare in the crust, partly because, like other members of the group, it is siderophile and much of the Earth's inventory of these elements is dissolved in the iron core (as also in meteoritic iron). It is not as rare as most other members of the group and has a range of uses for which the expense can be justified. It is occasionally found naturally as metal, usually alloyed with other members of the group and with iron, but most of the production is from sulphide ores of Ni and Cu, in which it occurs as a mixed sulphide, cooperite (Pt, Pd, Ni)S. The major producer is South Africa, with smaller outputs in Russia, Canada and a few other countries. The Canadian production is from the arsenide, sperrylite $PtAs_2$. Two properties of platinum have led to its principal uses, usually with iridium as a minor alloying constituent, corrosion resistance and catalytic action. It is favoured for specialised laboratory ware and surgical instruments, and is used in jewellery. For use as a catalyst, the surface is treated to produce a skin of platinum black, which is metallic but has a very fine roughness that effectively exposes a large surface area. This is applied to vehicle exhaust systems to control emissions by oxidising unburnt fuel. Platinum is not known to have any biological role and is not significantly toxic, although its compounds can be irritants and a few are used in chemotherapy for some tumours.

79. Gold (Au). A dense, soft and malleable metal characterised by a bright orange–yellow colour, to which the colour name 'gold' is applied. This distinction from the grey of almost all other metals and an untarnishing shiny surface have given it pride of place in jewellery, with the colour controlled by alloying, which is how most people see it. Its density is 19,300 kg m^{-3} and the melting point is 1064°C. Gold is widely but very sparsely distributed, a remarkable fraction of it as native metal, either reasonably pure or alloyed with silver, copper and other metals. It is also found as telluride. Traditionally, it was obtained from placer deposits by laborious hand extraction from large volumes of gravel in situations such as stream beds. Discoveries in newly explored territories, especially in North America and Australia, prompted 'gold rushes'. Longer-term, industrial-scale mining for gold began in a major way in South Africa, with deep mines exploiting ancient placer deposits. Now gold mining has extended to many places, China, Australia, USA, Russia and Peru, as well as South Africa. The extraction methods are energy intensive because very large volumes of material are processed to isolate small amounts of gold. They are also seriously polluting, involving mercury and cyanide and exposing large masses of fragmented material from which poisonous solutes are readily leached.

Once used as currency, gold is still treated as a tradable commodity that is more secure than currency, hoarded by national banks, as well as by and for individuals as secure savings. A major use continues to be in jewellery, usually hardened by alloying, and widely accepted as a safe investment as well as a reference standard for wealth. Most industrial uses of gold rely on its combination of high electrical conductivity and corrosion resistance, but account for less than 15% of total production. The principal use is for electrical contacts in electronic circuitry and although it constitutes only a small fraction of the circuit material, that fraction is still larger than that in most ores, making it a prime target for recycling. Gold is non-toxic, being, in any case, inert to body fluids, and has no biological role; claimed medical uses are, at best, questionable, but it serves well in dentistry as tooth capping.

80. Mercury (Hg). Once known as quicksilver, this is a liquid metal under ambient conditions (melting point −38.8°C), with a density of 13,534 kg m^{-3}. It is not abundant in the crust but occurs in some useful concentrations, notably at Almaden in Spain, and is readily extracted, and therefore there is a long history of its use. The most common mineral and the only one exploited is cinnabar, HgS. The unique combination of physical properties of mercury under normal environmental conditions, that of a dense, opaque, fluid, electrical conductor, have given it a range of uses in thermometers, barometers, manometers and zenith tubes (telescopes accurately focussed on the local vertical, which is perpendicular to the precisely horizontal reflecting liquid surface). It is used as an electrode in the chloralkali process for chlorine production by electrolysis of brine. The major use is in fluorescent lights and mercury vapour lamps, widely favoured for lighting busy roads because they have high efficiency, with the UV emissions from an electrical discharge converted to visible (blue) light by phosphorescence of the lamp lining. Mercury readily dissolves other metals, forming amalgams, and this is the basis of its use in extracting traces of gold from crushed rock. Mercury has no known biological role, but dental amalgam is the common filling for cavities in teeth. Since mercury became generally recognised as a dangerous environmental contaminant, demand has decreased to the point that mines, including that at Almaden, have closed and most production is now in China. For more than a century, it has been known as a cumulative poison, identified with particular trades, such as millinery, in which mercury was used in felt preparation, causing 'hatters' disease, even recognised in the 'mad hatter' character of the Alice in Wonderland story. A tragic mass poisoning occurred over many years in the town and surroundings of Minamata in Japan, where thousands of residents, many of whom died, were incapacitated or deformed as a result of eating fish and other sea food in which mercury from a discharge of industrial waste was concentrated. 'Minamata disease' alerted the world to the danger of mercury poisoning, which is not limited to ingestion as it vaporises sufficiently readily for inhalation to be a danger.

81. Thallium (Tl). A soft, grey metal, which rapidly discolours in moist air by formation of a hydroxide layer. Its density is 11,850 kg m^{-3} and the melting point is 304°C. It is found in potassium-bearing minerals but extraction is as a by-product of processing sulphide or selenide ores of Cu, Pb or Zn. It has a small range of specialised uses as a trace dopant in semiconductors and sodium iodide scintillation detectors of nuclear radiation, and is under investigation as a constituent of high-temperature superconductors, but use and demand are limited by concern over its toxicity. It is an acute poison, reputedly more dangerous than arsenic. Once used as rat poison, it is not being used now because it is indiscriminate. It is soluble in water, redistributed through soil and readily absorbed through the skin.

82. Lead (Pb). A soft, malleable metal of density 11,340 kg m^{-3} and a melting point of 327.5°C. It is bright and silvery when fresh but rapidly tarnishes to dull grey with an oxide, hydroxide and carbonate coating. With this protection, it is corrosion resistant and survives from its extensive use in the Roman Empire. Lead is the most abundant of the heavy metals. It is mainly extracted as sulphide, galena PbS, commonly associated with Cu, Zn and Ag. There are many sources, Australia, China and USA being the biggest producers, and recycling methods are also well established. Three of the naturally occurring isotopes occur as end products of the radioactive decays of thorium and uranium and there are also short-lived lead isotopes in all of the decay schemes, the longest lived being ^{220}Pb, with a 20-year half-life, a product of ^{238}U. Freshly produced lead from ore that has not had a prolonged isolation from uranium is slightly radioactive and lead screening of sensitive nuclear detectors is made from ancient lead in which the ^{220}Pb has decayed. The non-radiogenic isotope, ^{204}Pb, is itself radioactive but with a half-life so long (1.4×10^{17} years) that for most purposes it is regarded as stable.

Lead has numerous uses, but some have been discontinued because of its toxicity. It is a cumulative poison, especially harmful to the development of young children. It is no longer used in paint and, in particular, lead-painted toys are now discarded. The demand for lead decreased when it was no longer used as a fuel additive (tetraethyl lead), which had the effect of distributing lead widely in the environment, or for water pipes. The lead–acid batteries, almost universal in the automotive industry, remain a major use and the one in which the recycling of lead is best developed. The softness and ease of melting make lead a large component of most solders, and it continues as the main component of small arms ammunition, although that use is under threat. The high atomic number (82) makes lead an ideal material for nuclear and X-ray screening and its incorporation in glass imparts some X-ray opacity.

83. Bismuth (Bi). This is a silvery-white, brittle metal with a density of 9780 kg m^{-3} and a melting point of 271.5°C. It exists as a single isotope, ^{209}Bi, and this is radioactive, although, with a half-life of 1.9×10^{19} years, it is effectively stable for almost all purposes. The metal does not corrode, either wet or dry, and it is one of the few materials for which the liquid is denser than the solid, making it easy to cast. Bismuth occurs as sulphide, oxide and carbonate minerals. The sulphide, bismuthinite Bi_2S_3, is the one of economic interest, being extracted as a by-product in the processing of lead, and also zinc and copper sulphide ores. Principal producers are China, Peru and Mexico. Bismuth has some use in alloys, especially those with low melting points, such as solder, and in semiconductor materials, particularly bismuth telluride. It is a component of high-temperature superconducting materials, and manganese bismuthide is a magnetic material with an extraordinarily high coercive force that may give it wide application if and when the stability of that compound can be improved. Although, in the periodic table bismuth appears in a series of poisonous heavy metals, it is non-toxic and non-cumulative, even having a medicinal use in the treatment of digestive disorders. Bismuth has no known biological role. The non-poisonous nature has made it a substitute for lead in applications such as bird-shot and fishing sinkers.

90. Thorium (Th). A silvery metal that rapidly tarnishes to black. Its density is 11,724 kg m^{-3} and the melting point is 1750°C. Like uranium, it is strongly concentrated in the crust, relative to the mantle (by a factor exceeding 100 for the average in the upper continental crust), with the result that it is not regarded as rare. It is radioactive, with a half-life three times the age of the Earth, and is sufficiently abundant to account for almost half of the total radiogenic heat in the Earth. It is widely distributed in a range of minerals and is extracted, along with the heavier rare earths, from phosphate minerals, monazite, found as sand from eroded granite in several places, including Australia and Brazil, and xenotime in Brazil and USA. Its uses and hence production are limited. It is still seen in the luminous mantles of gas lanterns and is a standard component of welding rods, maintaining reliable electric arcs by radioactive emissions that ionise adjacent air. The possibility of widespread use of thorium as a nuclear fuel is attracting interest in several countries, especially India and China. Thorium itself, ^{232}Th, is not fissile but neutron bombardment converts it to an isotope of protoactinium, ^{233}Pa, which beta-decays with a 27-day half-life to ^{233}U, a fissile uranium isotope. Advocates point out that advantages, relative to uranium/plutonium reactors, include less highly active waste and greater safety, not least because no bomb-making material would be produced. Thorium has no biological role and is toxic.

92. Uranium (U). This is a hard and dense (19,100 kg m^{-3}) metal, silvery when fresh but spalling off a black oxide when exposed to air. It melts at 1132°C. It is widely distributed in many minerals and is extracted as a mixed oxide, uraninite (pitchblende), UO_2 and U_3O_8, in several places, particularly Kazakhstan, Canada and Australia. Apart from the fact that it is the central element of the nuclear industry, its uses are limited. It has been used in colouring glass and ceramics, to which it imparts a red-yellow colour in white light and a strong green in UV light, but its radioactivity discourages applications. Uranium occurs naturally as two radioactive isotopes that both decay to isotopes of lead via series of intermediate radioactive daughter elements. The more abundant isotope, ^{238}U, which comprises 99.27% of the natural element, decays with a half-life, 4.5 billion years, close to the age of the Earth. The other isotope, ^{235}U, is now only 0.72% of the element, being more active, with a half-life of 0.7 billion years. It is the nuclear properties of this rarer isotope that have spawned the nuclear industry and the demand for uranium. ^{235}U is referred to as fissionable, meaning that if bombarded with neutrons, the atoms split into smaller fragments (fission products) and release more neutrons, allowing either a self-sustaining fission process, as in a power station reactor, or a runaway amplification, as in a weapon. The neutrons produced in a reactor convert the non-fissionable ^{238}U to a fissionable isotope of the element plutonium, ^{239}Pu, via short-lived neptunium, ^{239}Np. Neither of these elements has any natural occurrence, but the production of plutonium is an industrial-scale process, with uranium as the feedstock. A sideshow to these processes is uranium enrichment, that is production of ^{235}U in more concentrated form and this leaves depleted uranium, with only a third of the normal content of ^{235}U, in need of a use. As a hard and very heavy metal, it found a military use in armour-piercing ammunition, which, when fragmented, immediately catches fire, because uranium is pyrophoric, making the ammunition incendiary as well as armour-piercing and spreading uranium in an environmentally problematic way. Independently of its radioactivity, uranium is chemically poisonous.

22.2 PRODUCTION OF SELECTED METALS

Production of metals from ores is economic only if they are naturally concentrated relative to the crustal average. The necessary concentration factors for key metals are listed in Table 22.1. Tables 22.2 through 22.5 present production rates and reserves for countries reporting more than 1% of the global total for each of the most important metals.

TABLE 22.1 APPROXIMATE ORE CONCENTRATION FACTORS, THE RATIOS BY WHICH THE CONCENTRATIONS OF SELECTED METALS MUST EXCEED THE CONTINENTAL CRUSTAL AVERAGE TO PRODUCE ECONOMICALLY EXPLOITABLE DEPOSITS WITH CURRENTLY AVAILABLE TECHNIQUES

Metal	Concentration Factor	Upper Continental Crust Concentration (ppm by mass)
Fe	4–14	36,000
Al	4	80,000
Cu	90	27
Ti	100	3000
Zn	130	67
Pt	600	6×10^{-4}
Ag	1000	0.05
Pb	2000	17
Au	3000	1.5×10^{-3}
U	11,000	4
Hg	100,000	0.05

Note: Reported values vary according to accessibility and deposit size.

TABLE 22.2 COPPER, LEAD, NICKEL, TIN AND ZINC PRODUCTION IN THOUSANDS OF TONNES/YEAR, WITH RESERVES IN THOUSANDS OF TONNES (2015) BY COUNTRIES ACCOUNTING FOR MORE THAN 1% OF THE GLOBAL TOTAL

	Copper		Lead		Nickel		Tin		Zinc	
Country	Production	Reserves	Production	Reserves	Production	Reserves	Production	Reserves	Production	Reserves
Australia	960	88,000	633	35,000	234	19,000	7	370	1580	63,000
Bolivia			82	1600					430	4600
Brazil					110	10,000	17	700		
Burma							30			
Canada	695	11,000							300	6200
Chile	5750	210,000								
China	1750	30,000	2300	15,800	102	3000	100	1500	4900	38,000
Colombia					73	1100				
Congo	990	20,000					6.4			
Cuba					57	5500				
India									830	10,000
Indonesia					170	4500	50	800		
Ireland									230	1100
Kazakhstan									340	4000
Madagascar					49	1600				
Malaysia							3.8	250		

(Continued)

TABLE 22.2 *(Continued)* COPPER, LEAD, NICKEL, TIN AND ZINC PRODUCTION IN THOUSANDS OF TONNES/YEAR, WITH RESERVES IN THOUSANDS OF TONNES (2015) BY COUNTRIES ACCOUNTING FOR MORE THAN 1% OF THE GLOBAL TOTAL

	Copper		Lead		Nickel		Tin		Zinc	
Country	Production	Reserves	Production	Reserves	Production	Reserves	Production	Reserves	Production	Reserves
Mexico	550	46,000	240	5600					660	15,000
New Caledonia					190	8400				
Peru	1600	82,000	300	6700			22.5	130	1370	25,000
Philippines					530	3100				
Poland			40	1700						
Russia	740	30,000	90	9200	240	2900				
South Africa					53	3700				
Sweden			76	1100						
Turkey			54	860						
USA	1250	33,000	385	5000					850	11,000
Vietnam							5.40			
Zambia	600	20,000								
World total	18,700	720,000	4710	89,000	2530	79,000	294	4800	13,400	200,000

TABLE 22.3 URANIUM, COBALT AND MOLYBDENUM PRODUCTION BY COUNTRIES ACCOUNTING FOR MORE THAN 1% OF THE GLOBAL TOTAL IN TONNES FOR URANIUM, THOUSANDS OF TONNES FOR COBALT AND MOLYBDENUM

	Uranium		Cobalt		Molybdenum	
Country	**Production**	**Reserves**	**Production**	**Reserves**	**Production**	**Reserves**
Armenia					7.3	150
Australia	6689 (2015)	1,706,000	6	110,000		
Brazil			2.6	78		
Canada	9134 (2014)	494,000	6.3	240	9.3	260
Chile					49	1800
China			7.2	80	101	4300
Congo			63	3400		
Cuba			4.2	500		
Iran					4	43
Kazakhstan	23,127 (2014)	679,000				
Mexico					13	130
Namibia	3255 (2014)	382,000				
New Caledonia			3.3	200		
Niger	4057 (2014)	405,000				
Peru					18.1	450
Philippines			4.6	250		

(*Continued*)

TABLE 22.3 *(Continued)* **URANIUM, COBALT AND MOLYBDENUM PRODUCTION BY COUNTRIES ACCOUNTING FOR MORE THAN 1% OF THE GLOBAL TOTAL IN TONNES FOR URANIUM, THOUSANDS OF TONNES FOR COBALT AND MOLYBDENUM**

	Uranium		Cobalt		Molybdenum	
Country	**Production**	**Reserves**	**Production**	**Reserves**	**Production**	**Reserves**
Russia	2990 (2014)	505,900	6.3	250	4.8	250
South Africa			2.8	31		
USA	1919 (2014)	207,400			56.3	2700
Uzbekistan	2400 (2014)	91,000				
Zambia			5.5	270		
World total (year)	59,370 (2014)	5,500,000	124 (2015)	7100	267 (2015)	11,000

TABLE 22.4 PALLADIUM AND PLATINUM PRODUCERS IN TONNES/YEAR OR TONNES (2015) FOR COUNTRIES ACCOUNTING FOR MORE THAN 1% OF THE GLOBAL TOTAL

Country	Palladium Production	Platinum Production	Palladium and Platinum Reserves
Canada	24	9	310
Russia	80	23	1100
South Africa	73	125	63,000
USA	12.5	3.7	900
Zimbabwe	10	12.5	
World total	208	178	66,000

TABLE 22.5 IRON ORE AND BAUXITE[a] PRODUCTION IN MILLIONS OF TONNES/YEAR WITH RESERVES IN MILLIONS OF TONNES BY COUNTRIES ACCOUNTING FOR MORE THAN 1% OF THE GLOBAL TOTAL

	Iron ore		Bauxite	
Country	Production	Reserves	Production	Reserves
Australia	824	54,000	80	6200
Brazil	428	23,000	35	2600
Canada	39	6300		
China	1380	23,000	60	830
Guinea			17.7	7400
India	129	8100	19.2	590
Iran	33	2700		
Kazakhstan			5.2	160
Malaysia			21.2	40
Russia	112	25,000	6.6	200
South Africa	80	1000		
Ukraine	68	6500		
USA	43	11,500		
World total	3320	190,000	274	28,000

[a] For these minerals in particular, the ore production data give a better idea of where material comes from than iron and aluminium production because much of the ore is sent to other countries for processing.

Chapter 23

Rocks and Minerals as Resources

The elements are treated as resources in Chapter 22, although they are obtained as compounds in geological materials from which they are separated by processes involving heat and chemical treatments. The products have little resemblance to the starting materials. This applies most obviously to the metals. However, many rocks have uses that require only mechanical treatment, such as crushing or cutting, leaving them still identifiable as rock. The difference is relevant to resource use, exhaustion and recycling, as considered in Chapter 26, although recycling and primary resource conservation are equivalent only in special situations, such as the use of helium.

23.1 THE IRREVERSIBILITY OF RESOURCE USE

'Resource industries' has become a term for the enterprises that obtain materials from the Earth and convert them to conveniently usable forms. It is almost synonymous with 'extractive industries' and conveys a sense that their products are expendable and that use of them is irreversible. In cases such as the fossil fuels this seems obvious. However, by starting our discussion of resources with the catalogue of elements in Chapter 22, we are implying that permanent destruction is not a universal consequence of resource use. Constituent elements are not destroyed (except in nuclear reactors), they are redistributed in less useful ways. A. Valero and Al. Valero (2015) argue that, fundamentally, this is a thermodynamic problem, with the scattering and dilution of resources causing an increase in entropy that could, in principle, be reversed with sufficient expenditure of energy. Their point is that the calculated cost of using our resource inheritance should include the effort and energy expenditure that would be required to restore it to a condition of usability equivalent to its original state. In many cases, it is not obvious just what this means, or whether it is possible, in principle. Table 22.1 presents a perspective on the problem by comparing the concentrations of industrially important elements, relative to their average crustal concentrations, that are required

TABLE 23.1 ROCKS AND MINERALS THAT ARE USED, AT LEAST PARTLY, WITHOUT THERMAL OR CHEMICAL PROCESSING

Type	Material	Uses and Comments
Native elements	Diamond, graphite, sulphur, gold	See carbon, sulphur and gold in Chapter 22.
Salts, solutes, evaporites	Halite, NaCl	Common salt. The most abundant of the salts, obtained from all salt sources. Culinary and food preservation. Some use on icy roads but $CaCl_2$ is preferred. (Basic chemical feedstock used for caustic soda and derivatives from it.)
	Sylvite, KCl Potassium sulphate, K_2SO_4 Carnallite, $KMgCl_3.6H_2O$	The dominant use of potassium salts is as fertiliser. KCl is most abundant but the chlorides are unsuitable for some crops and sulphate is preferred.
	Gypsum, $CaSO_4$	Used as building plaster, but also as hydrated material, $CaSO_4.2H_2O$, which is familiar as plaster of Paris.
	Borax, $Na_2B_4O_7$	Generally obtained as decahydrate, $Na_2B_4O_7.10H_2O$, in evaporite deposits in California, eastern Turkey and elsewhere. Direct uses include as a cleaning agent, fungicide/pesticide, fire retardant and fertiliser [borosilicate glass (Pyrex), fibreglass, carbide/nitride abrasives].
	Others	Some naturally occurring salts are claimed to have health benefits, particularly from the Dead Sea and in dilute forms in various spas. (Numerous salts are used as chemicals)
Carbonates	Limestone, chalk, $CaCO_3$	High-quality stone used for buildings, notably Portland Stone from Dorset, UK. Powdered material is used as a soil conditioner, especially for acid soils. (Cement manufacture)

(Continued)

TABLE 23.1 *(Continued)* ROCKS AND MINERALS THAT ARE USED, AT LEAST PARTLY, WITHOUT THERMAL OR CHEMICAL PROCESSING

Type	Material	Uses and Comments
Carbonates	Dolomite, $CaMg(CO_3)_2$	Uses similar to limestone but not for cement.
	Marble (metamorphosed limestone)	High-quality, hard limestone favoured for sculpture and decorative stonework in buildings.
Igneous/metamorphic	Basalt	Very widely distributed. Dark, hard rock widely used as aggregate, for road base and railway ballast. Not widely used as building stone. (Converted to fibre for various uses including flameproof clothing.)
	Greenstone	A low-grade metamorphic basic/ultrabasic rock with attractive textures favoured for monuments and interior decorative stonework.
	Granite	A hard, granular rock with a range of interesting colours used in construction, monuments, paving and kerbing and as boulders for retaining walls.
Granular sediments, placer deposits	Mixed sand	General landscaping, play pits, pavement foundations. (Concrete)
	Separated sand grains	Individual minerals extracted from sand by sieving, washing and cycloning methods, including quartz, rutile and zircon, in some cases concentrated by wave action or river flow. (Most of the products are subsequently processed industrially.)
	Eroded kimberlite	Placer deposits of water-transported debris have yielded diamonds.

(Continued)

TABLE 23.1 *(Continued)* **ROCKS AND MINERALS THAT ARE USED, AT LEAST PARTLY, WITHOUT THERMAL OR CHEMICAL PROCESSING**

Type	Material	Uses and Comments
Fine sediments	Clay	Very fine grained sediment, notable for water retention and resistance to water flow. Added to very porous soils and used to seal dams. (Raw material for pottery and bricks.)
	Shale	Fine-grained sediment with clear sedimentary layering. May host 'unconventional' oil. Used as road base.
Metamorphic	Slate	Metamorphosed clay/mudstone/shale. Very fine grained, strongly foliated and readily separated into sheets. Water resistant and used as roofs and as cladding of exposed buildings. Large sheets favoured as bases of billiard tables.
Residual	Bauxite	Mixed oxide/hydroxide of aluminium, a residue of very prolonged, slow water erosion of a range of primary rocks or soil. Used as landscaping material. (Primary use is in aluminium production.)

Note: Uses of the same materials that do involve such processing are listed in parentheses.

for economically exploitable deposits. Recovery and reuse is generally simplest for the least-processed materials, not all of which are high volume ones. Table 23.1 lists materials with uses that do not require compositional modification by heat or chemical action.

23.2 MINERALISATION ASSOCIATED WITH TECTONIC SETTINGS

23.2.1 Crust–Mantle Separation

The term 'tectonics' means the combination of effects that are caused by mantle convection, starting with the separation of the continental crust from the mantle. This means chemical fractionation. Some elements, colourfully termed incompatibles, are not favoured by mantle mineral structures and are volcanically transferred to the surface, forming a chemically distinct veneer. It is characterised by a greater abundance of Al and lower abundance of Mg, relative to the mantle, giving rise to the terms 'sial (Si–Al)' and 'sima (Si–Mg)' as shorthand expressions for the crust and mantle compositions. The comparison of the elemental compositions of the crust and mantle in the right-hand column of Table 8.3 shows that the majority of elements are relatively more abundant in the continental crust than in the mantle and could be described as, at least mildly, incompatible. The big difference is in the Mg concentrations, because Mg is a common element and the absolute values of its concentrations strongly influence the statistics of element distributions. Some of the element concentration ratios, such as the strong crustal abundances of radioactive elements (K, Rb, Th, U), in Table 8.3 have been recognised for many years but others are less obvious. Very high crust/mantle ratios, such as for tungsten, invite closer searches for evidence of fractionation processes that could also occur within the crust and lead to economically interesting mineral concentrations.

23.2.2 Tectonic Situations Identified with Mineralisation

Some metamorphic products or accumulations of minerals can be formed in a variety of tectonic settings depending on heat, pressure, composition and circulating fluids. Meteoric water circulating close to the surface causes redistribution of minerals, with oxidation and chemical weathering, known as supergene processes. Hot brines circulating at depth within oceanic or continental crust may mobilise and deposit other suites of minerals. The geodynamic and geological settings associated with deposit genesis are commonly considered in exploration practices. Table 23.2 presents examples of deposit types and mineral assemblages that are found in various tectonic settings.

TABLE 23.2 MINERAL ASSOCIATIONS FOUND IN DIFFERENT TECTONIC SETTINGS

Tectonic Setting	Mineral System Group	Deposit Type (Type Location)	Metal Association	Fluids and Magma
Continental backarc	Basin-related fluid flow, with active magmatism	Carlin-type (Nevada)	Au-Ag-As-Sb-Hg	Meteoric, magmatic-hydrothermal
		Broken Hill–type/Sullivan-type Zn-Pb-Ag (Australia and British Columbia)	Zn-Pb-Ag-Cu-Au	basinal brines
		Kuroko-type VAMS (Japan)	Zn-Cu-Pb-Ag-Au	Evolved sea water[a]; magmatic-hydrothermal
		Algoma-type BIF (Canada)	Fe	Evolved sea water[a]
	Magmatic-related hydrothermal	Intrusion-related Au	Au-Ag-Sb-Cu-Pb-Zn-Sn-W-Mo-Bi-Te	Magmatic-hydrothermal; metamorphic
		Cloncurry-type IOCG (Australia)	Fe-Cu-Au-Ag-As-Co-Mo-P	
		Olympic Dam–type IOCG (Australia)	Fe-Cu-Au-Ag-U-REE-Co-Mo-P-Nb	Magmatic-hydrothermal, meteoric
Island backarc	Basin-related fluid flow, with active magmatism	Kuroko-type VAMS (Japan)	Zn-Cu-Pb-Ag-Au	Evolved sea water; magmatic-hyrdrothermal
		Cyprus-style VAMS (Cyprus)	Cu-Zn-Co-Ag-Au	
		Algoma-type BIF (Canada)	Fe	Evolved sea water[a]
	Ortho-magmatic	Podiform[b] chromite	Cr	tholeiitic ultramafic magmas

(Continued)

TABLE 23.2 *(Continued)* MINERAL ASSOCIATIONS FOUND IN DIFFERENT TECTONIC SETTINGS

Tectonic Setting	Mineral System Group	Deposit Type (Type Location)	Metal Association	Fluids and Magma
Continental arc	Magmatic-related hydrothermal	Porphyry[c]	Cu-Au-Ag-Mo	Magmatic-hydrothermal
		Epithermal[d] [adularia-sericite (K-feldspar and fibrous muscovite)]	Au-Ag	Meteoric; magmatic-hydrothermal
		Epithermal[d] [advanced argillic (clays including kaolinite, dickite and alunite)]	Au-Cu-Ag	Magmatic-hydrothermal
		Skarn[e]	Fe-Cu-Zn-Pb-Sn-W-Mo	
		Candelaria style IOCG (Chile)	Fe-Cu-Au-Ag-U-As-Co-Mo-P-Nb-Ni-REE	
	Ortho-magmatic	Intrusion-hosted Ni-Cu-PGE	Ni-Cu-Pt-Pd-Au-Co	Tholeiitic mafic magmas
		Strata-bound Cr-PGE	Cr-Pt-Pd	Tholeiitic mafic-ultramafic magmas
		Merensky Reef–type Ni-PGE (South Africa)	Ni-Pt-Pd	
		Podiform[b] chromite	Cr	Tholeiitic ultramafic magmas

(Continued)

TABLE 23.2 (*Continued*) MINERAL ASSOCIATIONS FOUND IN DIFFERENT TECTONIC SETTINGS

Tectonic Setting	Mineral System Group	Deposit Type (Type Location)	Metal Association	Fluids and Magma
Island arc	Magmatic-related hydrothermal	Epithermal[d] [adularia-sericite (K-feldspar and fibrous muscovite)]	Au-Ag	Meteoric; magmatic-hydrothermal
		Epithermal[d] [advanced argillic (clays including kaolinite, dickite and alunite)]	Au-Cu-Ag	Magmatic-hydrothermal
		Porphyry[c]	Cu-Au-Ag-Mo	
		Skarn[e]	Fe-Cu-Zn-Pb-Sn-W-Mo	
Forearc basin	Magmatic-related hydrothermal	Intrusion-related Au	Au-Ag-Sb-Cu-Pb-Zn-Sn-W-Mo-Bi-Te	Magmatic-hydrothermal; metamorphic
	Weathering and regolith, physical concentration, reduced (low O) atmosphere	Palaeo-placer Au-U	Au-U-Pt-Pd	Meteoric
Retro-foreland	Basin-related fluid flow, without active magmatism	Mississippi Valley–type Pb-Ag-Zn (USA)	Zn-Pb-Ag	Basinal brines

(Continued)

TABLE 23.2 *(Continued)* MINERAL ASSOCIATIONS FOUND IN DIFFERENT TECTONIC SETTINGS

Tectonic Setting	Mineral System Group	Deposit Type (Type Location)	Metal Association	Fluids and Magma
Collisional	Deformation and metamorphism	Lode[f] Au	Au-Ag-As-Sb-Te-W-Bi	Metamorphic; magmatic-hydrothermal
		Mt. Isa–type Cu (Australia)	Cu	Metamorphic
		Cobar-style Pb-Zn-Cu-Au (Australia)	Pb-Zn-Ag-Cu-Au	
		Tennant Creek–type IOCG (Australia)	Cu-Au-Bi-Se-Pb-Zn-U	Magmatic-hydrothermal, basinal brines and metamorphic
Mid-oceanic ridge	Basin-related fluid flow, with active magmatism	Cyprus-style VAMS (Cyprus)	Cu-Zn-Co-Ag-Au	Evolved sea water[a]; magmatic-hydrothermal
	Ortho-magmatic	Ophiolite[g]-hosted Cr, Ni	Cr-Ni	Mafic magmas
Continental rift	Basin-related fluid flow, with active magmatism	Broken Hill–type/Sullivan-type (Australia and British Columbia)	Zn-Pb-Ag-Cu-Au	Basinal brines
	Ortho-magmatic	Carbonatite-hosted REE	REE-P-F-Mo-Cu-Pb-Zn	Alkaline carbonatitic magmas
		Diamonds	diamond	Alkaline ultramafic magmas
		REE and P-rich nepheline syenite	REE-P-F	Alkaline felsic magmas

(Continued)

TABLE 23.2 *(Continued)* MINERAL ASSOCIATIONS FOUND IN DIFFERENT TECTONIC SETTINGS

Tectonic Setting	Mineral System Group	Deposit Type (Type Location)	Metal Association	Fluids and Magma
Continental rift	Ortho-magmatic	Anorthosite-hosted Fe-Ti-V	Fe-Ti-V	Tholeiitic mafic magmas
		Intrusion-hosted Ni-Cu-PGE	Ni-Cu-Pt-Pd-Au-Co	
		Komatiitic-hosted nickel sulphide	Ni-Cu-Pt-Pd-Au	Komatiitic mafic magmas
Rifted arc	Basin-related fluid flow, with active magmatism	Kuroko-type VAMS (Japan)	Zn-Cu-Pb-Ag-Au	Evolved sea water[a]; magmatic-hydrothermal
		Algoma-type BIF (Canada)	Fe	Evolved sea water[a]
	Magmatic-related hydrothermal	Epithermal[d] (adularia-sericite)	Au-Ag	Meteoric; magmatic-hydrothermal
Oceanic	Basin-related fluid flow, with active magmatism	Cyprus-style VAMS (Cyprus)	Cu-Zn-Co-Ag-Au	Evolved sea water[a]; magmatic-hydrothermal
Continental	Ortho-magmatic	Flood basalt–associated Ni-Cu-PGE	Ni-Cu-Pt-Pd-Au-Co	Mafic magmas
Ocean basin	Sedimentary	Mn-Ni-Co nodules	Mn-Ni-Co	Oxidised sea water
		Sedimentary sulphate	SO_4-Ca-Ba	
		Sedimentary phosphate	P	

(Continued)

TABLE 23.2 *(Continued)* MINERAL ASSOCIATIONS FOUND IN DIFFERENT TECTONIC SETTINGS

Tectonic Setting	Mineral System Group	Deposit Type (Type Location)	Metal Association	Fluids and Magma
Passive margin	Sedimentary	Sedimentary manganese	Mn-Ni-Co	Reduced sea water
		Hammersley-type BIF (Australia)	Fe	
	Basin-related fluid flow, without active magmatism	Mt. Isa–type Zn-Pb-Ag (Australia)	Zn-Pb-Ag	Basinal brines
Distal contraction (occurring in the outer part of the affected area)	Basin-related fluid flow, without active magmatism	Mississippi Valley–type Pb-Ag-Zn (USA)	Zn-Pb-Ag	Basinal brines
		Irish-style Pb-Zn	Zn-Pb-Ag	
		Sediment-hosted Cu-Co	Cu-Co-Ag	
		Unconformity U	U-P-REE-Cu-Au	
		Laisvall-type Pb (Sweden)	Pb	
		Kipushi-type Cu-Zn-Pb (Democratic Republic of Congo)	Cu-Zn-Pb	*Currently unknown*

(Continued)

TABLE 23.2 (*Continued*) MINERAL ASSOCIATIONS FOUND IN DIFFERENT TECTONIC SETTINGS

Tectonic Setting	Mineral System Group	Deposit Type (Type Location)	Metal Association	Fluids and Magma
Inactive continental	Basin-related fluid flow, without active magmatism	Rollfront-palaeo-channel U	U-P-REE	Basinal brines
	Weathering and regolith, chemical concentration	Lateritic Ni	Ni-Au	Meteoric
		Lateritic bauxite	Al	
		Calcrete Au-U	Au, U	
		Supergene enrichment	Au-Cu-Pb-Zn	
		Placer Ti-Zr-Th-Hf	Ti-Zr-Th-Hf	
		Placer Au	Au	
		Placer Sn-Ta	Sn-Ta	
	Weathering and regolith, physical concentration	Placer Ti-Zr-Th-Hf	Ti-Zr-Th-Hf	Meteoric
		Placer Au	Au	
		Placer Sn-Ta	Sn-Ta	
Variable	Ortho-magmatic	Pegmatite[h]	Ta-Sn	Felsic magmas
Pull-apart basin	Basinal fluid flow	Salton Sea–type Zn-Pb-Cu (USA)	Zn-Pb-Cu	Basinal brines

(Continued)

TABLE 23.2 *(Continued)* MINERAL ASSOCIATIONS FOUND IN DIFFERENT TECTONIC SETTINGS

Tectonic Setting	Mineral System Group	Deposit Type (Type Location)	Metal Association	Fluids and Magma
Pop-up (compression, reverse faults and thrusts)	Deformation and metamorphism	Lode[f] Au	Au-Ag-As-Sb-Te-W-Bi	Metamorphic; magmatic-hydrothermal

Note: BIF, banded iron formation; IOCG, iron oxide, copper and gold; PGE, platinum group elements; VAMS, volcanic-associated massive sulphide deposit.

[a] Evolved sea water is sea water enriched in minerals through the process of hydrothermal circulation.

[b] Podiform, a pod-shaped mass or lens (sometimes elongated) of ore typically of chromite in dunite or peridotite at ophiolite complexes.

[c] Porphyry, an igneous rock consisting of large crystals that provide a stockwork of veins through which ore minerals have been disseminated by hydrothermal fluid.

[d] Epithermal deposits form from warm water at shallow depth.

[e] Skarn, typically in the contact area between a carbonate sedimentary rock and an igneous intrusion – the reaction of the limestone to the acid (silica and iron rich) hydrothermal fluid leads to formation of specific minerals including diopside and tremolite in association with metal ore.

[f] Lode is a metal deposit that has filled a crack within or between layers of rock.

[g] Ophiolite is a sequence of uplifted rocks that include oceanic crust and upper mantle material with associated hydrothermal fluid deposits on and through the rocks.

[h] Pegmatite, a plutonic, igneous rock with exceptionally large grains and crystals, often from late-stage water- and metal- or mineral-enriched magma.

Chapter 24

Fossil Fuels

24.1 FUEL TYPES AND BASIC CHEMISTRY

Fossil fuels are fossils in the sense of being derived from buried organic material. As with virtually all living material, the central element is carbon and this is the essential component of all of the fuels. The biological matter that decomposes in the process of its conversion to fossil fuel does so under anoxic conditions, buried at various depths and prevented from oxidation by isolation from the atmosphere. The carbon compounds contain a variety of elements, of which all but one are incidental to fuel generation. That one is hydrogen, and it is a sufficiently large fraction of the decomposing material to become an important, even a major, ingredient of most fuels. These are basically hydrocarbons with proportions of H/C varying from almost zero for the highest rank of coal to 4:1 (atomic ratio) for methane, CH_4, the lightest hydrocarbon, which is the dominant constituent of oil-related natural gas as well as the 'unconventional' coal seam gas and shale gas. The formation and classification of the three fuel types, coal, oil and gas, are considered in Sections 24.2 to 24.4. Data on fuel production and reserves are presented in Sections 24.5 and 24.6. Section 24.7 lists the principal uses and the CO_2 released by combustion. Inevitably, some specialist jargon has developed, especially in the oil industry, and is referred to in Section 24.8.

The development processes of coal and oil are similar. The initial phase is compaction with very moderate heating, expulsion of volatiles and rearrangement of the chemical bonding of the organic materials. In the case of algae and bacteria-rich sea floor sediments, the consolidated organic material, recognised as the precursor of oil, is termed kerogen. The plant constituents in the peat formed from land vegetation remain more distinct. In both cases, this is referred to as the diagenetic phase. The second, catagenic, phase involves heating past 100°C and occurs with burial to greater depths. Kerogen starts to decompose, forming liquid oil and, with increasing heat, gas. Coal forms a progressively more compact solid with increased expulsion of volatiles, and the various plant constituents of peat assume different granular features (macerals, discussed in Section 24.2). The third identified phase, metagenesis, at temperatures above 150°C, is characterised by further increases in gas production, leaving coal or crude oil that is more carbon rich.

24.2 COAL

Coal is a product of long-term, and often deep, burial of land vegetation (grasses, trees, ferns, other plants and pollen), usually after preliminary decomposition in low oxygen (anoxic) environments, such as bogs and stagnant lakes. A normal first stage is peat, which can be regarded as the lowest rank of coal. As peat is converted to the higher ranks of coal, volatiles are lost with compaction to a harder solid. The amount of peat that has developed in the last 10,000 years amounts to 500 gigatonnes (5×10^{14} kg) of stored carbon, globally. As a carbon reservoir, this corresponds to a 1-mm-thick global layer of coal. The recognised reserves of coal, of various grades, would amount only to a 7-mm-thick global layer, although if all deposits that are too deep, thin or low quality to be of present commercial interest are included, the layer would be more like 4-cm thick. Since this has developed over 350 million years, these numbers indicate that only a small fraction of the peat that develops eventually becomes coal.

In considering the categories and descriptions of coal, it is helpful to have explicit meanings for the words that are applied. Although they are often used interchangeably, there are three keywords, italicised in the following quotation from Schopf's (1956) definition of coal:

> Coal is a readily combustible rock containing more than 50 percent by weight and more than 70 percent by volume of carbonaceous material, formed from compaction or induration of variously altered plant remains similar to those of peaty deposits. Differences in the kinds of plant materials (*type*), in degree of metamorphism (*rank*), and range of impurity (*grade*), are characteristic of the varieties of coal.

These keywords are used to distinguish three methods of classifying coals in Table 24.1, according to the purposes of the classifications. The classes are not mutually exclusive so that, for example, a coal classed as a bright type will generally also have a high rank, for example, anthracite.

Coal types are identified on a macroscopic scale by the proportion of bright (reflective) bands contained within the seams. Brightness profiles (Table 24.2) are logged to describe the overall general character of coal and may be used to assess variations in texture and composition within or between coal seams and coal measures. Coal type may also be identified on a microscopic scale by its maceral composition (Table 24.3).

TABLE 24.1 COAL CLASSIFICATION METHODS

Classification	Meaning	Significance
Type (see Table 24.2)	Describes the character of a coal as determined by its depositional origin and maceral composition (see Table 24.3)	Identified in seams or hand specimens by bright (reflecting) and dull bands and in thin sections by maceral content
Rank (see Table 24.4)	The metamorphic stage reached. Thermal maturity of coal determined by age, depth (pressure), temperature and length of time of burial	Indicates chemical and physical properties
Grade (see Table 24.5)	Refers to amount of non-organics, minerals and water present in the coal	Specifies industrial use and usefulness

TABLE 24.2 COAL TYPES CLASSIFIED BY BRIGHTNESS

Type	% of Bright Content
Bright	>90%
Bright banded	60–90%
Inter-banded dull and bright bands	40–60%
Dull banded	10–40%
Dull with minor bright bands	1–10%
Dull	<1%

With increasing temperature and pressure of burial, over geologic time, coal is converted to successively higher ranks (degrees of metamorphic processing). Hydrogen-containing volatiles, including methane, are lost, leaving an increasing proportion of carbon. Lignite or brown coal, the lowest rank (after peat), is typically 40% to 65% water and only 30% to 45% carbon, whereas the highest rank used as a fuel is anthracite with 90% to 95% carbon. It is the hardest coal with the highest energy output and has low moisture, volatiles and non-combustible particles (ash). The end product in the process of coal formation is graphite (effectively pure carbon), which is too valuable to be used as fuel. Graphite traces are found in metamorphic rocks, providing evidence that

TABLE 24.3 CHARACTERISTICS OF MACERALS

Maceral/Particle Type	Description and Formation	Reflectance (Defining the Brightness Profile)
Vitrinite	'Vitreous' or glassy in appearance, often in bands through coal seams. Formed from woody material at high temperatures. Low ash and volatile contents.	Moderate (medium bright). Anisotropic (two maxima and two minima in most orientations). The reflectance of vitrinite (VRo) is characteristic of coal rank (see footnote b to Table 24.4).
Liptinite	Formed from waxy and resinous spores and cuticles from leaf material. High in volatiles, including hydrogen. Causes bands of lower reflectance than vitrinite.	Lowest (dull). Reflectance varies little through most ranks.
Inertinite	A product of advanced fungal decomposition of woody material, incorporating charcoal. Unresponsive to heat, as in the coking process. Low in volatiles.	Bright. Higher reflectance than vitrinite but with a range of reflectance within this classification.
Mineral matter (non-maceral) also known as 'ash'	Inorganic particles, such as quartz and clay. Non-combustible material that becomes ash after combustion of coal.	Reflectance of non-organic matter not assessed in coal grain analysis.

organic carbon is widely distributed and more abundant than the concentrated forms in fossil fuels. Coal rank is generally judged by a microscopic assessment of its constituent macerals. These are granular constituents derived from different plant components, summarised in Table 24.3. They are analogues of the minerals in inorganic rocks but are all of biological origin. Routine microscopic examination measures their reflectances, as in footnote b to Table 24.4.

24.2.1 Coal Grade and Non-Carbonaceous Constituents

Although coal is essentially composed of carbon, hydrogen and oxygen, there is a wide range of minerals present in most coals, in various concentrations.

TABLE 24.4 COAL RANKS: MEASURES OF THERMAL MATURITY

Rank	Carbon Content[a] (mass %)	Volatile Matter[a] (mass %)	Vitrinite Reflectance[b]	Calorific Content (MJ/kg)	Transformation Phase	Max Burial (m)	Hardness	Relative Burial Depth and Time
Peat	<60%	63%–69%	<0.3%	14.6–16.7	Diagenesis		Softest	Shallowest burial and most recently deposited
Lignite (brown coal)	60%–71%	53%–63%	0.3%–0.38%	16.7–18.8		1500		
Sub-bituminous	71%–76%	42%–53%	0.4%–0.6%	17.5–29.3		2100		
High-volatile bituminous	76%–80%	31%–46%	0.5%–1.15%	23.0–29.3	Catagenesis			
Medium-volatile bituminous	82%–86%	23%–31%	1.2%–1.5%	29.3–33.5				Deepest burial for longest, from around 350 million years ago

(*Continued*)

TABLE 24.4 (*Continued*) COAL RANKS: MEASURES OF THERMAL MATURITY

Rank	Carbon Content[a] (mass %)	Volatile Matter[a] (mass %)	Vitrinite Reflectance[b]	Calorific Content (MJ/kg)	Transformation Phase	Max Burial (m)	Hardness	Relative Burial Depth and Time
Low-volatile bituminous	86%–90%	14%–23%	1.5%–1.9%	33.4–36.2		6000		
Semi-anthracite	90%–91%	8%–14%	1.9%–2.6%	32.6–33.4	Metagenesis			
Anthracite	91%–92%	<8%	2.6%–5%	31.8–32.6		7600		
Meta-anthracite	>92%	<8%	>5%	31.8			Hardest	

Note: Listed values refer to dry ash-free samples.

[a] These are not mutually exclusive categories because volatiles contain carbon.

[b] Reflectivity is observed under precisely specified, automated conditions to ensure clear physical significance of differences between observations on different samples and compatibility of measurements in different laboratories. A narrow (2 to 15 μm) beam of filtered green light, of wavelength 546 nm, is directed at 45° incidence on to the carefully polished surface of a sample that is held under a specified oil (density 1515 kg m^{-3}). The reflected beam is observed with a microscope fitted with an oil-immersion objective lens and a photomultiplier, monitored by a computer-controlled microprocessor. Reflections from standard reference materials, such as sapphire, are used for repeated calibrations. Average reflectivities of 20 to 100 grains may be recorded or, if polarised light is used, maximum, minimum and mean reflectivities are recorded as the specimen is rotated through 360°. The reflectivities, as normally observed for specimens under oil, are all very low. Reflections in air are of order 10 times stronger.

The most common inorganic minerals are quartz, clays (kaolinite, illite, montmorillonite and chlorite), pyrite, calcite and siderite. Coals that have high concentrations of these materials are classed as low grade. Most commercial coals are of bituminous compositions and a general overview of a bituminous coal's elemental content is given in Table 20.10, but it varies widely within and between coal formations. Table 24.5 summarises coal grades.

The commercial value and potential uses of coal are determined in a number of ways. Coking coal for example is selected by rank, physical properties

TABLE 24.5 TYPICAL COAL GRADE DIVISIONS

General Description	Other Terms Used		Character	Non-Organic/ Mineral Material
Low grade[a]	Steaming, non-coking, thermal		Low proportion of organic matter, high proportion of mineral matter compared with high-grade coal	30 to 55%
Medium grade	metallurgical	Semi-coking	Intermediate proportions of organic and mineral matter	Not exceeding
		Semi-coking, grade I		19%
		Semi-coking, grade II		19 to 24%
High grade[b]		Coking	Low proportion of mineral matter, high organic matter, low sulphur and phosphorus	
		Steel grade I		<15%
		Steel grade II		15% <18%
		Washery grades[c] I–IV		18 to 35%

Note: Coal grade describes the proportion of inorganic (mineral) material present and helps to identify the economic value and usefulness for particular industrial applications.

[a] Low-grade coal is typically considered to be high ash, high moisture and low energy, although some low-grade coals are reported as having as little as 18% non-organics and up to 55% for the lowest energy coal.

[b] High-grade coal is typically considered to be low ash, low moisture and high energy, with the highest grade coking coals containing less than 10% ash, although ash contents up to 35% may occur before crushing and washing. Coking coals are high-grade coals that devolatilise and swell on heating and resolidify to produce coke.

[c] Washery grade is coal that would be washed after crushing to reduce dust and non-organic materials to improve grade.

(coke strength, swelling index, caking and plasticity) and organic properties, and can be hard or soft. Pulverised Coal Injection (PCI) coal (used in steel-making blast furnaces) may be softer than high-energy thermal coal. Both soft coking coals and PCI coals are cheaper than hard coking coal and are used to reduce the cost of steel-making operations. Thermal or 'steaming' coal is classified as that which is burnt to produce electrical power and this is also the term used for all except coking and PCI coals. However, the thermal properties (specific energy, volatile and carbon contents), moisture content, abrasion and grindability indices, in addition to ash and sulphur contents, are attributes used to assess the values of different thermal coals. The amounts and types of minerals present also impact on a coal's suitability for use, as the ash must be removed after combustion. Coals that have significant amounts of calcium and iron-bearing minerals, such as calcite and siderite, are satisfactory for use in bottom-ash furnaces, the ash being removed as a molten slag, and minerals such as calcite and those with high melting temperatures, such as quartz and clay, can be used in boilers that produce fly ash (ash driven from the boiler with flue gases). Additionally, some coal uses are sensitive to small quantities of particular elements. For example, in selecting coking coal, used in steel production, elements such as sulphur and phosphorus, which may be present in minerals such as pyrite and apatite, are avoided or else the coal is processed before use to reduce their abundances.

Since nearly all mined coal is ultimately destined to be burnt, there is a particular concern with the constituents that become gaseous combustion products (CO_2, H_2O, SO_2, N_2, N_2O and NO_x) of atmospheric or other environmental concerns (Table 24.6) (see also Table 24.13).

24.2.2 Variations in Coal Formation with Time, Latitude and Environmental Conditions

Coal started forming almost as soon as land plants became available, about 350 million years ago. It has been forming throughout subsequent geological history, as shown in Figure 24.1a, but the earliest coal is by far the most abundant. Possible contributory causes include the later development of microorganisms that digest lignin, a constituent of wood that resists decomposition. It may also be significant that the Carboniferous period appears to have been particularly favourable to the growth of abundant vegetation. Although the variation in abundance with the latitude of formation (Figure 24.1b) indicates formation at very high latitudes, this may be, in part, a consequence of global climate variations, with a particularly warm period during the time of Carboniferous coal formation (Table 24.7).

TABLE 24.6 ELEMENTS WITH VOLATILE COMBUSTION PRODUCTS IN A TYPICAL COMMERCIAL COAL

Constituent	Typical Content Range (mass % dry, ash-free)
Carbon	60% to 95%
Hydrogen	<3.75% to 6%
Oxygen	<2.5% to 34%
Sulphur	0.5% to 3.0% (higher Sulphur coals can contain up to 10%)[a]
Nitrogen	~ 1.5%
Phosphorus	<0.03% for coking coal; thermal coal can contain more phosphorous
Mercury	0.10 ± 0.01 ppm
Arsenic[b]	1.4 to 71 ppm
Selenium[b]	3 ppm

[a] Some higher sulphur coals are used for power generation.
[b] Some coals have much higher arsenic and selenium contents.

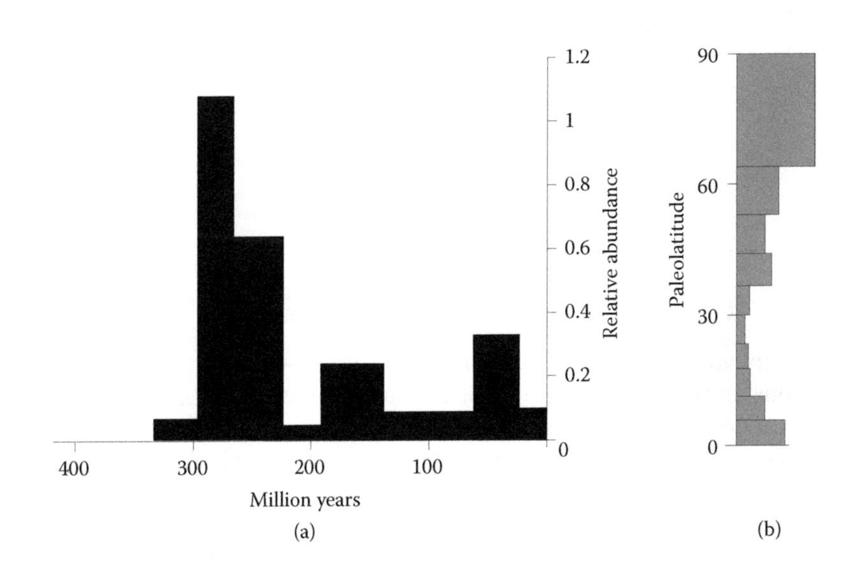

Figure 24.1 (a) Development of global coal reserves over time, represented by shaded areas proportional to the totals for different time intervals and (b) palaeolatitudes at which coal deposits formed, many of which are now found at very different latitudes. Based on palaeomagnetic data by Tarling (1983).

TABLE 24.7 CONDITIONS AND VEGETATION DURING COAL-FORMING PERIODS

Age	Character	Example Basins
Carboniferous (360–290 Ma)	Warm, moist, tropical to sub-tropical, gymnosperms, cycadophytes	UK and French coal fields, Appalachian, USA, Donets (Ukraine) and Kuzbass (Russia)
Permian (290–251 Ma)	Warm wet summers, freezing winters, gymnosperms	Eastern Australia, India, South America, South Africa, China
Triassic (251–205 Ma)	Similar to Permian although warmer, conifer forests, ferns, gymnosperms	Australia (Callide, Tarong)
Jurassic (205–141 Ma)	Warm, generally wet (Pangaea) but with large dry areas, angiosperms, gymnosperms, cycads	Australia (Gunnedah, Walloon, Milmerran) and Russia (Yakutia, Pechora)
Cretaceous (141–65 Ma)	Cool to warm climate, mainly angiosperms	Canada, USA (Wyoming, Colorado), Spitsbergen, New Zealand, Venezuela
Tertiary (65–17.78 Ma)	Warm, mainly angiosperms	Indonesia, New Zealand, Australia, China, Germany, Japan, USA

24.3 OIL

24.3.1 Generation, Migration and Trapping

Most exploitable oil is derived from marine organisms which rain onto the sea floor as they die and are buried by accumulating sediment. The fragments of decaying plankton and other debris accumulate algae and bacteria that become a significant component of kerogen, the insoluble solid material formed by burial and compression of organic matter. This is the source material that develops into oil. It remains a stable component of the sediment until it is deep enough to reach a temperature of 100°C, when the carbon bonds in the complex molecules start breaking, releasing the lighter hydrocarbon molecules of oil and gas. This process is termed maturation. At this stage, the gas remains dissolved in the oil. With further maturation, at temperatures up to about 150°C, increasing fragmentation produces more gas, which may be too abundant to remain dissolved and separates out. The gas molecules are identified as those with five or fewer carbon atoms, the heavier ones remaining as liquid. Up to 50% of the kerogen may be converted, the rest being unproductive and described as 'inert'. The light molecules, especially methane, have higher H/C ratios than the liquid, leaving the residual oil with larger, more carbon-rich molecules. What may start as a light oil becomes progressively heavier and, for maturation proceeding that far, the end point is conversion of virtually all of the oil to gas, leaving a residual tar layer (asphaltene).

Being derived from complex organic materials, oil is not just a mix of simple hydrocarbons. Only for the light gases are the molecules just linear (or branched) chains of carbon bonds. Double bonds and ring (aromatic) structures are important in the heavier material. Since they give lower H/C ratios than in the light gases, as H-rich gas is expelled, these structures accommodate the increasingly C-rich composition of the remaining liquid. There are also other elements that influence the properties and one, sulphur, is of particular concern and is extracted at an early stage in the processing of crude oil. This is now the source of all of the sulphur used in the chemical industry. Oil that is naturally low in sulphur is described as sweet.

Oil and gas are lighter than the matrix of their sedimentary source rocks and pore water, so their buoyancy drives them into any neighbouring porous material that is at a higher level, even if only slightly so. Although the driving force is upwards, sedimentary basins are normally more or less horizontally layered and the easiest path may be close to horizontal, so that some oils are found at considerable distances from their source rocks. They are trapped if they reach a porous rock that is so capped by impervious material that further migration is blocked and an oil or gas reservoir is formed. Otherwise, the oil and gas leak into the atmosphere or ocean and this appears to be the end result for most of what has been produced over the last several hundred million years.

To be useful, a reservoir rock must be not just porous but permeable enough to allow flow within it. In referring to permeability, the oil industry has adopted a measurement unit, the darcy $= 0.987 \times 10^{-12}\ \mathrm{m}^2 \approx (1\ \mu\mathrm{m})^2$, originally introduced for Henri Darcy's 1856 study of the flow of water through sand and still used in hydrogeology as well as oil reservoir engineering. Although the definition (Section 24.8) makes it appear remote from modern practice, it gives convenient numerical values. Some reservoir rocks have permeabilities of several darcys and values below 1 millidarcy make oil extraction slow and expensive.

24.3.2 Variations in Oil Formation with Time and Latitude

The earliest oil appeared 200 million years before the first coal (compare Figures 24.2a and 24.1a). This is consistent with the earlier appearance of life at

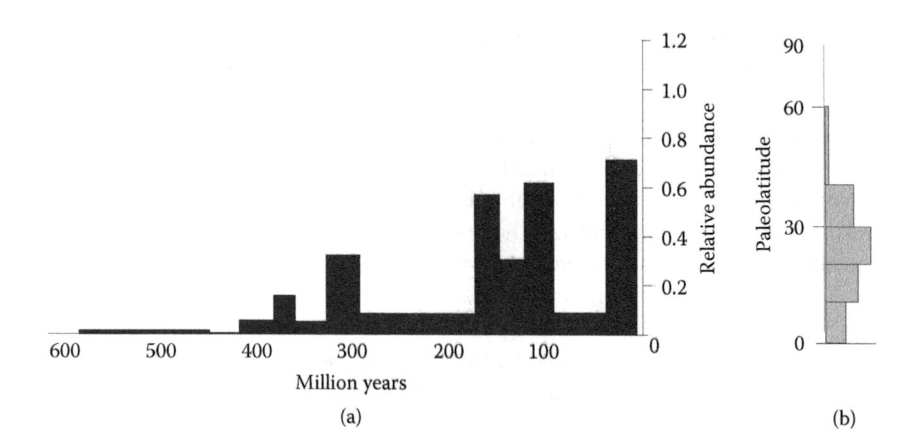

Figure 24.2 (a) Ages of source rocks of exploitable oil. (b) Palaeolatitudes (latitudes of formation) of oil deposits, from palaeomagnetic data by Tarling (1983).

sea than on land. However, the very early oil is rare and the trend of Figure 24.2a is a general increase with time. This can be understood from the tendency of oil (and even more so gas) to leak and to be lost to the atmosphere and ocean with time and with geological disturbance. The latitude distribution of oil generation in Figure 24.2b is also quite different from the corresponding figure for coal. It is not obvious why there is no high-latitude oil generation, but see Section 24.4.2.

24.3.3 Unconventional Oil

Oil that does not naturally flow from reservoirs, but requires unconventional means to extract it from the reservoir rock, has become an economically viable resource over the past few decades, and the term used is unconventional oil. Tight oil (also known as shale oil) is the oil that is trapped in rock that has extremely low permeability and requires hydraulic fracturing (fracking) techniques to allow the oil to flow. Not to be confused with shale oil, another resource is oil shale, which is a collective term for a range of sedimentary rocks, not all being shale, that contain marine-derived kerogen and are immature oil source rocks found at or near the surface. They do not contain significant oil but can be made to yield synthetic oil by heating (retorting) mined rock, effectively accelerating the maturation process. Before development of the now conventional oil industry, shale oil was produced in several countries, including Scotland, where production was continuous for 100 years starting early in the 19th century. Although not competitive now without a particularly rich source rock, for strategic or political reasons production continues in a few countries, notably Estonia, where it is the only locally derived energy source, and production exceeds domestic demand making it an export product. Countries reporting current or recent production are Brazil, China and Estonia with outputs of 0.44, 0.04 and 0.46 GW, respectively. These are small numbers compared with those in Table 24.8. But oil shales are widely distributed around the world and in most places their existence is noted, without significant steps towards exploitation, waiting for two developments: a sufficient rise in the price of conventional oil to make shale oil competitive and hoped-for technical advances that will reduce the serious environmental impact of shale oil processing. Tar sands, notably those in Alberta, Canada, present opportunities and problems similar to those of shale oil.

The kerogens in almost all coals, even of low rank, have lower hydrogen contents than the algal kerogen in marine sediments. The vitrinite in coal is richer in hydrogen than the coal average and can yield some liquid with maturation, but it is not expelled, remaining held in the coal until further maturation converts it to gas. Coal seams are a source of gas but not oil, and coal kerogen is often not recognised as kerogen, which is regarded as an oil source. Nevertheless, the production of synthetic oil from coal has been pursued when strategic or political considerations over-ride the economics, as in South Africa during the sanctions of the apartheid era. The economics are most favourable for countries with abundant coal but little or no oil, which include South Africa and also China, where there is a strong interest. Several industrial-scale processes have been used, a basic requirement being the need to introduce hydrogen. The relative merits of shale oil and coal liquefaction as eventual substitutes for petroleum are argued by their proponents, but conclusions appear to be dictated by the availability of raw material.

24.4 GAS

24.4.1 Sources

Gas consists of hydrocarbons with five or fewer carbon atoms per molecule and is normally dominated by methane. It can be regarded as an end product in the maturation of both oil and coal, but most of the natural gas that is used comes from oil fields. It is also recovered from coal and from sedimentary rocks with high biological contents that have not been subjected to the temperatures and pressures that lead to oil and coal (gas and oil shales). It is one of the three major fuels with production rates and reserves detailed in Tables 24.8 and 24.9. For many years, natural gas was treated as a waste product by the oil industry and flared off, burnt in spectacular flames over oil fields. Although its value as a fuel is now well recognised, and it has become an exploration target for its own sake, flaring off continues where inertia inhibits abandonment of old industrial practices. In addition to gas extracted from conventional reservoirs, unconventional sources are available, including coal seam gas and tight or shale gas. Gas that is associated with coal but is insufficient as a resource is released by ventilation of coal mines, in which its mixture with air is dangerously explosive. This fugitive gas is not as easily controlled or collected as is oil field gas, but coal seam gas is now extracted from seams that are not also being mined for coal. Exploitation methods for tight or shale gas require artificial stimulation through cracking (fracturing or fracking) of rock to mobilise the gas and for management of groundwater, which can be problematic and environmentally contentious. Some coal seam gas resources also require artificial fracturing to allow the gas to flow. For both, commercial activity is accelerating, with shale gas developing faster globally.

For both coal seam gas and shale gas production, control of methane leakage (fugitive gas) is a problem that draws attention to atmospheric methane and the natural sources that release it. The global perspective is indicated by the historical record of its atmospheric concentration in Table 12.3. Since methane survives in the atmosphere for about 10 years, the natural (pre-industrial and pre-intensive animal husbandry) rate of leakage to the atmosphere was about 200 M tonnes/year, 5% of the present rate of industrial gas production. The total rate of increase in atmospheric methane is now three times this, with the digestive systems of cattle often blamed, although they cannot plausibly be the major contributor. The natural leakage from the Earth is part of the global carbon cycle and the magnitude indicates how significant it is. Much plant material is short-lived, or even seasonal, and most of that reverses its photosynthesis by releasing CO_2 as it decays. But an important fraction decomposes under anoxic conditions, yielding methane, without accumulating as large deposits and going through the extended processes of burial, compression and heating that produce the fossil fuels. Marsh gas, gently bubbling from stagnant ponds, is a familiar example, but methane clathrate is another one of particular interest.

24.4.2 Methane Clathrate

This is an ice-like solid consisting of a framework of water molecules encaging methane molecules. It is formed when methane under moderate pressure meets water at or close to its freezing point. It is unstable when released from pressure or warmed, allowing the methane to escape, but it is stable in high-latitude marine sediment and in permafrost. It is widely distributed and estimates of its total abundance vary, some suggesting that it is a store of methane comparable to the fossil fuel reserves. It appears to be too widely distributed for economic harvesting but is fundamentally very significant. It is of biological origin, a decay product of high-latitude biota, presumably marine life-forms, raw material for oil that took a different decay path, possibly answering the question raised by Figure 24.2b, why is there no high palaeolatitude oil?

24.5 FUEL AND ENERGY

24.5.1 Physical Units and Conversion Factors

Quantities of different fuels are reported in a variety of units; conversion factors presented here allow comparisons. The essential measure is the energy released by complete combustion and the fundamental significance is most obvious when SI units are used, allowing direct comparison with the energies of the natural processes considered in Chapter 25. This means writing energies in joules or, for the large quantities handled by the energy industries, gigajoules (1 GJ = 10^9 J) or petajoules (1 PJ = 10^{15} J). For rates of consumption, the fundamental unit is the watt (1 W = 1 joule/second) but for the geophysical scale, the practical units are the gigawatt (1 GW = 10^9 W) and the terawatt (1 TW = 10^{12} W = 31,557 petajoules/year). The British thermal unit (1 BTU = 1055.06 J) has often appeared in American usage, with its multiple, the quad (quadrillion BTU = 10^{15} BTU = 1055.06 PJ). Primary data sources for the rates of fuel production and reserves are the International Energy Agency's (IEA) Key World Energy Statistics and the BP Statistical Review. Differences between these data sets serve to emphasise that the numbers are approximate. Reasons for this mainly arise from variability in the compositions of fuels. There is also a difference in reporting energies involved in electricity generation. The BP report assumes that generation of electric power from fossil fuels is 38% efficient, a value appropriate for modern power stations, but the IEA data assume 33%, which appears to be a better estimate of the global average. This problem arises when electricity generation by means other than fossil fuels (e.g. nuclear and hydro) is multiplied by a factor that converts it to the fuel energy that would be required for the same generation. The confusion is avoided in data reported here, in which fuel energy is total thermal energy and electricity generation is reported as electric power. The BP data are generally more recent but reuse some earlier IEA data.

In reports of the amounts of fuels produced, oil is frequently used as a reference standard, with the tonne of oil equivalent (toe), or million tonnes of oil equivalent (Mtoe), as a unit of fuel energy. Since oil is variable, this is not really directly useful and an assumed standard average is applied, commonly

1 toe = 41.868 GJ of fuel energy, but slightly lower values are also used. Even for oil itself, quality can be assessed to assign a toe value that does not coincide with the actual mass. Similarly, for coal, which is more variable, toe values can be applied, although, in the absence of quality control, a notional average rating may be assumed: 1 toe = 1.428 tonnes of coal, or higher values for higher ranks. Natural gas production is recorded in cubic feet, cubic metres or millions or billions of these volumes at standard pressure. 1 cu ft = 0.02832 m^3 and the energy rating is 39,100 cu ft = 1107 m^3 = 1 toe. All of these can be translated to joules with the conversion ratio 1 Mtoe = 41.868 PJ and, in referring to annual production, 1 Mtoe/year corresponds to an average rate of 1.327 GW.

In spite of the adoption of the toe as a reference standard, production of oil is generally reported in barrels. Although the barrel is precisely defined (158.99 L, originally 42 American gallons), the density of crude oil is quite variable and barrels do not immediately translate to tonnes. But since a further conversion to toe is required to account for oil quality, in marketing oil the problem is avoided by reporting quantities in barrels of oil of specific types, such as (in USA) West Texas crude and (in Europe) Brent (North Sea) light. A reference average valuation is 1 barrel = 0.146 toe = 6.113 GJ, but lower values are also reported.

In the listing of the production of fuels in the following section (Table 24.8), all values are converted to average rates in GW, to simplify comparisons not just with one another but with natural energy dissipations and non-fossil fuel alternative energies in Chapter 25. Proven reserves in GW years are listed in Tables 24.9 and 24.10.

24.5.2 Rates of Production

TABLE 24.8 FOSSIL FUELS PRODUCED BY COUNTRIES REPORTING 1% OR MORE OF THE GLOBAL TOTAL FOR ANY FUEL

Country	Coal	Oil	Natural Gas
Global total	5220	6274	4163
Algeria	–	108	100
Angola	–	121	–
Argentina	–	–	43
Australia	373	–	67
Azerbaijan	–	60	–
Brazil	–	166	–
Canada	49	304	195

(Continued)

TABLE 24.8 *(Continued)* FOSSIL FUELS PRODUCED BY COUNTRIES REPORTING 1% OR MORE OF THE GLOBAL TOTAL FOR ANY FUEL

Country	Coal	Oil	Natural Gas
China	2448	300	162
Colombia	76	70	–
Egypt	–	–	59
Germany	58	–	–
India	323	63	–
Indonesia	374	60	88
Iran	–	256	208
Iraq	–	232	–
Kazakhstan	73	120	–
Kuwait	–	221	–
Mexico	–	197	70
Netherlands	–	–	67
Nigeria	–	167	46
Norway	–	134	131
Oman	–	67	–
Pakistan	–	–	51
Poland	73	–	–
Qatar	–	134	213
Russia	227	767	696
Saudi Arabia	–	814	130
South Africa	200	–	–
Thailand	–	–	51
Trinidad	–	–	51
Turkmenistan	–	–	83
United Arab Emirates	–	263	70
United Kingdom	–	60	44
United States	674	824	876
Uzbekistan	–	–	69
Venezuela	–	192	–

Note: Values are average rates of production in 2014 in gigawatts (GW). Some reports list numbers differing from these, primarily because they include still developing (unconventional) sources from shales.

24.6 FOSSIL FUEL RESERVES

TABLE 24.9 CONVENTIONAL FUELS: 'PROVEN RESERVES' AT THE END OF 2014, AS CITED IN THE BP STATISTICAL REVIEW, CONVERTED TO GW-YEARS, AS IN THE FOOTNOTE, AND DIVIDED BY THE RATE OF USE TO YIELD A NOTIONAL EXHAUSTION TIME

	Coal		Oil		Gas	
Country	GW-years	Years	GW-years	Years	GW-years	Years
Global total	828,314	159	329,331	52	224,212	54
Algeria	–	–	–	–	5400	54
Australia	70,983	190	–	–	4480	67
Canada	–	–	33,493	110	2433	12
China	106,381	43	3584	12	4147	26
Egypt	–	–	–	–	2213	38
Germany	37,673	650	–	–	–	–
India	56,303	174	–	–	–	–
Indonesia	26,588	71	–	–	3445	39
Iran	–	–	30,568	119	40,774	196
Iraq	–	–	29,057	125	4300	2950
Kazakhstan	31,217	428	5811	48	–	–
Kuwait	–	–	19,662	89	–	–
Libya	–	–	9376	266	–	–
Nigeria	–	–	7187	43	6112	133
Qatar	–	–	5172	39	29,397	138
Russia	145,877	643	19,991	26	39,124	56
Saudi Arabia	–	–	51,721	64	9788	75
South Africa	28,018	140	–	–	–	–
Turkmenistan	–	–	–	–	20,950	252
United Arab Emirates	–	–	18,945	72	7300	104

(Continued)

TABLE 24.9 *(Continued)* CONVENTIONAL FUELS: 'PROVEN RESERVES' AT THE END OF 2014, AS CITED IN THE BP STATISTICAL REVIEW, CONVERTED TO GW-YEARS, AS IN THE FOOTNOTE, AND DIVIDED BY THE RATE OF USE TO YIELD A NOTIONAL EXHAUSTION TIME

	Coal		Oil		Gas	
Country	**GW-years**	**Years**	**GW-years**	**Years**	**GW-years**	**Years**
Ukraine	31,471	753	–	–	–	–
USA	220,469	327	9395	11	11,709	13
Venezuela	–	–	57,785	301	6689	196

Note: These are almost all lower bounds, being based on very conservative estimates of total resources, but represent indicative time scales. The countries listed are those reporting reserves amounting to 1% or more of the global total for each of the fuels.
With conversion factors in Section 24.5:
1 gigawatt-year (GW-year) = 31.557 petajoules (PJ).
1 million tonnes of coal corresponds to 0.9291 GW-years.
1 billion barrels of oil corresponds to 193.7 GW-years.
1 trillion cu ft of gas equates to 33.96 GW-years.

TABLE 24.10 SHALE OIL AND GAS RESERVES

Country	Shale Oil (GW-years)[a]	Shale Gas (GW-years)[a]
Global total	81,000	257,000
Algeria	1100	24,000
Argentina	5200	27,000
Australia	3000	14,600
Brazil	1000	8300
Canada	1700	19,400
Chad	3100	–
China	6200	38,000
Colombia	1300	–
Egypt	900	3400
France	900	4600

(Continued)

TABLE 24.10 *(Continued)* SHALE OIL AND GAS RESERVES

Country	Shale Oil (GW-years)[a]	Shale Gas (GW-years)[a]
India	–	3300
Indonesia	1500	–
Kazakhstan	2100	–
Libya	5000	4100
Mexico	2500	18,500
Oman	1200	–
Pakistan	1800	3600
Paraguay	–	2600
Poland	–	5000
Russia	14,500	9700
South Africa	–	13,000
Turkey	900	–
Ukraine	4300	–
United Arab Emirates	4400	7000
USA	15,200	21,000
Venezuela	2600	5700

Note: These estimates are less secure than those in Table 24.9. Values are given for countries reporting 1% or more of the global total for each fuel.

[a] 1 GW-year is equivalent to 5.16 million barrels or 0.75 million tonnes of oil equivalent or 29.4 billion cubic feet of gas.

24.7 USES OF THE FOSSIL FUELS AND CO_2 EMISSION

The industrial revolution was driven by coal, which became not just the almost universal fuel in industrialising countries for steam-driven transport, both on land and at sea, as well as general heating, but also feedstock for large-scale steel production and for the developing chemical industry. In established industrial economies, the balance has shifted towards an increasing use of oil and gas but globally coal remains dominant in electricity generation (Tables 24.11 and 24.12), steel making and cement production. Particularly in China, its cost competitiveness with imported oil and gas is reviving the use of coal gasification and liquid fuel production, as well as a

TABLE 24.11 GLOBAL PERCENTAGE USES OF THE FOSSIL FUELS

Use	Coal	Oil	Gas
Electricity	68%	0.7%	31.7%
Steel	14%	–	–
Cement	8%	–	–
Chemical feedstock	7%	22%	32%
Transport	1%	72%	3%
Heating	1%	5%	34%
Other	1%		

TABLE 24.12 AVERAGE ELECTRIC POWER GENERATION IN TERAWATTS FROM EACH OF SEVERAL SOURCES

Fuel Type	Fuel		% of Total Electricity Generation
Fossil fuels	Coal	1.152	41.3
	Oil	0.129	4.6
	Gas	0.648	23.2
	Total	1.929	69.1
Non-fossil sources	Nuclear	0.230	8.2
	Hydro	0.419	15
	Other	0.215	7.7
	Total	0.864	30.9
Total		2.793	

Note: Global data for 2014.

range of industrial chemical production for which oil and gas are competitors. Most coal is referred to as steaming coal, being used in power stations, but different qualities are required for other applications, in particular the coking coal, with low sulphur and phosphorus, used in steel making. The use of coal continues to evolve and to increase, in spite of its reputation as a CO_2 emitter. Oil has become the almost universal transport fuel and gas is now the favoured fuel for general heating, with an increasing role in electricity generation. Both are important feedstock for the chemical industry. The CO_2 released by fossil fuel burning depends on the fraction of the fuel that is carbon and so differs between samples even of the same fuel. Table 24.13 gives approximate averages of the mass of CO_2 released per unit of energy generated by complete combustion and the total CO_2 emissions from each of the fuels in 2014.

In spite of efforts to increase the supply of energy from the 'alternative' sources summarised in Chapter 25, the world's electricity generation continues to be dominated by fossil fuels (69%), with coal the biggest contributor. Oil is contributing a diminishing fraction of the total as gas takes over, with 'unconventional' gas (shale gas and coal seam gas) supplementing the supply of natural gas. The generation, in 2014, is presented here in terawatts of average power derived from each of the fuels, with the contribution by non-fossil sources added for comparison. The non-fossil generation is considered in more detail in Chapter 25.

The fossil fuel generation is by thermal power stations, operating at efficiencies between 30% and 40%, so that the total thermal energy expended in generating 1.9 TW is about 5.5 TW. The lower efficiencies are identified with older power stations that are mostly coal fired and about 68% of the 5.22 TW of the energy use attributed to coal in Table 24.8 is consumed in electricity generation.

TABLE 24.13 CO_2 EMITTED FROM FOSSIL FUEL USE AND A COMPARISON WITH VOLCANIC EMISSION

Fuel	kg of CO_2 per GJ	CO_2 Release in 2014 (megatonnes)
Natural gas	52	6656
Oil	57	11,200
Coal	87	14,390 (coal + peat)
Peat	106	
Total	Average 68	33,250[a]
Volcanoes	–	65 (annual average)

[a] The total release amounts to 6.47 ppm by mass or 4.26 ppm by volume of the atmosphere, close to twice the increase in atmospheric CO_2 (without consideration of cement production or deforestation). The difference is attributed primarily to solution in the oceans.

24.8 A NOTE ON TERMINOLOGY AND UNITS USED IN OIL/GAS RESERVOIR ENGINEERING AND HYDROGEOLOGY

24.8.1 Darcy's Law of Fluid Flow in Porous Media

The quantitative treatment of the flow of fluids through porous media, as applied to oil and gas reservoirs and to aquifers containing accessible groundwater, is an application of classical hydrodynamics to situations in which fine details of the flow are not resolved (and do not need to be). Consider the simple case of laminar (non-turbulent) flow of a viscous fluid in a tube or pipe (of circular cross-section). Laminar flow is readily satisfied by slow flow in a narrow tube, and this is relevant to what follows. Poiseuille's formula gives the rate of flow, q (m^3/s if SI units are used), in terms of the area, a, and length, l, of the tube, the pressure difference, P, between its ends and the viscosity, η, of the fluid (often written as μ in the literature of the subject):

$$q = a^2/(8\pi\eta).P/l \tag{24.1}$$

Now, we consider not a physically distinct tube but a large number, n, of narrow flow paths in a porous medium, treated simplistically as a bundle of n tubes. Its total cross-sectional area is $A = na/f$, where $f << 1$ is the fraction of the cross-section that is open pores (or tubes), not occupied by solid material. The total flow, expressed as flow per unit area of the medium, termed darcy flux, is

$$Q/A = nq/A = (fa/8\pi).(1/\eta).P/l = (\kappa/\eta).P/l \tag{24.2}$$

where P/l is the pressure gradient and $\kappa = (fa/8\pi)$ is the permeability of the medium. It is a physical property of the medium, representing the geometry of the pores and their interconnectedness. In layered rock, particularly sediments that have remained more or less horizontally layered, permeability can be much greater in horizontal directions than vertically.

Equation 24.2 is Darcy's law and the unit of κ is the darcy. Its original definition appears antiquated in the age of SI units:

1 darcy is the permeability of a medium that allows a flow of 1 cm^3/second of fluid with dynamical viscosity of 1 centipoise (1milliPascal-second) through a cross-section of 1 cm^2 under a pressure gradient of 1 atmosphere/cm,

but its value, $\sim 1\ (\mu m)^2$, makes it the practical unit of the subject.

Permeability, κ, is a property of a porous material, independent of the viscosity of fluid flowing in it. In the case of oil in reservoirs, there is a wide range of viscosities, but in hydrogeology, concerned only with water, the viscosity is almost a constant (1.8×10^{-3} Pa s at 0°C to 0.9×10^{-3} Pa s at 25°C) and the need to consider it is generally avoided by using the concept of hydraulic conductivity

$$K = (Q/A)/(-\mathrm{d}h/\mathrm{d}l) \tag{24.3}$$

in which the driving pressure gradient is represented by the hydraulic head, h, with a negative sign to indicate that pressure decreases downstream. κ depends only on the structural properties of the porous material, but K is controlled by the fluid viscosity as well as its density, ρ, and gravity, g, so that

$$K = (\rho g/\eta)\kappa \tag{24.4}$$

Since, in hydrogeology the fluid is water, $(\rho g/\eta) \approx 5.8 \times 10^{6}$ kg m^{-2} s^{-1} is effectively a constant.

24.8.2 Some Technical Jargon

Some of the terminology of hydrogeology and reservoir engineering is introduced in Section 24.8.1. Although the physical principles of these disciplines are similar, the different applications have led to quasi-independent technical terms. Meanings of some additional ones (italicised) follow.

Specific storage is the volume of water extractable from an aquifer per unit decline in hydraulic head, usually expressed as a fraction of the aquifer volume. The same term is used for the hydrocarbon output from a reservoir per unit drop in reservoir pressure, applied particularly to reservoirs with free gas.

Storativity is an alternative measure, being the water extracted per unit decline in hydraulic head per unit area of aquifer. It is the product of specific storage and aquifer thickness. The word is not used for oil reservoirs for which the term *productivity index* has essentially the same meaning.

Specific yield is the fraction of the pore volume of a saturated aquifer that would free drain if it were unconfined. There is no corresponding term for oil reservoirs.

Transmissivity is a measure of the rate at which water can be transferred horizontally in an aquifer. Although this word is not used in oil reservoir engineering, the concept is applied in assessing reservoir performance.

Chapter 25

Energy Sources Alternative to Fossil Fuels

25.1 ENERGIES OF SOME NATURAL PHENOMENA

Several natural processes dissipate energy that can be exploited. The abundances, accessibility and consequences of use have been discussed for decades, not always with consistent conclusions. The fundamental considerations emphasised here draw attention to two general principles that appear self-evident when they are recognised. The total energy of any natural process imposes an extreme upper bound on the energy that may be extracted from it, in principle, and the environmental consequences of use increase with the fraction that is used. That fraction will, in any case, be limited by technical considerations. These are inescapable reasons for concentrating on the phenomena with the most abundant energies. Inevitably that means, first and foremost, solar energy. As numbers in Table 25.1 show, wind power ranks second and all the other sources may have particular uses but are also-rans in the quest for power on the scale of the present human use, 16 TW. Harnessing any of these sources has no influence on the energy balance of the Earth, because all of the energy is converted to heat, whether naturally or by diversion to human use. Subsections 25.1.1 to 25.1.8 present comments and explanations of the entries in Table 25.1.

25.1.1 Solar Energy

The solar entries in Table 25.1 give the total power of sunlight reaching the surface and include the fraction that is directly reflected, which is as readily intercepted as is the radiation that is absorbed by the surface. Allowing for latitude and day–night variations, the average radiation at the surface is 190 Wm^{-2} and if this is to be converted to usable power with 10% efficiency, the generation of 16 TW requires an area of 840,000 km^2, 0.5% of the land area.

TABLE 25.1 AVERAGE RATES OF ENERGY DISSIPATION BY SOME NATURAL PROCESSES, WITH ALL VALUES IN TERAWATTS

Energy Sources		TW
Solar	Land	28,000
	Sea	68,000
Wind	Land	200
	Sea	500
Tides	Solid	0.5
	Marine	3.2
Geothermal	Land	10
	Ocean	37
Tectonics		7.7
River flow		6.6
Waves		4
Atm. electricity		0.001

Note: Entries in this table are not necessarily exploitable but are included for comparison.

25.1.2 Wind Power

The total dissipation of the kinetic energy of wind in the atmosphere is much greater than the entries in Table 25.1, which refer only to the consequence of wind friction with the Earth. Energy dissipated in the bulk of the atmosphere is presumed not to be harnessable. Only wind in a boundary layer within ~100 m of the surface, that is accessible by wind turbines, is considered. However, the driving force is convection in the atmosphere as a whole. Rotation of the Earth diverts the motion into cyclonic and anticyclonic horizontal circulations, but the energy is derived from the buoyancy forces of vertical circulation over tens of kilometres in elevation. Some of the energy is fed downwards into the boundary layer, where it interacts with the Earth and can be intercepted by wind turbines, but there would be no significant wind in a thin layer isolated from the rest of the atmosphere. The magnitude of the accessible wind energy can be estimated from the strength of its coupling to the Earth that is apparent from astronomically observed short-term variations in the rate of the Earth's rotation, but it is also the subject of more direct observations of the distribution of usefully strong winds. To some extent, wind and solar power are complementary. Wind is generally stronger at high latitudes where solar power is weak and is as effective at night as during daylight. There is a case for combining solar and wind farms.

25.1.3 Tides

Tidal deformation extends over the depth of the mantle (2900 km) and there is no useful concentration of strain energy near to the surface. The only useful tidal energy is in the marine tides. It is derived from the rotation of the Earth in the gravity fields of the Moon (83%) and Sun (17%), which have a global effect, so that the tidal forces are generated in the oceans and are not significant in small bodies of water. However, most of the dissipation occurs in shallow, marginal seas, where the tide is driven by connections to the open oceans and may be amplified by local resonances. This means that the upper bound on the energy that can be derived from tides is the energy put into the marginal seas by the open ocean drivers and cannot be inferred from the tidal behaviour in the shallow areas themselves, where the tide is actually harnessed. The locally resonant tidal amplitudes would collapse if the motion were damped by harnessing the energy. Although it may appear possible that extraction of tidal energy would be additional to the natural dissipation, this would decrease as a response, and the tidal energy in Table 25.1 must be regarded as a global upper bound of extractable energy. For practical reasons, that cannot be even remotely approached. Nevertheless, tidal power is harnessed, notably in the Rance estuary on the Atlantic coast of France, where an average of 60 MW (6×10^{-5} TW) of electrical power has been generated for 50 years. There is a number of other installations around the world, especially in Korea, but mostly with limited operating cycles, and the reported output capacities are maxima that greatly exceed the total or average output.

25.1.4 Geothermal Power

Heat from the Earth has been used in restricted locations (as water heated by igneous activity) for millennia. However, the numbers in Table 25.1 give no indication of the opportunities offered by such localised heat and report only the average background heat from the Earth as a whole. This is less than 1 kW per hectare and over most of the Earth it simply diffuses from the crust, driven by a temperature gradient of 20 to 25 K/km. The direct use of local igneous heat now extends to domestic and even community heating systems, notably in Iceland, but this is mostly low-grade heat, meaning that it is at temperatures that are too low for useful thermodynamic efficiency in applications such as electricity generation. That requires wells drilled to obtain superheated water that becomes steam when brought to the surface. The geological requirement is porous material, generally sediment, naturally charged with groundwater and overlying very hot rock or even magma. These conditions are found in several countries (Section 25.2.4) but the requirements are too restrictive to anticipate that geothermal power can become a major contributor to global energy use. The possibility of expanding the method to areas of hot but dry rock has been under investigation for several decades, but success has been very limited. It involves fracturing rock at a few kilometres depth to allow flow of water from an injection well to an extraction well. Favourite targets are large granite batholiths that retain their igneous heat for hundreds of thousands of years. Although granite is the most radioactive among common rock types, its radiogenic heat ($\sim 10^{-9}$ W kg^{-1}) is not significant. Globally, radioactivity is the largest continuing heat source (~30 TW), but is too weak and widely dispersed to be useful in the present context.

25.1.5 Tectonics

This entry in Table 25.1 is the energy dissipated by mantle convection. Although it is concentrated in near surface seismically active subduction zones, it is difficult to envisage a method of harnessing it. The energy released in fault zones is not apparent as local heat, presumably because they are fragmented and porous, allowing free flow of groundwater.

25.1.6 River Flow

In many ways, hydroelectric installations are ideal power sources. Energy can be stored indefinitely, as the gravitational energy of impounded water, and, on very short notice, converted to electric power with high efficiency. The nominally available power in Table 25.1 is simply calculated as the gravitational energy that is dissipated by all of the surface water that flows into the sea, assuming that it does so from the average elevation of all land. Although it is an extreme limit that supposes that all of the available water could flow through turbines of 100% efficiency all the way, the use of hydropower (Section 25.2.1) has reached 6.9% of this limit and continues to increase. That is a commentary on its accessibility. In a few countries, Norway being a striking example, most of the electric power is obtained from hydro schemes, but hydropower is used more widely to smooth irregularities in the mismatch of demand and the supply from other sources that cannot respond quickly to changes. This use is extended in pumped storage systems, in which water is pumped to an upper reservoir when supply exceeds demand and used when demand peaks.

Use of this ideal energy storage method will need to increase with increases in irregularly generated solar and wind power.

The report card for hydropower has some negative remarks. Large dams displace populations and agricultural activity, modify groundwater levels and dehydrate downstream ecology and communities. They silt up, imposing limits on their useful lives which, in the extreme case of the Aswan Dam on the Nile, may be as short as 100 years. Also, dam failures that have occurred were calamitous. Increasingly popular mini- and micro-hydro installations can reduce these problems.

25.1.7 Waves

Taking waves of 2 m amplitude and 10 s period as normal on wave-exposed coastlines, the dissipation, about 40 kW per metre of coast, appears to present an inviting engineering challenge and numerous teams (and companies) have taken up the challenge with a wide variety of devices. A difficulty is that there is not one obvious way to proceed (or even a very few ways) and the research effort is fragmented. It has not reached the stage at which large-scale projects could be contemplated and enthusiasm is damped by recognition of the length of coastline that would be required for waves to make a major contribution. Ignoring the question of efficiency in converting wave power, the entry in Table 25.1 assumes wave harnessing on 10^5 km of shoreline, an estimate of the global total length of suitably wave-exposed coast. Locally useful power generation has been demonstrated, but on a global scale, wave energy cannot be regarded as a serious contender.

25.1.8 Atmospheric Electricity

The only seriously proposed application was to use lightning strikes on wires strung across valleys in Switzerland as a source of extreme electrostatic potentials for particle acceleration in nuclear physics. This predated modern particle accelerators. It appears not to have been used.

25.2 GENERATION OF NON-FOSSIL FUEL POWER

25.2.1 Major Contributions: Nuclear and Hydropower

Nuclear power is really in a category of its own and is not a comfortable partner of the other energy sources in this chapter. In Table 25.2, it is coupled with hydropower because the global scales are comparable and are much larger than are the sources in subsections 25.2.3 and 25.2.4. Average generation in

TABLE 25.2 GLOBAL DISTRIBUTION OF NUCLEAR AND HYDROPOWER GENERATION

	Nuclear		Hydro	
Country	**GW**	**Domestic %**	**GW**	**Domestic %**
Global total	300	–	428	–
Argentina	–	–	4.6	30.6
Australia	–	–	1.7	8.2
Austria	–	–	5.0	67
Brazil	–	–	47	75
Canada	11	15	43	60
Chile	–	–	22.8	29.5
China	11	2	100	18
Colombia	–	–	5.4	81
Ecuador	–	–	1.4	55
Egypt	–	–	1.5	8.5
Finland	–	–	1.9	25
France	49	76	6.6	11
Germany	11	16	2.4	3.6
Greece	–	–	0.5	7
Iceland	–	–	1.4	70
India	–	–	14	12
Indonesia	–	–	1.5	7
Iran	–	–	1.4	5
Iraq	–	–	0.6	9
Italy	–	–	4.7	5
Japan	2011: 35 2014: 0	2011: 30 2014: 0	8.5	8
Korea	17	28	–	–
Kyrgyzstan	–	–	1.6	94
Laos	–	–	1.3	92
Mexico	–	–	3.6	11
Mozambique	–	–	1.7	99.9
New Zealand	–	–	2.6	52
Norway	–	–	16	97

(Continued)

TABLE 25.2 *(Continued)* GLOBAL DISTRIBUTION OF NUCLEAR AND HYDROPOWER GENERATION

	Nuclear		Hydro	
Country	**GW**	**Domestic %**	**GW**	**Domestic %**
Pakistan	–	–	3.4	32
Paraguay	–	–	6.8	99.9
Peru	–	–	2.5	56
Philippines	–	–	1.2	15
Portugal	–	–	0.45	13
Romania	–	–	1.4	21
Russia	20	17	11	16
Spain	–	–	2.3	7
Sweden	7.3	38	9	48
Switzerland	–	–	4.4	60
Tajikistan	–	–	1.9	95
Thailand	–	–	1.0	5
Turkey	–	–	6.9	25
UK	7.3	19	0.6	1.6
Ukraine	10	45	1.2	5.5
USA	91	18	32	6.8
Uzbekistan	–	–	1.3	22
Venezuela	–	–	9	66
Vietnam	–	–	6.0	45
Zambia	–	–	1.3	99.7

GW (10^{-3} TW) is listed for countries with the biggest contributions and for smaller countries with hydroelectric generation that is a significant, or even dominant, component of their domestic electricity supplies. This is indicated by the percentages of domestic power consumptions. The length of the list indicates how widespread the use of hydroelectricity is.

25.2.2 Hydroelectric Installations

There are more than 3000 stations with output capacities exceeding 1 MW and a much larger number of smaller ones. Table 25.3 lists those with average

TABLE 25.3 TYPICAL OR ANNUAL AVERAGE VALUES OF HYDROELECTRIC GENERATION FOR RECENT YEARS (2015 IF DATA ARE AVAILABLE)

Location	Capacity GW	Average Electricity Production (GW)
Itaipu Dam, Brazil, Paraguay	14	10.18 (2015)
Three Gorges Dam, China	22.5	10.02 (2015)
Xiluodu, China	13.86	6.30 (2015)
Guri, Venezuela	10.225	6.10
Tucuruí, Brazil	8.370	4.73
Churchill Falls, Canada	5.428	4.00
Xiangjiaba, China	6.448	3.50 (2015)
Sayano–Shushenskaya, Russia	6.4	3.06
Robert-Bourassa, Canada	5.616	3.03
Bratsk, Russia	4.500	2.58
Ust Ilimskaya, Russia	3.840	2.48
Yacyretá, Argentina, Paraguay	3.1	2.29
Grand Coulee, USA	6.809	2.28
Xiaowan Dam, China	4.2	2.17
Longtan Dam. China	6.426	2.13
Xingó Dam, Brazil	3.162	2.13
Ilha Solteira Dam, Brazil	3.444	2.04
Boguchany Dam, Russia	2.997	2.01
Gezhouba Dam, China	2.715	1.94
Jinping-I, China	3.6	1.94
Ertan Dam, China	3.3	1.94
Macagua, Venezuela	3.167	1.74
Krasnoyarsk, Russia	6	1.71
Pubugou Dam, China	3.3	1.67
Guanyinyan Dam, China	3	1.55
W.A.C. Bennett Dam, Canada	2.876	1.50
Tarbela Dam, Pakistan	3.478	1.48
Caruachi, Venezuela	2.160	1.48
Volzchskaya, Russia	2.639	1.47

(Continued)

TABLE 25.3 *(Continued)* TYPICAL OR ANNUAL AVERAGE VALUES OF HYDROELECTRIC GENERATION FOR RECENT YEARS (2015 IF DATA ARE AVAILABLE)

Location	Capacity GW	Average Electricity Production (GW)
Chief Joseph Dam, USA	2.620	1.43
Zhiguliovskaya, Russia	2.383	1.34
Iron Gates-I, Romania, Serbia	2.254	1.29
Nurek Dam, Tajikistan	3.015	1.28
Aswan Dam, Egypt	2.1	1.26
Son La Dam, Vietnam	2.4	1.17
Laxiwa Dam, China	4.2	1.16
Goupitan Dam, China	3	1.10
Atatürk Dam, Turkey	2.4	1.02

outputs exceeding 1 GW. The differences between these numbers and the rated capacities (maximum outputs) indicate the fractions of downtime arising from not only fluctuating electricity demand but also seasonal and other variations in available water flow. Many of the installations in Table 25.3 have attached pumped storage facilities and there are also pumped storage hydroelectric stations operating independently of the conventional stations that use river flow. They act as generators at times of peak demand and as pumps to reverse the flow by using 'surplus' power during quieter times, particularly at night, allowing thermal power stations to operate at steadier power levels. Unlike the thermal stations, including nuclear stations, which require substantial lead times for changes in output, pumped storage stations can respond very quickly (even within seconds) to changes in demand. Table 25.4 lists the pumped storage installations with capacities (maximum power generation) exceeding 1.5 GW and there are at least 50 others with capacities exceeding 1.0 GW. They are an ideal complement to solar and wind generation, which is intermittent, and, at least partly for this reason, plans for more pumped storage facilities are increasing with the expansion of solar and wind farms. The maximum duration of full power use varies from a few hours to a few days. Although the global total of pumped storage capacities exceeds 100 GW, this is only a small fraction of what will be needed to complement the non-hydropower demand, 2400 GW, if that is to come from solar and wind sources. Major dams are listed in Section 11.2.

TABLE 25.4 PUMPED STORAGE FACILITIES WITH CAPACITIES EXCEEDING 1.5 GW

Installation	Max. Output (GW)
Bath County, Virginia, USA	3.00
Huizhou, China	2.45
Guangdong, China	2.40
Okutataragi, Japan	1.93
Ludington, Michigan, USA	1.87
Tianhuangping, China	1.84
Grand'Maison, France	1.80
Dinorwig, N Wales	1.73
Raccoon Mt., Tennessee, USA	1.65
Mingtan, Taiwan	1.60
Castaic, California, USA	1.57
Tumut-3, Australia	1.50

25.2.3 Rapidly Expanding Sources: Solar and Wind

Table 25.5 gives annual average production of electricity from solar and wind sources. The solar power entries are totals for photovoltaics and solar concentrators combined. Heat produced from sunlight and used directly as heat, for space heating and in roof panels for water heating, is not easily assessed and is not included in this table but exceeds 400 GW (thermal) globally. Values less than 0.1 GW are not listed. Many of the values date from 2012. Global rates of increase are about 35% per year for solar and 26% per year for wind.

25.2.4 Geographically Restricted Sources: Geothermal, Tides and Waves

The geothermal power entries in Table 25.6 denote electrical power generation and do not include heat that is used directly for heating. Except for the pioneering installation on the Rance estuary in France, estimates of tidal generation are very approximate, being based on reported generation capacity without clear statements of operating cycles. Separate identification of energy derived from waves is not considered here because, although there is a variety of operating systems in several countries, they are still at the stage of being demonstration projects.

TABLE 25.5 ANNUAL AVERAGE ELECTRICITY GENERATION BY SOLAR AND WIND SOURCES FOR COUNTRIES PRODUCING MORE THAN 0.1 GW

	Solar		Wind	
Country	**GW**	**Domestic %**	**GW**	**Domestic %**
Australia	0.56	2.8	1.1	5.5
Austria	0.54	7.3	0.28	3.8
Belgium	0.25	2.8	0.32	3.6
Brazil	–	–	0.58	0.94
Canada	–	–	1.3	1.8
China	0.73	0.13	11.0	2.0
Denmark	–	–	1.2	36
Egypt	–	–	0.14	0.81
France	0.49	0.80	1.7	2.8
Germany	3.0	4.5	5.8	8.7
Greece	0.19	2.9	0.44	6.7
India	0.24	0.20	3.2	2.7
Ireland	–	–	0.46	16
Italy	2.2	6.7	1.5	4.8
Japan	0.79	0.72	0.55	0.50
Mexico	–	–	0.42	1.3
Netherlands	–	–	0.57	5.3
New Zealand	–	–	0.24	4.8
Norway	–	–	0.18	1.1
Pakistan	0.11	1.1	–	–
Poland	–	–	0.54	3.1
Portugal	–	–	1.2	24
Romania	–	–	0.30	4.7
Spain	1.4	4.3	5.6	18
Sweden	–	–	0.82	4.4
Turkey	–	–	0.67	2.6
UK	0.14	0.4	2.2	5.8
USA	0.49	0.11	16	3.5

TABLE 25.6 GEOTHERMAL, TIDAL AND WAVE POWER GENERATION

	Geothermal		Tides and Waves	
Country	**GW**	**Domestic %**	**GW[a]**	**Domestic %**
Australia	–	–	(7.8×10^{-5})	–
Canada	–	–	(0.02) 0.0036	0.005
Costa Rica	0.15	13	–	–
El Salvador	0.17	2.4	–	–
France	–	–	(0.24) 0.061	0.10
Iceland	0.54	28	–	–
Indonesia	1.1	5.2	–	–
Italy	0.65	2.0	–	–
Japan	0.31	0.2	–	–
Kenya	0.17	1.8	–	–
Korea	–	–	(0.254) 0.063	0.10
Mexico	0.74	2.3	–	–
New Zealand	0.70	14.3	–	–
Philippines	1.1	13.8	–	–
Spain	–	–	(3×10^{-4})	–
Turkey	0.10	0.3	–	–
UK	–	–	(0.012) 0.007	0.018
USA	1.8	0.37	(2×10^{-4}) 2×10^{-5}	–
Global total	7.6	–	~0.1	–

[a] Values in parentheses give peak or maximum generation.

25.2.5 Ethanol and Biodiesel

Ethanol is a fermentation product of crop material used either directly as a substitute for petrol or mixed with it. Biodiesel is a product of processed vegetable oils, and also recycled oils, which can be used directly for some sources. Both are used as transport fuels. A total of 1 billion litres/year converts to average power of 0.8 GW for ethanol or 1.1 GW for biodiesel. Production of each is listed in Table 25.7.

TABLE 25.7 ETHANOL AND BIODIESEL PRODUCTION IN BILLION LITRES/YEAR FOR THE LARGER PRODUCERS

Country	Ethanol	Biodiesel
Global	94	30
Argentina	0.7	2.9
Belgium	0.6	0.7
Brazil	27	3.4
Canada	1.8	0.3
Colombia	0.4	0.6
China	2.8	1.1
France	1.2	2.5
Germany	0.9	3.4
Indonesia	0.1	3.1
Italy	0.3	0.5
Netherlands	0.4	1.7
Poland	0.2	0.8
Spain	0.4	0.8
Thailand	1.1	1.2
UK	0.3	0.4
USA	54	4.7

Chapter 26

Land Degradation, Waste Disposal and Recycling

26.1 A HISTORICAL PERSPECTIVE

Ancient rubbish dumps, mine workings, soil disturbances and old building foundations are combed through over by archaeologists seeking evidence of the activities of our ancestors. What are now interesting artefacts may be obscured by natural processes, but they were not deliberately destroyed, except incidentally, as in the reuse of favourable building sites and scavenging of materials. Societies moved on, from one phase or activity to the next, with little or no attention to what was (or was not) left behind. Population expansion has gradually led to recognition that this cannot continue indefinitely because the Earth and its resources are finite. Given a clear measure of the scale of effects of human activity on the environment, an assessment of their reversibility is possible using evidence of consequential changes in the activity. This feedback loop has attracted increasing attention because of climatic effects. Earth scientists with an interest in the geology of the last 10,000 years, the period referred to as the Quaternary (Figure 18.1), shared with archaeologists one of the targets of their studies, the control exerted by climate and changes in it, on human activity. A new recognition of the significance of these studies has arisen with the realisation that human modifications of the environment are not restricted to local effects but that we are modifying the global climate. Other activities in this feedback loop include land clearing, mining developments, industrial and city expansions, waste disposal, intensive fishing and resource exhaustion generally, with global effects that are effectively instantaneous when viewed on the time scale of Earth history.

26.2 LAND USE AND USABILITY

Following is a list of the global total areas of various land uses in units of 1000 km^2 (10^9 m^2) with percentages of the total land area. The categories are not mutually exclusive, so the total exceeds 100%. Hooke et al. (2012) presented a review of global land use and modification, as in 2007, with some numbers differing from those presented here, essentially not only because definitions of the various categories depend on the emphasis that is put on them but also because multiple land uses were disallowed by constraining the total to 100%.

Global land area 148,000 (100%)

'Reclaimed' land (formerly under water, not including drained wetlands) ~15 (10^{-2}%)

Ice caps, glaciers and permanent snow 17,000 (11.5%)

Permafrost 24,000 (16.2%)

Deserts and arid areas (rainfall <250 mm/year, not including Antarctica) 20,000 (13.5%)

Mountain terrain 35,000 (24%)

Water covered areas (lakes, rivers, impoundments) 5600 (3.8%)

Wetlands (marshes, bogs, permanent or seasonal) 5500 (3.7%)

Area that would be inundated per metre of sea level rise 300 (0.2%)

Area required to produce 16 TW of average power from sunlight at 10% efficiency 840 (0.5%); see Section 25.1.1

Natural forest and scrub 39,000 (26.4%)

National parks 3380 (2.3%)

Plantation forest 1900 (1.3%)

Grazing land 33,600 (22.7%)

Cropland 15,500 (10.5%)

Total urban area (55% of world population) 3600 (2.4%)

Roads and margins ~6000 (4%)

Railways and margins 140 (0.1%)

Airports, airfields (42,000 total) and adjacent facilities 600 (0.4%)

Military facilities (rough estimate) 300 (0.2%)

Extractive industries (mines, quarries, oilfields) 3000 (2%)

Other industrial sites 2000 (1.4%)

Land degraded by human activities:

(i) Bad agricultural practice (salinisation, desertification) 8000 (6%)
(ii) Abandoned mine sites and extractive industry residua ~4400 (3%)
(iii) Industrial wastelands (include Chernobyl and Fukushima) ~2500 (1.7%)

Landfill sites and dumps (expanding with 1.9 billion tonnes of garbage per year) 450 (0.3%)

26.3 MATERIAL RECYCLING

Recycling, as considered here, is the extraction of material constituents from discarded items for uses that generally differ from the original uses. This excludes the recovery of materials or manufactured items for direct reuse, perhaps after repair or minor modification. An example of the distinction is in the treatment of building materials. The recovery of quality building stone from demolished buildings for use in new buildings, perhaps with only minor recutting, is direct reuse, whereas we classify as recycling the conversion to rubble for uses such as road base. Reuse is, almost by definition, economic, but recycling may be driven as much by the need to reduce waste disposal, to extract environmentally harmful materials or even to recover increasingly rare materials for which processing is still only marginally economic.

26.3.1 High-Volume Recyclables

Consideration of the recycling of specific elements focusses attention on some of the basic problems and, in some cases, this is referred to in Chapter 22. It is generally well developed for two types of element, those that are concentrated in their products and therefore easy (cheap) to process and those for which extraction from ores makes a high energy demand. It is not always necessary, or even desirable, to extract pure elements but to separate for reuse alloys or compounds in which they occur. The economic viability of a recycling process also depends on the scale of operation, and partly for this reason, iron and steel provide the largest volume of recycled material in the form of motor vehicles and white goods, such as refrigerators, for which the fraction of discarded material that is recycled exceeds 90% in industrialised communities. However, a global assessment is needed because in some cases, waste material is traded internationally for processing. An interesting example is the development of an industry dismantling ships for scrap metal in Bangladesh. The vehicle industry accounts for other elements with high recycling fractions and a particularly significant one is lead. About 85% of the current global lead production is used in the manufacture of vehicle batteries and they are virtually all recycled. Two other high-volume metals, aluminium and copper, have well-developed recycling industries, in these cases approaching 50% recycling, with the incentive of a major power saving. All aluminium is produced by electrolysis of molten oxide and although not all copper requires electrolysis, it is a necessary final stage in the production of the high-conductivity copper used in the electrical industry. Some other important metals (Ni, Cr, V and Mn) are used as alloying ingredients in steel, especially stainless steel, and are conveniently processed in that form. The same applies to much of the zinc and tin, which are alloyed with copper, to form brass and bronze, respectively, and magnesium, which strengthens most of what we refer to generically as aluminium.

26.3.2 Rarer Elements

A reason for reviewing the recycling of less abundant elements is to draw attention to those for which continued supply appears least secure or for which a sharply increasing demand may be incompatible with availability. Section 26.4 lists obvious ones. Helium is an interesting example. Its use, as liquid at very low temperature, in cooling the superconducting coils in instruments such as magnetic resonance imaging (MRI) machines, has been increasing rapidly. The collection and compression of evaporating gas for reuse is an obvious route, but far from universally practised and not generally 100% effective. Although the increasing demand has been met by extraction of He from new natural gas supplies, not all natural gas has a useful concentration of it and the possibilities of continued expansion of the supply appear limited. Although He is continuously leaking from the Earth, it escapes too rapidly to space to leave a sufficient atmospheric concentration for economic extraction. It appears just possible that the development of high-temperature superconductors, operating at liquid nitrogen temperature, will reduce the demand for He. The other inert gases, Ne, Ar, Kr and Xe, are all obtained from the atmosphere by partial distillation in the processes of liquid oxygen and nitrogen production. Argon is abundant (1% of the atmosphere), but the others are rarer and require greater expenditure of energy. This is particularly serious for the rarest of them all, xenon, which may have increasing use as a general anaesthetic, creating a demand that will be difficult to satisfy or make its use prohibitively expensive.

Two light elements that are rare are Li and Be. Although there are several uses of Li, the supply problem is drawn to attention by its use in high-performance batteries. It appears unlikely that the supply could, even in principle, suffice for Li batteries to become storage devices for much of the world's electrical power, as has been suggested. It appears prudent to make the recycling of Li batteries mandatory. In the case of Be, the combination of lightness and strength gives it special uses. It is unrivalled as window material for X-ray machines, but this does not cause a supply problem. Of greater concern is the more widespread use in small, high-performance mechanical components, such as Be–Cu alloy springs, that are widely distributed in devices in which their compositional significance may not be recognised when the devices are discarded.

26.3.3 Precious Metals

Gold, silver and the platinum group metals (Ru, Rh, Pd, Os, Ir, Pt) are regarded as precious in the sense of being permanent and incorruptible and so worth saving for their own sake. Gold, in particular, is regarded as secure material for the storage of wealth by both governments and individuals, and both gold and silver are the classical coinage metals. They are also the most favoured metals for jewellery, but share that role with Pt and the other members of the Pt group. There is no significant role for recycling in these 'uses'. That becomes important in the industrial applications that may lead to wide dispersion and effective loss. Three of these metals are prime targets for recycling, Au, Pt and Pd. A major use of gold is now as thin coatings on electrical contacts in electronic equipment. The accumulating mountains of obsolete and unwanted equipment contain traces of gold that are still concentrated relative to its abundance in many of the ores used for its primary extraction. Economic salvaging requires manual extraction of the coated contacts and has become an activity in some low-income countries, especially Ghana. However, it is not very effective, as little more than 10% of the 300 tonnes/year of the 'electronic gold' is actually recovered and Ghana has become an indiscriminate dump, not only of electronic discards but also of unprocessed European white goods. The gold that is 'lost' is nearly 10% of global production. The platinum group metals, particularly platinum itself and palladium, are catalysts applied as coating metals in vehicle exhausts to reduce the emission of unburnt fuel. The intrinsic value of the metals is high and the technology for their recovery is straightforward but the logistics of collection and processing are not well developed and there is still limited progress. Palladium has another use in the separation of hydrogen from other gases. It reversibly absorbs hydrogen (as much as 1 atom per 1.5 atoms of Pd at atmospheric pressure) and is used as a screen, allowing hydrogen to pass through but blocking other gases. Palladium is one of the elements for which the supply is rated as insecure.

26.3.4 Rare Earths and Other Elements with Special Uses

Elements 57 to 71, 21 and 39 are categorised as rare earths. Although not all rare, as mentioned in Chapter 22, they are chemically very similar, making them difficult (expensive and environmentally problematic) to separate. Their physical properties give them very particular uses, making it desirable to maintain isolation from one another, once refined, even when they are incorporated as minor constituents in devices with a range of other elements. The wide distribution and diversity of devices in which they are used make this difficult to do systematically and will require a selection of those identified as more critical or more easily recognised as constituents of common devices before indiscriminate reprocessing. An important minor element of which a significant fraction (~30%) is reprocessed is niobium because it occurs (mostly as Nb–Ti alloy) in readily identified and reasonably large quantities as winding wire in superconducting magnets.

26.4 A WATCH LIST OF ELEMENTS WITH SUPPLY INSECURITY

A. Valero and Al. Valero (2015) presented a discussion on the resource exhaustion problem from a physical perspective, and Henckens et al. (2016) considered the supply–demand approach, with a selection of critical elements differing somewhat from the list presented here. Elements for which supply problems appear most likely to arise in the short term are antimony, indium, lithium, niobium, palladium, phosphorus and xenon.

26.5 SOME PROBLEMS

Plastics probably present the most serious waste disposal problems. Recycling is made difficult by the variety of plastics with incompatible compositions that are common in waste material and their low value as recycling products. They are a major component of landfill and an estimated 8 million tonnes per year end up in the sea. Biodegradable variants have been developed for some purposes, particularly wrapping material and plastic bags, but have been adopted by only a tiny fraction of the potential market. Heating of mixed plastics to produce compacted briquettes reduces the volume of waste, but such briquettes have limited use and even less demand and are really no more than condensed waste. A plastic of particular interest in the recycling context is polytetrafluoroethylene (PTFE or Teflon). Its composition is 76% fluorine by mass, which is neither abundant nor easily extracted from minerals, but, as with many technical materials, it is widely dispersed in small components, and even as coatings, making selective recovery uneconomic. Some waste processing plants have optical sensing of major plastic types to allow sorting, but they are rare and, in any case, only some thermoplastics can be reused as is. This is a problem still awaiting a satisfactory solution. Another material with a low recycling value is glass, but this is quite extensively recycled, particularly as bottles, which are reasonably homogeneous in shape and readily sorted mechanically into different types (uses, colour). Automation is not difficult but in many places this is still done by hand. Reprocessing of paper and cardboard is increasing, not always enthusiastically, with the aim of reducing the need for wood pulp. The recycled material mostly becomes cardboard, but there is some use of higher quality paper that is at least partly recycled material. Some of these processes are not driven by economics, but by legal requirements or public subsidies, a familiar case being vehicle tyres.

The retrieval of useful material from the global total of 1.9 billion tonnes/year of municipal waste, dubbed urban mining, is not systematically practiced and recovers only about 2% of the material that reaches official dumps. Half of this is organic compost or methane from decaying biomaterial in sealed dumps. The management of these dumps requires particular attention to the mobilisation and redistribution of heavy metals, of which the mercury in fluorescent light fittings is a special concern. The heavy metals problem arises also with mine tailings dams and water pumped from mines and gas wells, which commonly compromise the quality of both surface water and groundwater over wide areas.

References

Anderson, O. L. and Isaak, D. G. 1995. Elastic constants of mantle minerals at high temperature. In *Mineral Physics and Crystallography: A Handbook of Physical Constants, Volume 2,* ed. T. J. Ahrens, 64–97. Washington, DC: American Geophysical Union.

Bass, J. D. 1995. Elasticity of minerals, glasses and melts. In *Mineral Physics and Crystallography: A Handbook of Physical Constants Volume 2,* ed. T. J. Ahrens, 45–63. Washington, DC: American Geophysical Union.

Beckley, B. D., Zelensky, N. P., Holmes, S. A., et al. 2010. Assessment of the Jason-2 extension to the TOPEX/Poseidon, Jason-1 sea-surface height time series for global mean sea level monitoring. *Mar. Geo.* 33(Suppl 1): 447–471. doi: 10.1080/01490419.2010.491029.

Bird, P. 2003. An updated digital model of plate boundaries. *Geochem. Geophy. Geosy.* 4(3): 1027. doi: 10.1029/2002GLO16002.

Bond, D. P. G. and Wignall, P. B. 2014. *Large Igneous Provinces and Mass Extinctions: An Update.* Geological Society of America Special Paper 505, The Geological Society of America.

Bullard, E. C., Freedman, C., Gellman, H. and Nixon, J. 1950. The westward drift of the earth's magnetic field. *Phil. Trans. Roy. Soc. Lond.* A243: 67–92.

Dlugokencky, E. J., Lang, P. M., Crotwell, A. M., Masarie, K. A. and Crotwell, M. J. 2014. *Atmospheric Methane Dry Air Mole Fractions from the NOAA ESRL Carbon Cycle, Cooperative Global Air Sampling Network, 1983–2013,* Version: 2014-06-24.

Dunlop, D. J. and Özdemir, Ö. 1997. *Rock Magnetism: Fundamentals and Frontiers.* Cambridge, UK: Cambridge University Press.

Dziewonski, A. M. and Anderson, D. L. 1981. Preliminary reference earth model. *Phys. Earth Planet. Inter.* 25: 297–356.

Etheridge, D. M., Steele, L. P., Francey, R. J. and Langenfelds, R. L. 1998. Atmospheric methane between 1000 AD and present: Evidence of anthropogenic emissions and climatic variability. *J. Geophys. Res.* 103: D13, 15979–15993.

Fegley, B. 1995. Properties and composition of the terrestrial oceans and of the atmospheres of the earth and other planets. In *Global Earth Physics: A Handbook of Physical Constants Volume 1,* ed. T. J. Ahrens, 320–345. Washington, DC: American Geophysical Union.

GSFC. 2013. *Global Mean Sea Level Trend from Integrated Multi-Mission Ocean Altimeters TOPEX/Poseidon Jason-1 and OSTM/Jason-2 Version 2*. San Diego, CA: PO.DAAC.

Gubbins, D. 2003. Thermal core-mantle interactions: Theory and observations. In *Earth's Core: Dynamics, Structure, Rotation*, eds. V. Dehant, K. C. Creager, S.-I. Karato and S. Zatman, 163–179. Washington, DC: American Geophysical Union.

Gutenberg, B. and Richter, C. F. 1954. *Seismicity of the Earth and Other Recent Phenomena*. Princeton, NJ: Princeton University Press.

Henckens, M. L. C. M., van Ierland, E. C., Driessen, P. P. J. and Worrell, E. 2016. Mineral resources: Geological scarcity, market price trends, and future generations. *Resour. Policy* 49: 102–111.

Hodgkinson, J. H., McLoughlin, S. and Cox, M. E. 2006. The influence of structural grain on drainage in a metamorphic sub-catchment: Laceys Creek, southeast Queensland, Australia. *Geomorphology* 81: 394–407.

Hooke, R. L., Martin-Duque, J. F. and Pedraza, J. 2012. Land transformation by humans: A review. *GSA Today* 22(12): 4–10. doi: 10.1130/GSAT151A.I.

Horton, R. E. 1945. Erosional development of streams and their drainage basins: Hydrophysical approach to quantitative morphology. *Geol. Soc. Am. Bull.* 56(3): 275–370. doi: 10.1130/0016-7606(1945)56[275:EDOSAT]2.0.CO;2.

Jaeger, J. C. 1971. *Elasticity, Fracture and Flow: With Engineering and Geological Applications*. London: Methuen & Co.

Jourdan, F., Hodges, K., Sell, B., et al. 2014. High-precision dating of the Kalkarindji large igneous province, Australia, and synchrony with the Early-Middle Cambrian (Stage 4–5) extinction. *Geology* 42: 543–546.

Kanamori, H. and Brodsky, E. E. 2004. The physics of earthquakes. *Rep. Prog. Phys.* 67: 1429–1496.

Keating, P. N. 1966. Effect of invariance requirements on the elastic strain energy of crystals with application to the diamond structure. *Phys. Rev.* 145: 637–645.

Keeling, C. D., Piper, S. C., Bacastow, R. B., et al. 2001. *Exchanges of atmospheric CO_2 and $^{13}CO_2$ with the terrestrial biosphere and oceans from 1978 to 2000, I. Global aspects*. SIO Reference Series, No. 01-06. San Diego, CA: Scripps Institution of Oceanography.

Kokubu, N., Mayeda, T. and Urey, H. C. 1961. Deuterium content of minerals, rocks and liquid inclusion from rocks. *Geochim. Cosmochim. Acta* 21: 247–158.

Krumbein, W. C. 1937. The sediments of Barataria Bay. *J. Sediment. Petrol.* 7: 3–17.

Lane, C. S., Chom, B. T. and Johnson, T. C. 2014. Ash from the Toba supereruption in Lake Malawi shows no volcanic winter in East Africa at 75 ka. *Proc. Natl. Acad. Sci. U. S. A.* 110(20): 8025–8029.

Lodders, K. and Fegley, B., Jr. 1998. *The Planetary Scientist's Companion*. Oxford: Oxford University Press.

Mason, B. and Moore, C. B. 1982. *Principles of Geochemistry*. New York: Wiley.

McDonough, W. F. and Sun, S.-s. 1995. The composition of the Earth. *Chem. Geol.* 120: 223–253.

Mei, S. and Kohlstedt, D. L. 2000. Influence of water on plastic deformation of olivine aggregates 2: Dislocation creep regime. *J. Geophys. Res.* 105: 21471–21481.

Merrill, R. T., McElhinny, M. W. and McFadden, P. L. 1996. *The Magnetic Field of the Earth: Paleomagnetism, the Core and the Deep Mantle.* San Diego, CA: Academic Press.

Mooney, W., Laske, G., Masters, T. G. 1998. CRUST 5.1: A global crustal model at $5^0 \times 5^0$. *J. Geophys. Res.* 103(B1): 727–747.

Newell, D. B. 2014. A more fundamental international system of units. *Phys. Today* 67(7): 35–41.

Newsom, H. E. 1995. Composition of the solar system, planets, meteorites and major terrestrial reservoirs. In *Global Earth Physics: A Handbook of Physical Constants Volume 1*, ed. T. J. Ahrens, 159–189, Washington, DC: American Geophysical Union.

Oppel, A. 1856. *Die Juraformation Englands, Frankreichs und des südwestlichen Deutschlands: nach ihren einzelnen Gliedern eingetheilt und verglichen.* Stuttgart: Verlag von Ebner & Seubert.

Rudnick, R. L. and Gao, S. 2014. Composition of the continental crust. In *Treatise on Geochemistry*, 2nd edition, eds. Holland, HD and Turekian, KK, Vol. 4, 1–51, Amsterdam, The Netherlands: Elsevier.

Schopf, J. M. 1956. A definition of coal. *Econ. Geol.* 51: 521–527.

Schweinfurth, S.P. and Finkelman, R.B., Coal—A complex natural resource: an overview of factors affecting coal quality and use in the United States. *USGS Circular* 1143, USGS, Denver.

Sella, G. F., Stein, S., Dixon, T. H., et al. 2007. Observations of glacial isostatic adjustment in 'stable' North America with GPS. *Geophys. Res. Lett.* 34: LO2306. doi: 10.1029/2006GL027081.

Shirey, S. B., Hauri, E. H., Thomson, A. R., et al. 2013. Water content of inclusions in superdeep diamonds. In *Goldschmidt 2013 at Florence*, pp. 2205. doi: 10.1180minmag.2013.077.5.19.

Shreve, R. L. 1966. Statistical law of stream numbers. *J. Geol.* 74: 17–37.

Shreve, R. L. 1967. Infinite topologically random channel networks. *J. Geol.* 75: 178–186.

Stacey, F. D. and Davis, P. M. 2008. *Physics of the Earth*, 4th edition. Cambridge: Cambridge, UK: Cambridge University Press.

Stacey, F. D. and Hodgkinson, J. H. 2013. *The Earth as a Cradle for Life, the Origin, Evolution and Future of the Environment.* Singapore: World Scientific Publishing Co.

Strahler, A. N. 1952. Hypsometric (area-altitude) analysis of erosional topology. *Geol. Soc. Am. Bull.* 63(11): 1117–1142. doi: 10.1130/0016-7606(1952)63[1117:HAAOET]2.0.CO;2.

Strahler, A. N. 1957. Quantitative analysis of watershed geomorphology. *Trans. Am. Geophys. Un.* 38(6): 913–920. doi: 10.1029/tr038i006p00913.

Takahashi, F., Matsushima, M. and Honkura, Y. 2005. Simulations of a quasi-Taylor state geomagnetic field including polarity reversals on the Earth simulator. *Science* 309: 459–461.

Tanimoto, T. 1995. Crustal structure of the Earth. In *Global Earth Physics: A Handbook of Physical Constants*, Vol. 1, 214–224. Washington, DC: American Geophysical Union.

Tarling, D. H. 1983. *Palaeomagnetism: Principles and Applications in Geology, Geophysics and Archaeology.* Chapman and Hall.

Tsuya, H. 1955. On the 1707 eruption of volcano Fuji. *Bull. Earthq. Res. Inst. Tokyo* 33: 341–393.

Valero, A. and Valero, Al. 2015. *Thanatia: The Destiny of the Earth's Mineral Resources; A Thermodynamic Cradle-to-Cradle Assessment.* Singapore: World Scientific.

Wentworth, C. K. 1922. A scale of grade and class terms for clastic sediments. *J. Geol.* 33: 377–392.

White, R. S., McKenzie, S. D. and O'Nions, R. K. 1992. Oceanic crustal thickness from seismic measurements and rare earth element inversions. *J. Geophys. Res.* 97: 19683–19715.

White, W. M. and Klein, E. M. 2014. Composition of the oceanic crust. In *Treatise on Geochemistry*, 2nd edition, eds., H. Holland and K. Turekian, 457–496. Amsterdam: Elsevier.

Yardley, B. W. D. 1989. *An Introduction to Metamorphic Petrology.* Harlow: Longman.

Zhang, L., Meng, Y., Yang, W., et al. 2014. Disproportionation of (Mg,Fe)SiO_3 perovskite in the Earth's deep lower mantle. *Science* 344: 877–882.

Index

A

D

H

L

M

Q

R

S

T

U

Z